Minerals and Rocks   18

Editor in Chief
P.J.Wyllie, Chicago, IL

Editors
A.El Goresy, Heidelberg
W. von Engelhardt, Tübingen · T. Hahn, Aachen

G. Gottardi   E. Galli

Natural Zeolites

With 218 Figures

Springer-Verlag
Berlin Heidelberg New York Tokyo

Professor Dr. GLAUCO GOTTARDI
Professor Dr. ERMANNO GALLI
Istituto di Mineralogia
e Petrologia
Università di Modena
Via S. Eufemia 19
I-41100 Modena

Volumes 1 to 9 in this series appeared under the title
Minerals, Rocks and Inorganic Materials

ISBN 3-540-13939-7 Springer-Verlag Berlin Heidelberg New York Tokyo
ISBN 0-387-13939-7 Springer-Verlag New York Heidelberg Berlin Tokyo

Library of Congress Cataloging in Publication Data. Gottardi, Glauco, 1928–
Natural zeolites. (Mineral and rocks; 18) Bibliography: p. 1. Zeolites. I. Galli,
Ermanno, 1937– II. Title. III. Series. QE391.Z5G67 1985   549′68   84-26874

Typesetting: Fotosatz GmbH, Beerfelden
Printing and binding: Brühlsche Universitätsdruckerei, Giessen
2131/3130-543210

To Laura and Maura

Preface

Zeolites form a family of minerals which have been known since the 18th century, but they remained a curiosity for scientists and collectors until 60 years ago, when their unique physicochemical properties attracted the attention of many researchers. In the past 30 years there has been an extraordinary development in zeolite science; six International Conferences on Zeolites have been held every 3 years since 1967, and a large number of interesting contributions have been published in their proceedings. Many books, written either by individual authors or by several authors under a leading editor, have been published on these interesting silicate phases, but none has been devoted specifically to natural zeolites, even though this theme may be of interest not only to earth scientists, but also to chemists, as the information obtained from natural samples completes and integrates the characterization of many zeolites. We are trying to fill this gap on the basis of 20 years of research on natural zeolites, which we performed at the University of Modena together with many friends and colleagues.

If it is in general difficult to write a scientific book without upsetting somebody, this is particularly true for a book on natural crystals, because mineralogy is an interdisciplinary science which covers some fields of physics, chemistry, petrology, geology, and it is almost impossible to meet every requirement.

To begin with crystallography, a partially new structural classification has been proposed without any pretension to having it generally adopted; as a matter of fact, the structural classification of minerals (and of inorganic crystals) is a still open problem. Obviously enough, most of the crystallographic chapters are devoted to crystal structure, the information on morphology being more restricted.

Although the same scheme has been chosen for the description of all species, for rare minerals the information given represents nearly all that is at present available, whereas for common minerals a representative spectrum of data has been selected. We have done our best to give a complete list of the occurrences of rare zeolites.

Most thermal curves and X-ray powder patterns have been specially recorded and interpreted by the authors, in order to guarantee the homogenity of all these data.

Modena, 1985   GLAUCO GOTTARDI
                                               ERMANNO GALLI

Acknowledgements

We wish to thank all the members of the Modena group, A. Alberti, R. Rinaldi, E. Passaglia (now at Ferrara University), D. Pongiluppi, and G. Vezzalini for their advice, and for their permission to publish some new data. Special thanks are due to Laura Gottardi, who kindly typed the manuscript, and to Roy Phillips, who critically read and revised it (although we wonder if he really succeeded in transforming the text as we had written it in a language which only vaguely resembled English, into proper English). Thanks are also due to Maria Franca Brigatti and Luciano Poppi for their assistance in recording the thermal curves. Preparation of some of the drawings and pictures were made possible by the financial support of the Consiglio Nazionale delle Ricerche, Roma. Philips S.p.A., Milano, are also acknowledged for their substantial contribution to the preparation of drawings and the correction of the manuscript. We also wish to thank W. Lugli for drawing many diagrams, V. Casodi for a photograph of phillipsite, and P. Dunn, E. Tillmanns and G. Hentschel for the permission to reproduce some of their pictures. The facilities of the Centro Strumenti and Centro di Calcolo of Modena University were used for preparing the SEM pictures, for indexing the powder patterns and for drawing crystals with the program of Kanazawa and Endo (1981). The structure drawings were made by "Studio grafico Dimitri Nicola" in Modena.

Modena, 1985   GLAUCO GOTTARDI
                                                            ERMANNO GALLI

Contents

The descriptions of individual zeolite genera are arranged systematically as follows:
1   History and nomenclature
2   Crystallography
3   Chemistry and synthesis
4   Optical and other physical properties
5   Thermal and other physico-chemical properties
6   Occurrences and genesis
7   Uses and applications

0 General Information on Zeolites

0.1 Introduction and Historical Remarks

The term "zeolite" was created by Cronstedt (1756) from the Greek $\zeta\varepsilon\iota\nu$ = "to boil" and $\lambda\iota\vartheta o\varsigma$ = "stone", for minerals which expel water when heated and hence seem to boil. They are very common and well known as fine crystals of hydrothermal genesis in geodes and fissures of eruptive rocks, or as microcrystalline masses of sedimentary origin.

A natural zeolite is a framework alumino-silicate whose structure contains channels filled with water and exchangeable cations; ion exchange is possible at low temperature (100°C at the most) and water is lost at about 250°C and reversibly re-adsorbed at room temperature.

As is well known, the primary building units (PBU) of the structures of silicates are the TO_4 tetrahedra, where T is mainly Si. In framework silicates, these PBU are linked so as to form a 3-dimensional framework and (nearly) all oxygens are shared by two tetrahedra; the sharing coefficient (Zoltai 1960) is hence 2 or a little smaller. If all tetrahedra were centred by Si, and the sharing coefficient were 2, the chemical formula of the framework would be Si_nO_{2n}; if there is some Al in the tetrahedra, the formula is $[Al_mSi_{n-m}O_{2n}]^{m-}$, the m negative charges being balanced by the extraframework cations, usually K, Na, Ca, less frequently Li, Mg, Sr, Ba.

In synthetic zeolites the composition may vary more widely, including Ga, Ge, Be, P, as tetrahedral cations, and all alkali, alkaline-earth and rare earth elements and also organic complexes (for instance tetramethylammonium) as extraframework cations.

The general formula for natural zeolites is hence

$$(Li, Na, K)_a (Mg, Ca, Sr, Ba)_d [Al_{(a+2d)} Si_{n-(a+2d)} O_{2n}] \cdot mH_2O$$

where the part in square brackets represents the framework atoms and the part outside the square brackets the extraframework atoms, cations plus water molecules. Please note that $m \leqslant n$, usually. This formula allows a wide variability, because the only real constraint is given by the condition $Si \geqslant Al$; this variability is true not only considering the zeolite family as a whole, but also within a single zeolitic species. In other words, the chemical composition of one zeolite is well represented by a definite chemical formula in some cases, but generally speaking this is not true.

The general formula contains a clue to check if a chemical analysis is reliable or not; in fact the balance error

$$E\% = \frac{(Al+Fe)-(Li+Na+K)-2(Mg+Ca+Sr+Ba)}{(Li+Na+K)+2(Mg+Ca+Sr+Ba)} \times 100$$

is 0 in a "perfect" analysis, in any case should be low ($<10\%$) in an acceptable one; analyses with higher E should be discarded.

The chemical composition of zeolites can be satisfactorily represented as a point in the tetrahedron

$$Si_2O_4 - Na_2Al_2O_4 - K_2Al_2O_4 - (Mg, Ca, Sr, Ba)Al_2O_4$$

because Li is normally absent, Ca normally prevails by far over the other divalent cations, and the water content may be neglected. For the sake of simplicity, the triangular plot

$$Si_2O_4 - (Na, K)_2Al_2O_4 - (Mg, Ca, Sr, Ba)Al_2O_4$$

is often adopted. Alberti (1979a) has shown that if we use a triangular diagram whose vertices each represent a linear function of all chemical components, it is possible to get a more complete separation of points in the triangle than in the simpler plot, where each chemical component influences the function of only one vertex.

The ability to loose water at high temperature (150 to 400°C) and re-adsorb it from the atmosphere at room temperature is one of the properties which qualifies a framework silicate mineral as a zeolite. Hence thermal analysis has great importance in the study of these minerals. Out of the four common types of thermal analysis (differential thermal analysis or DTA, thermo-gravimetry or TG, differential thermo-gravimetric analysis or DTGA, differential scanning calorimetry or DSC), we think that TG and DTGA of zeolites are more reproducible and less subject to instrumental conditions than DTA, whereas DSC data are still scanty.

A zeolitic phase, once dehydrated (or „activated") may re-adsorb not only water, but also gases, vapours and fluids, especially if their molecules are polar. This capacity to adsorb a variety of molecules is in turn related to the catalytic properties of zeolites; these physico-chemical qualities are widely described in detail in several books (Breck 1974; Rabo 1976; Barrer 1978, 1982) and will be discussed only briefly in this book.

To conclude this introduction, some historical remarks on the research on zeolites may be useful. As already said, this history began as early as 1756 with a publication by Cronstedt. 100 years later, Dufrénoy (1859) listed 17 zeolites, but in 1914 Dana recorded 19 zeolites, plus another 2 which were discredited later; today the zeolitic species number 32, but new species are discovered at the rate of one every three years (see Table 0.1 I). At the beginning, mineralogical studies were based mainly on the morphology of well developed crystals, so that only well shaped zeolites of hydrothermal genesis were considered for description as mineral species. In 1862 von Seebach found analcime crystals 1 mm in size in a clay; Murray and Renard (1891) found twinned

Table 0.1I. Year of proposal of the zeolite names with their present meaning

Chabazite: Bosc d'Antic (1792)
Stilbite: Haüy (1801)
Analcime: Haüy (1801)
Harmotome: Haüy (1801)
Natrolite: Klaproth (1803)
Laumontite: Werner (in Karsten 1808)
Scolecite: Gehlen and Fuchs (1813)
Mesolite: Fuchs and Gehlen (1816)
Gismondine: Leonhard (1817)
Thomsonite: Brooke (1820)
Heulandite: Brooke (1822)
Brewsterite: Brooke (1822)
Edingtonite: Haidinger (1825)
Phillipsite: Lévy (1825)
Gmelinite: Brewster (1825a)
Levyne: Brewster (1825b)
Epistilbite: Rose (1826)
Faujasite: Damour (1842)
Mordenite: How (1864)
Offretite: Gonnard (1890)
Gonnardite: Lacroix (1896)
Wellsite: Pratt and Foote (1897)
Erionite: Eakle (1898)
Dachiardite: D'Achiardi (1906)
Stellerite: Morozewicz (1909)
Ferrierite: Graham (1918)
Clinoptilolite: Schaller (1932b)
Viséite: Mélon (1943)
Yugawaralite: Sakurai and Hayashi (1952)
Wairakite: Steiner (1955)
Bikitaite: Hurlbut (1957)
Garronite: Walker (1962a)
Paulingite: Kamb and Oke (1960)
Roggianite: Passaglia (1969)
Tetragonal natrolite: Krogh Andersen et al. (1969) named tetranatrolite by Chen and Chao (1980)
Mazzite: Galli et al. (1974)
Barrerite: Passaglia and Pongiluppi (1975)
Cowlesite: Wise and Tschernich (1975)
Svetlozarite: Maleev (1976)
Merlinoite: Passaglia et al. (1977)
Amicite: Alberti et al. (1979)
Partheite: Sarp et al. (1979)
Paranatrolite: Chao (1980)
Goosecreekite: Dunn et al. (1980)
Gobbinsite: Nawaz and Malone (1982)
Willhendersonite: Peacor et al. (1984)

phillipsite crystals of the same size in deep sea sediments; these first findings of sedimentary zeolites were possible because of the exceptionally large size of the crystals, but normally sedimentary zeolites are microcrystalline ($\sim$ 10 µm), so that only when the X-ray powder technique came into common use was it possible to ascertain just how common sedimentary zeolites are too. Bramlette and Posnjak (1933) started this kind of research describing a pyroclastic sediment altered to clinoptilolite; from then on, evidence for the sedimentary genesis of at least a dozen zeolite species has been given.

0.2 The Topology of the Tetrahedral Frameworks of Natural Zeolites

0.2.0 Introduction

Following Wells (1954, 1977), the topology of a tetrahedral framework may be represented by its tetrahedral centres (nodes) and by oxygen bridges (linkages) only and, hence, is equivalent to the topology of a 4-connected 3-dimensional net, at least if, as normally, every oxygen is shared by two tetrahedra, i.e. if Zoltai's (1960) sharing coefficient is 2.00 (if the sharing coefficient is lower, one should consider a mixed 3- and 4-connected net).

The number of theoretical nets thus obtainable is infinite. However, in the frameworks found so far in silicates, the tetrahedra form quite a small number of complex structural units (chains, rings, cages) of low potential energy. A large number of existing frameworks may therefore be described as different ways of connecting the same complex structural units.

According to this principle, Meier (1968) defined a small number of complex structural units formed by a finite number of tetrahedra (Secondary Building Units = SBU) which may be assembled to form a complete framework; Meier excluded groups which are either (ideally) infinite or very large.

Careful consideration of all known zeolite structures, allows us to propose, as a basis for the framework classification, some complex structural units of tetrahedra, whether finite or infinite. They are (see Fig. 0.2 A):

The chain of fibrous zeolites
The singly connected 4-ring chain
The doubly connected 4-ring chain
The 6-ring, single or double
The hexagonal sheet with handles
The heulandite unit

These complex structural units mostly are simply connected to form the actual frameworks, but in some cases vertices, edges or also faces are shared with nearby units. As a consequence, the natural zeolites may be classified as follows:

1. Fibrous zeolites
2. Zeolites with singly connected 4-ring chains
3. Zeolites with doubly connected 4-ring chains

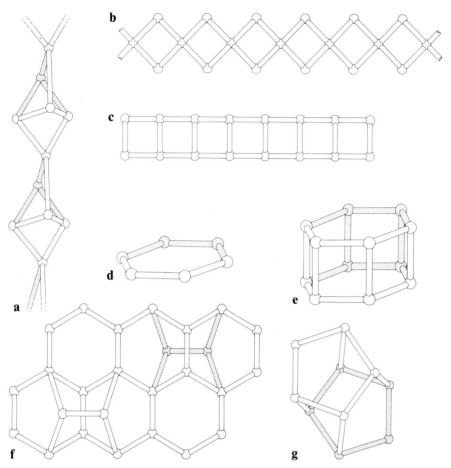

Fig. 0.2 A a – g. The structural units, finite or infinite, which may be used to assemble the frameworks of zeolites: **a** – The chain of fibrous zeolites; **b** – The singly connected 4-ring chain; **c** – The doubly connected 4-ring chain; **d** and **e** – The 6-ring, single and double; **f** – The hexagonal sheet with handles; **g** – The 4-4-1-1 heulandite unit

4. Zeolites with 6-rings
5. Zeolites of the mordenite group
6. Zeolites of the heulandite group

Different zeolite species with the same framework may exist, as order-disorder phenomena in the framework or the type of extraframework cations may have enough influence on the structure to justify a separate classification (Gottardi 1978). The set of all zeolites with the same framework may be called a genus (Gottardi 1978). Meier and Olson (1978) gave a complete list for all frameworks known for both natural and synthetic silicates, each with a couple of stereodrawings, its topological symmetry and its code, as proposed by the

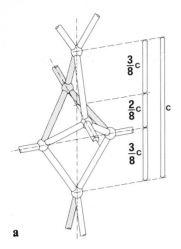

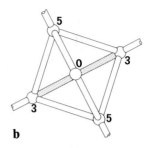

Fig. 0.2B a, b. The repeat-unit of the chain of fibrous zeolites. **a** – Clinographic view and distance along *c* between nodes; **b** – Orthogonal view along *c*, and height above the (001)plane as multiples of *c*/8

a

b

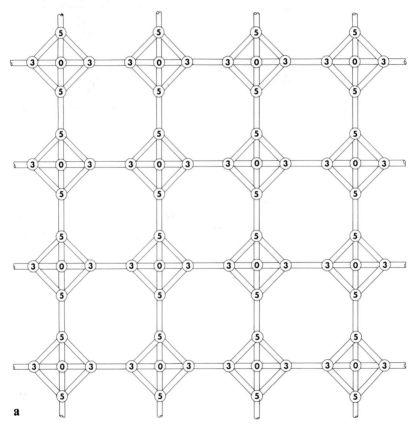

a

Fig. 0.2C a, b. The framework of the fibrous zeolites viewed along *c*; The chains are represented as in Fig. 0.2Bb. **a** – Edingtonite; **b** – Thomsonite

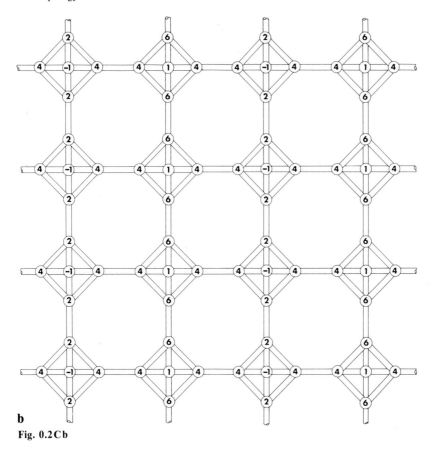

b

Fig. 0.2Cb

special I.U.P.A.C.-Commission. We recommend Meier and Olson's booklet to all those who are interested in the matter and like stereodrawings, which are not shown in this book. Our approach to the problem of understanding 3-dimensional objects like the frameworks will be used here: we feel that the best way to comprehend the matter is to build a 3-dimensional model using the easily available tetrahedral units and plastic tubing ("spaghetti" or similar material). So the following description could also be entitled "How to build models of zeolite frameworks using tetra-units and spaghetti" (please do not forget the authors are Italian).

Other recommended review articles on the tetrahedral frameworks are: Alberti and Gottardi (1975), Merlino (1976), Smith (1977, 1978, 1979, 1983), Alberti (1979b), Smith and Bennett (1981, 1984).

Detailed references on the structure determinations are given in the chapters devoted to the mineral species.

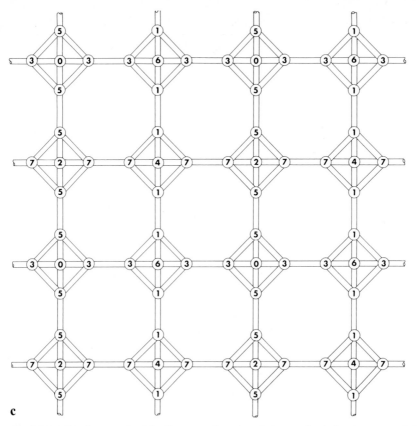

Fig. 0.2Cc. The framework of the fibrous zeolites viewed along *c*; the chains are represented as in Fig. 0.2Bb. **c** – Natrolite

0.2.1 The Frameworks of the Fibrous Zeolites

All the different ways of assembling the chain of fibrous zeolites have been described by Alberti and Gottardi (1975) and by Smith (1983). There are 4 side-linkages in each repeat-unit of the chain (Fig. 0.2B), two lower at $3/8c$ from the lowest node (at $0/8c$), and two higher at $5/8c$ from the lowest or at $3/8c$ from the highest node (at $8/8c$); let us take the lowest node of the repeat-unit as reference for the height in the chain. A second chain may be connected to the first so that both have the same height, or the second is $2/8c$ higher or lower. On the hypothesis that only one connection scheme exists throughout the structure, six frameworks are possible, but only three correspond to minerals, and are represented in Fig. 0.2C.

This is a homogeneous group, as a straightforward derivation of all structure from a single element is possible. Table 0.2I summarizes the essential data for this group.

Table 0.2I. Frameworks with the chains of the fibrous zeolites

IUPAC Code	Structure type	Topological symmetry	Type species	Symmetry of the type species
EDI	All chains at the same level	$P\bar{4}2_1m$	Edingtonite	$P\bar{4}2_1m$ or $P2_12_12_1$
THO	The chains are at two levels and form planes, in which the level is constant	Pmma	Thomsonite	Pcnn
NAT	The chains are at four levels around screw tetrads	$I4_1/amd$	Natrolite	Fdd2

Note: reviews of real and possible frameworks of this group are in Alberti and Gottardi (1975) and Smith (1983)

0.2.2 The Frameworks with Singly Connected 4-Ring Chains

This group contains some frameworks which are not strictly related to each other, namely those of analcime, laumontite, yugawaralite and roggianite.

A description of the framework of analcime, which is also present in leucite and wairakite, has been given by Galli et al. (1978). The chain (Fig. 0.2A b) is wrapped around a square prism of infinite length (Fig. 0.2D), thus forming an element with a screw-tetrad, right (4_1) or left (4_3), which may be called a right or left prism. These prisms are set side by side as shown in Fig. 0.2E, with a regular alternation of left and right. Two adjacent prisms are connected throught a glide plane, as shown in Fig. 0.2F. At this point the framework is completely connected. Figure 0.2G a shows a cage in the form of a square prism, the height of which is twice the side of the square base: this cage may be easily identified in Fig. 0.2F. The topological symmetry of this framework is cubic, hence there are equivalent left and right screw tetrads and

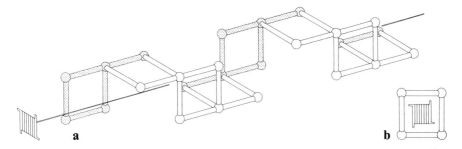

Fig. 0.2D a, b. The singly connected 4-ring chain is shown as rolled around a square prism, following a right-handed screw axis; in what follows, this prism with right-rolled chain is called a right prism. **a** – Clinographic view of the right prism; **b** – The same shown in orthogonal projection

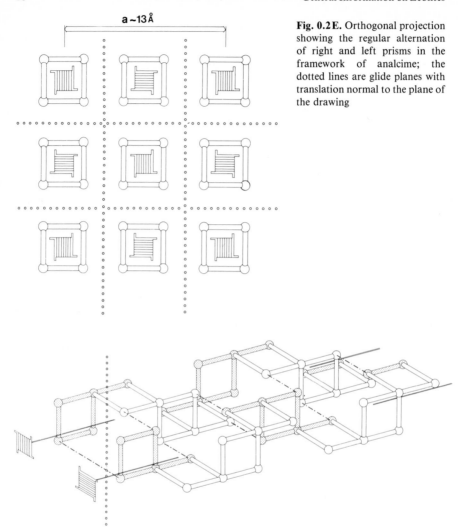

Fig. 0.2E. Orthogonal projection showing the regular alternation of right and left prisms in the framework of analcime; the dotted lines are glide planes with translation normal to the plane of the drawing

Fig. 0.2F. The connection (*dashed-dotted lines*) between two nearby prisms, a *right* and a *left* one; the *dotted line* is a reminder that a glide plane exists between the two prisms

left and right prisms in all three orthogonal directions, and each node is part of three orthogonal prisms and three orthogonal cages (Fig. 0.2Gb).

    The framework of laumontite may be built by assembling a 12-nodes unit (Fig. 0.2Ha and b) to form at first the sheets of Fig. 0.2Hc and d, in which the singly connected 4-ring chains may be distinguished. These sheets are further interconnected to form a framework as shown in Fig. 0.2 He.

    The framework of yugawaralite can be built by joining through a mirror plane the sheets of Fig. 0.2I. The chains run normal to the sheets and hence

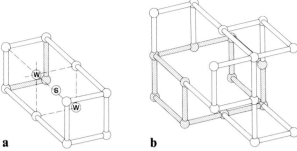

a   **b**

Fig. 0.2 G. a − The cage typical of the analcime framework; this cage may be easily seen in the connection of two prisms shown in Fig. 0.2F; the sites S (centres of the cages, multiplicity = 24) and W (midpoints of the long edges, which are void of nodes, multiplicity = 16) are possible positions for extraframework atoms. **b** − The relative positions of three different cages, which are parallel to the three crystallographic axes. Each nodal point pertains to two different cages, being at the same time in the middle of the long edge of one cage and in the base of another cage

Table 0.2 II. Frameworks with singly connected 4-ring chains

IUPAC Code	Structure type	Topological symmetry	Type species	Symmetry of the type species
ANA	The chains are wrapped around square prisms and interconnected so as to form typical cages	Ia3d	Analcime	Ia3d or $I4_1/acd$ or Ibca
LAU	The assembling of the framework using the chains is possible, but it is easier to do it with a special 12-nodes unit	C2/m	Laumontite	C2/m
YUG	The assembling of the framework is easy with a special 16-nodes unit, it is possible with the 4-ring chains only if some nodes are in common between two chains	C2/m	Yugawaralite	Pc
No code	Four chains around a tetrad form a column; nodes in the outer part of a column form linkages to other columns, but remain with a free vertex	I4/mcm	Roggianite	I4/mcm

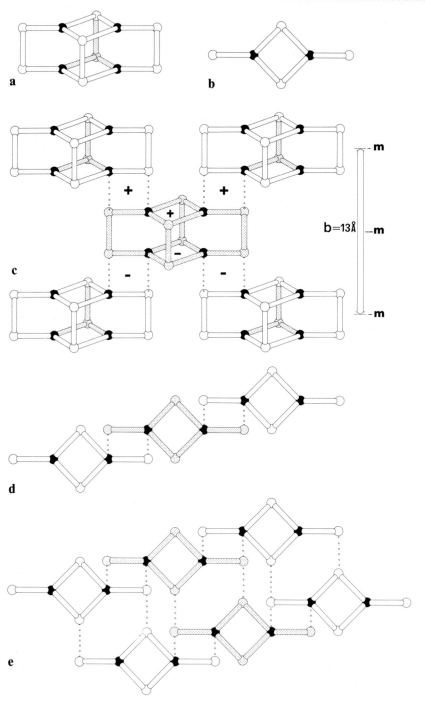

a

b

c

b=13Å

d

e

Fig. 0.2H a – e

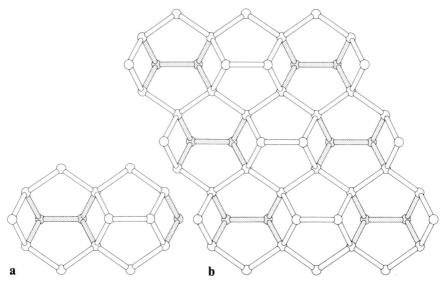

Fig. 0.2Ia, b. The framework of yugawaralite: **a** − The 16-nodes unit that generates the frame-work; **b** − The sheet formed by these units; some nodes are already 4-connected, other nodes have one free linkage; the sheets are linked to each other with mirror planes between them, using the free linkages. The singly connected 4-ring chains run normally to the drawing, but half of the nodal points pertain to two chains

normal to the mirror planes, but half the nodal points pertain to two chains: in other words it is impossible to build the yugawaralite framework by assembling a certain number of separate 4-ring chains, as, by contrast, is possible for analcime and laumontite.

Also the framework of roggianite may easily be built directly from separate chains, as shown in Fig. 0.2J. This framework is exceptional because it is the only one with some nodes which are only 3-connected. The sharing coefficient (Zoltai 1960) is 1.917 in this framework.

Table 0.2II summarizes the data of this group.

Fig. 0.2Ha − e. The framework of laumontite: **a** − Clinographic view of the 12-nodes unit; **b** − Orthogonal view along c of the same unit; white circles have a linkage normal to this drawing, black circles have no such linkage; **c** − Clinographic view of the sheet formed by interconnection of the above units; the sequence of squares marked + (or −) gives the singly connected 4-ring chain; the unit with gray rods only has its centre at $z/c = 1/2$, other units have their centres at $z/c = 0$ or 1; the *dotted line* marks the linkages between units; **d** − Orthogonal view of the same sheet; **e** − The framework built by interconnecting the sheets shown in **c** and **d**

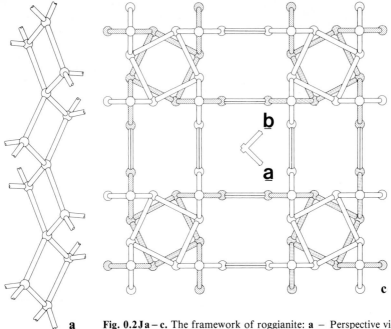

Fig. 0.2J a – c. The framework of roggianite: **a** – Perspective view of the singly connected 4-ring chain; **b** – The same chain viewed along the chain axis; **c** – The framework projected along c; half of the nodes are 4-connected, the others remain 3-connected: rods with a central line link just these 3-connected nodes; eight chains, out of the sixteen in the drawing, are shifted by $c/2$ with respect to the other eight chains, so that *white squares* are at $c = 0$ and the *gray ones* are at $c = 1/2$

0.2.3 The Frameworks with Doubly Connected 4-Ring Chains

The most common arrangement of the doubly connected 4-ring chain, usually known as the double crankshaft, is shown in Fig. 0.2K. Each node of this chain has one free linkage, which may be used to join these chains to each other to form frameworks. Smith (1978) has enumerated all 17 possible frameworks which may be obtained by interconnecting the double crankshaft; Sato (1979) gave a proof for the completeness of Smith's enumeration; Sato and Gottardi (1982) emphasized some relationships between the possible frameworks of this type. The three frameworks which correspond to the known zeolites merlinoite, gismondine, phillipsite, are shown in Fig. 0.2L. Double 8-rings (shaded in Fig. 0.2L a) are present in the framework of merlinoite. Another "double crankshaft" (shaded in Fig. 0.2L b) is formed in the framework of gismondine, so that in the plane normal to the arrows, there are two "crankshafts", one parallel and one normal to the plane of the drawing; in the direction of the arrows, a screw tetrad 4_1 is created. A wavy ribbon (shaded in Fig. 0.2L c), which is another kind of doubly connected 4-ring chain, is present in the framework of phillipsite. Another framework of

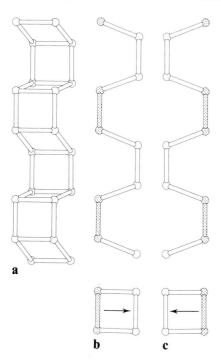

Fig. 0.2 K a – c. The doubly connected 4-ring chain in the arrangement known as "double crankshaft": **a** – Perspective view; **b** – Side view (*top*) and (*bottom*) a scheme for its view along the chain axis; **c** – (*top*) The same as in b but rotated by 180°, to favour a better understanding of its scheme (*bottom*)

Table 0.2 III. Frameworks with doubly connected 4-ring chains

IUPAC Code	Structure type	Topological symmetry	Type species	Symmetry of the type species
MER	4 double crankshaft chains around a tetrad	I4/mmm	Merlinoite	Immm
GIS	4 double crankshaft chains around a screw-tetrad	$I4_1/amd$	Gismondine	$P2_1/c$
PHI	Double crankshaft chains are set on planes, two nearby chains being alternately rotated by 180°; there is a mirror between successive planes	Cmcm	Phillipsite	$P2_1/m$
MAZ	Columns of gmelinite cages are linked around a hexad, a glide-plane being present between two columns	$P6_3/mmc$	Mazzite	$P6_3/mmc$
PAU	Double 8-rings and double 24-rings are joined in a complex way	Im3m	Paulingite	Im3m

Note: more data on real and possible frameworks of this kind are in Smith (1978)

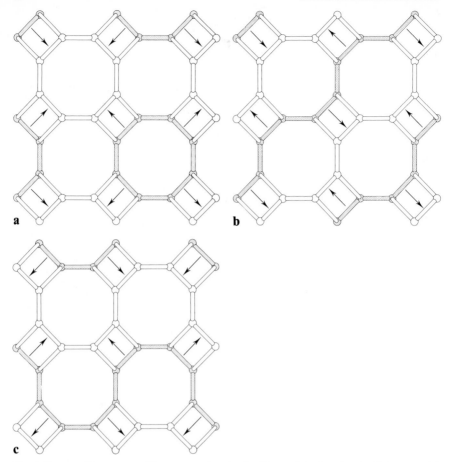

Fig. 0.2 L a – c. Three ways of interconnecting the double crankshaft, using the linkages which remain free in the isolated chain: **a** – The framework of merlinoite; **b** – The framework of gismondine, amicite, garronite; **c** – The framework of phillipsite, harmotome

this kind corresponds to minerals (paracelsian, slawsonite, danburite) which are not zeolites.

The framework of mazzite can be assembled by using the doubly connected 4-ring chain shaped in another way (Fig. 0.2 M a). If three such chains are connected around a ternary axis a complex unit is built (Fig. 0.2 M b and c), which can be considered as the piling up of gmelinite cages (see Chap. 0.3) to form columns. These columns are in turn connected through glide planes to form the whole framework (Fig. 0.2 M d).

Paulingite has a very complex cubic framework, which can be included in this group, because it may be considered as the assemblage of two kinds of closed 4-ring chains, one with 8-repeat (or double 8-ring) (Fig. 0.2 N a) and another with 24-repeat (or double 24-ring) (Fig. 0.2 N b); none of the double 8-rings

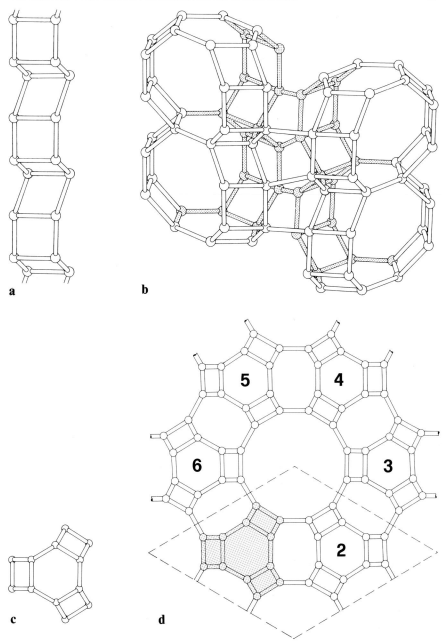

Fig. 0.2 M a – d. The framework of mazzite: **a** – The shape of the doubly connected 4-ring chain in mazzite; **b** – Three chains are connected around a triad (c-axis) to form a column of gmelinite cages (see Fig. 0.3 A h); two such columns are actually shown to explain how they are linked together, a glide plane being in between. **c** – One column viewed along c; **d** – The whole framework viewed along c

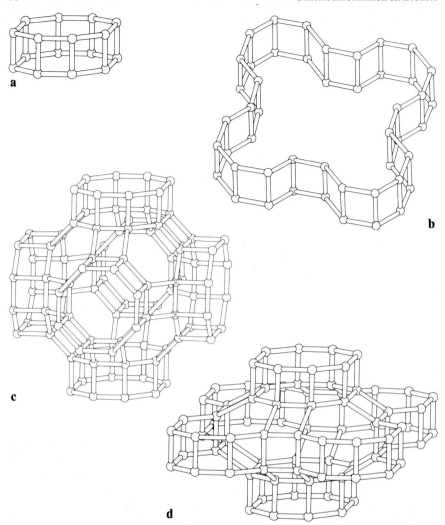

Fig. 0.2 N a – e. The framework of paulingite: **a** – The double 8-ring; **b** – The double 24-ring; **c** – The truncated cubo-octahedron (or α-cage, see Fig. 0.3 A l) connected with six double 8-rings; **d** – Two double 8-rings connected with a double 24-ring; **e** – The connection of the units shown in **c** and in **d**

share any node with other rings of the same kind, but every double 24-ring shares eight squares with similar rings. Figure 0.2 N shows how to assemble the paulingite framework from these two kinds of double rings.

Table 0.2 III summarizes the main features of these frameworks.

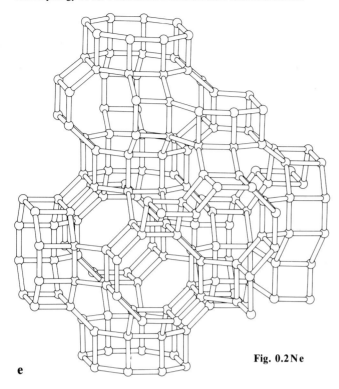

Fig. 0.2 N e

e

0.2.4 Frameworks with 6-Rings

Tetrahedra may be linked to form 6-rings, single or double. Let us consider first single rings only, namely rings which are not directly superposed; then three possible sets of positions are possible (Fig. 0.2 O), so that each structure of this kind can be represented as a succession of the three letters A, B, C in a given sequence. As double rings are possible, but triple not, each letter may be repeated twice successively (Fig. 0.2 Q) obeying Merlino's (1984) rule: besides sequences of only single or double rings, mixed sequences of single and double rings regularly alternating are also possible. The number of possible frameworks is infinite: Table 0.2 IV lists those corresponding to zeolites. All these frameworks contain characteristic cages, which are described in Chap. 0.3.

The 6-rings are lying in one and the same plane in all frameworks of this group described so far. One different zeolite framework can be assembled putting double 6-rings in four planes: that of faujasite. To describe it, let us first consider a cage usually known as a truncated octahedron, a sodalite cage or β-cage; a crystallographer would call it a combined form of a cube and an octahedron with such a mutual development that the eight octahedron faces are regular hexagons (see Fig. 0.3 A f and g). In this cage, every node has a free linkage and is part of two out of the eight regular hexagons: in other words,

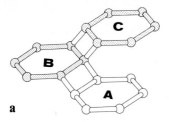

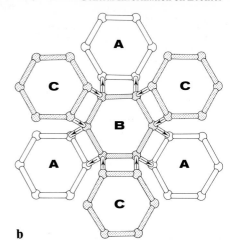

Fig. 0.2O. a – The way of connecting three parallel non-superposed 6-rings A – B – C; **b** – The same three 6-rings projected along the axis normal to the rings

b

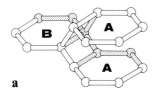

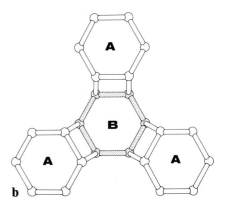

Fig. 0.2P. a – The way of connecting two parallel non-superposed 6-rings A – B; **b** – Projection of the same two 6-rings along the axis normal to the rings

b

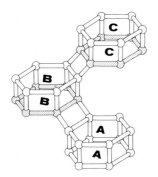

Fig. 0.2Q. The succession of rings AA – BB – CC with double 6-rings

Table 0.2 IV. Frameworks with 6-rings

IUPAC Code	Structure type	Topological symmetry	Type species	Symmetry of the type species
GME	Sequence AABB... of parallel 6-rings	$P6_3/mmc$	Gmelinite	$P6_3/mmc$
CHA	Sequence AABBCC... of parallel 6-rings	$R\bar{3}m$	Chabazite	$R\bar{3}m$
LEV	Sequence AABCCABBC... of parallel 6-rings	$R\bar{3}m$	Levyne	$R\bar{3}m$
ERI	Sequence AABAAC... of parallel 6-rings	$P6_3/mmc$	Erionite	$P6_3/mmc$
OFF	Sequence AAB... of parallel 6-rings	$P\bar{6}m2$	Offretite	$P\bar{6}m2$
FAU	Double 6-rings are linked so as to form sodalite cages in a diamond arrangement	Fd3m	Faujasite	Fd3m

Note: data on real and possible frameworks of this kind are in Smith and Bennett (1981)

the cage can be considered the union of four 6-rings, which correspond to the four faces of a tetrahedron. Now, it is possible to join two such cages by connecting the hexagons of two nearby cages and thus forming double 6-rings; the axes of the four double 6-rings are directed tetrahedrally; if the assembling scheme is that of diamond, the resulting framework is faujasite (Fig. 0.2 R). Other ways of connecting the sodalite cages have been described by Smith and Bennett (1981).

0.2.5 The Frameworks of the Mordenite Group

To build these frameworks, one must start with a sheet with a hexagonal pattern (Fig. 0.2 S): in this sheet each node has a free bond, which may be directed up- or downwards, this possibility giving rise to different kinds of sheets. A well known example is the mica sheet, in which all free bonds have the same direction, or the tridymite or cristobalite sheet, in which the free bonds are directed alternately up and down; the latter situation exists also in zeolites (Fig. 0.2 S), together with another two, the dachiardite and the mordenite sheets.

The second element which is present in these frameworks, is the "handle" or a pair of nodes, which has to be attached to four nodes (Fig. 0.2 T) on one or other face of the sheet; the attachment of the "handles" deforms the sheet

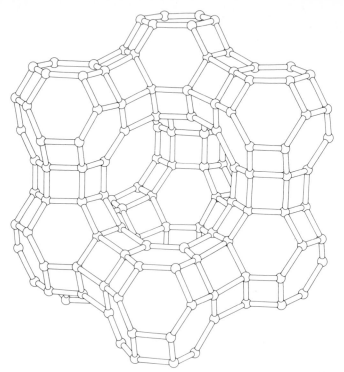

Fig. 0.2 R. The framework of faujasite: the sodalite cages (or truncated octahedron or β-cage, see Fig. 0.3 A f, g) are connected through 6-rings to form a cubic structure

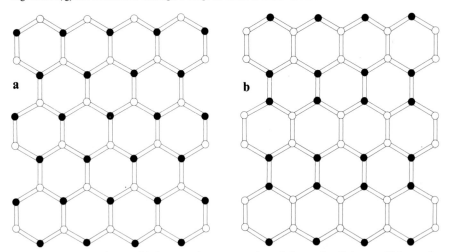

Fig. 0.2 S a – c. The sheets with hexagonal pattern; each node has a free linkage, which may be directed upwards (*black dots*) or downwards (*white circles*). **a** – The tridymite (or bikitaite) sheet, with a regular alternation of up and down; **b** – The dachiardite (also epistilbite and ferrierite) sheet, with a regular alternation up and down of pairs of linkages. **c** – The mordenite sheet, with a less simple rule of up and down

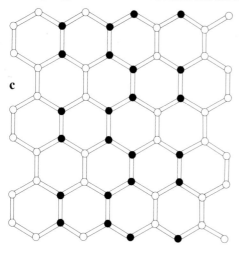

c

Fig. 0.2Sc

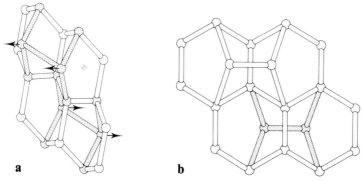

a   b

Fig. 0.2T a, b. A "handle", or pair of nodes, may be attached on the front side or on the rear side of a dachiardite or mordenite sheet. a – Perspective view; b – Orthogonal view

which becomes corrugated; two bonds of the handle remain free for further linkages.

The mordenite sheet is decorated with handles and the decorated sheets are connected through mirror planes to form the complete framework, as shown in Fig. 0.2U.

The dachiardite sheet may be decorated with handles in two ways; the sheets can be connected through binary axes to form the complete dachiardite framework (Fig. 0.2V) or connected through mirror planes to form the framework of epistilbite (Fig. 0.2W).

The dachiardite sheet may also be decorated with double handles, as shown in Fig. 0.2X; this attachment causes a more pronounced corrugation than a single handle does. These decorated sheets are connected through mirror planes (Fig. 0.2Y) to form the ferrierite framework.

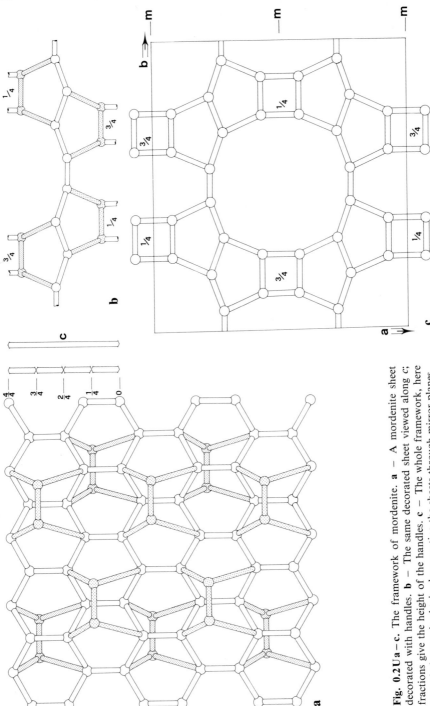

Fig. 0.2U a – c. The framework of mordenite. **a** – A mordenite sheet decorated with handles. **b** – The same decorated sheet viewed along **c**; fractions give the height of the handles. **c** – The whole framework, here viewed along **c**, is obtained connecting the sheets through mirror planes

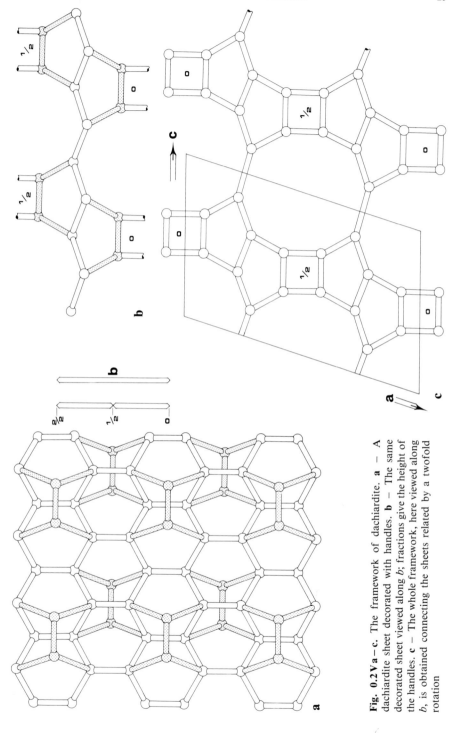

Fig. 0.2V a – c. The framework of dachiardite. **a** – A dachiardite sheet decorated with handles. **b** – The same decorated sheet viewed along *b*; fractions give the height of the handles. **c** – The whole framework, here viewed along *b*, is obtained connecting the sheets related by a twofold rotation

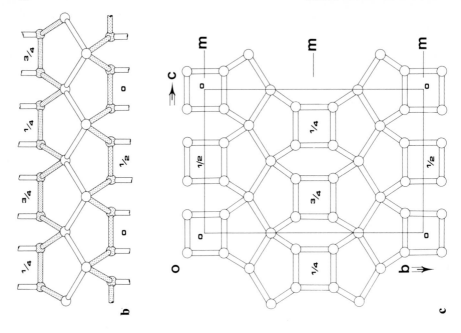

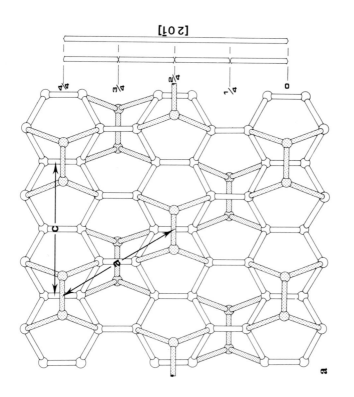

Fig. 0.2Wa–c. The framework of epistilbite. **a** – The dachiardite sheet decorated with handles in the epistilbite style. **b** – The same viewed along [20$\bar{1}$]; fractions give the height of the handles. **c** – The whole framework, here viewed along [20$\bar{1}$], is obtained connecting the sheets through mirror planes

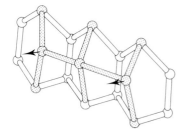

Fig. 0.2 X. A double handle attached to a dachiardite sheet

The tridymite sheet is present in the bikitaite framework, which does not contain handles but a pyroxene chain (also called "Zweierkette"). These chains are attached on both sides of the sheet (Fig. 0.2 Z) and the decorated sheets are interconnected; there is no symmetry element between the sheets but a c-translation superposes one sheet on the nearby one.

Several other hypothetical frameworks of this group have been described by Sherman and Bennett (1973), by Merlino (1975, 1976), and by Smith and Bennett (1984).

Table 0.2 V. Frameworks with decorated hexagonal sheets

IUPAC Code	Structure type	Topological symmetry	Type species	Symmetry of the type species
MOR	Hexagonal sheets of the mordenite type are decorated with handles which connect the sheets	Cmcm	Mordenite	Cmcm
DAC	Hexagonal sheets of the dachiardite type are decorated with handles which connect the sheets	C2/m	Dachiardite	C2/m
EPI	Hexagonal sheets of the dachiardite type are decorated with handles which connect the sheets	C2/m	Epistilbite	C2/m
FER	Hexagonal sheets of the dachiardite type are decorated with double handles which connect the sheets	Immm	Ferrierite	Immm
BIK	Hexagonal sheets of the tridymite type are connected by pyroxene-type chains	Cmcm	Bikitaite	$P2_1$

Note: data on real and possible frameworks of this type are in Merlino (1976) and in Smith and Bennett (1984)

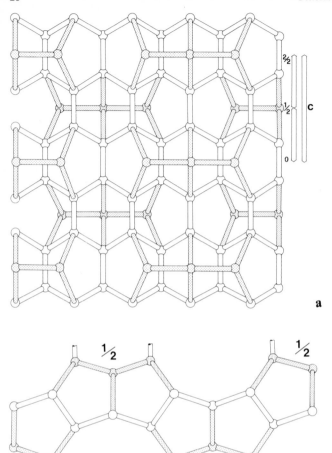

Fig. 0.2 Y a – c. The framework of ferrierite. **a** – A dachiardite sheet decorated with double handles. **b** – The same sheet viewed along c; fractions give the height of the double handles. **c** – The whole framework, viewed along c, is obtained by connecting sheets through mirror planes

0.2.6 The Frameworks of the Heulandite Group

The frameworks of this group feature the common unit shown in Fig. 0.2 AA which is built with 10 nodes and is also known as a 4-4-1-1 unit. In zeolites, these units are connected so as to share one or two nodes. Figures 0.2 AB, 0.2 AC, 0.2 AD show how these units are grouped to form the sheets in heulandite (and clinoptilolite), in stellerite (and stilbite, barrerite) and in brewsterite; these sheets in turn form frameworks by putting mirror planes between them and using the few nodes which are only 3-connected in the isolated sheets, for linkages through the mirror planes.

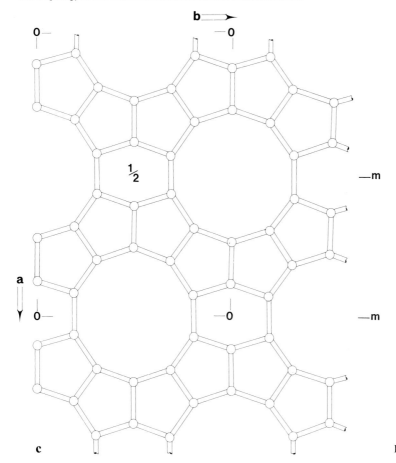

Fig. 0.2 Yc

Table 0.2 VI. Frameworks with the 4-4-1-1 heulandite unit

IUPAC Code	Structure type	Topological symmetry	Type species	Symmetry of the type species
HEU	The units sharing vertices form chains which by translation form sheets, which are connected through mirror planes	C2/m	Heulandite	C2/m
STI	The units sharing vertices form chains which by reflection form sheets, connected through mirror planes	Fmmm	Stilbite	C2/m
BRE	The units sharing edges form chains, which by translation form sheets connected through mirror planes	$P2_1/m$	Brewsterite	$P2_1/m$

Note: data on real and possible frameworks of this type are in Alberti (1979b)

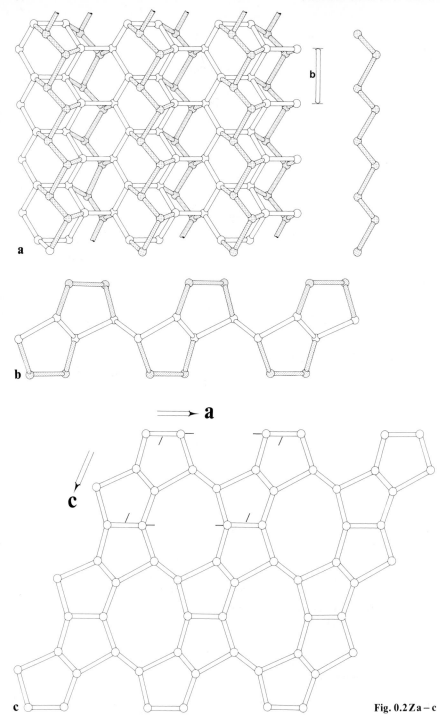

a

b

b

$\longrightarrow$ **a**

c

c

Fig. 0.2 Z a − c

Fig. 0.2 AA

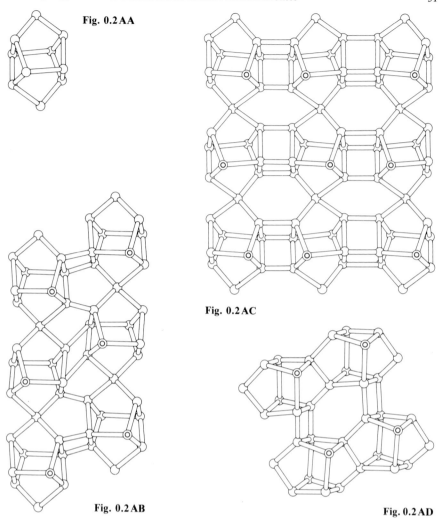

Fig. 0.2 AC

Fig. 0.2 AB

Fig. 0.2 AD

Fig. 0.2 AA. The 4-4-1-1 unit or heulandite unit

Fig. 0.2 AB. The framework of heulandite: the drawing actually shows a sheet in which some nodes have a free linkage, which is used for the connections to other sheets through mirror planes

Fig. 0.2 AC. The framework of stilbite: the drawing actually shows a sheet in which some nodes have a free linkage, which is used for the connections to other sheets through mirror planes

Fig. 0.2 AD. The framework of brewsterite: the drawing actually shows a sheet in which some nodes have a free linkage, which is used for the connections to other sheets through mirror planes

Fig. 0.2 Z. The framework of bikitaite. **a** – A tridymite sheet decorated with a pyroxene chain, shown as an isolated unit on the right. The chains decorating the sheet are shown as grey rods above or under the plane of the sheet. Both kinds of chains are shown a little shifted downwards to avoid superpositions. **b** – The same decorated sheet viewed along *b*. **c** – The whole framework, here viewed along *b*, is built up by connecting sheets related by a *c*-translation

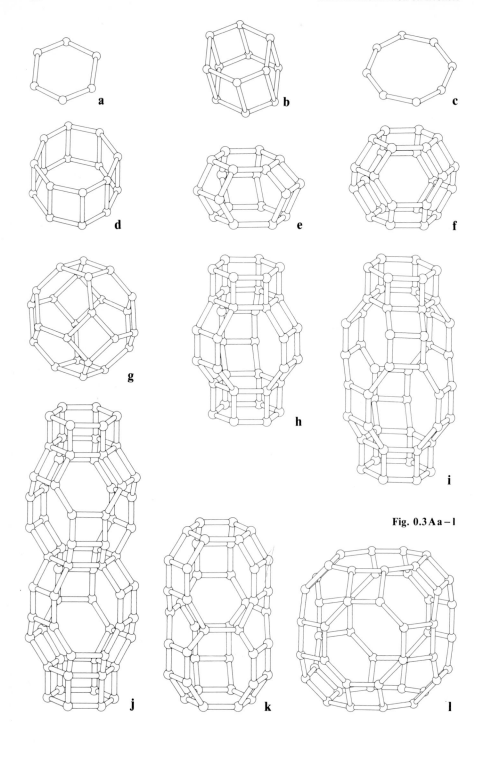

Fig. 0.3 A a – l

A thorough investigation on all possible frameworks built with the heulandite unit was done by Alberti (1979b), who found, with some constraints on the unit cell dimensions, 45 types: 4 correspond to known minerals (the three above-mentioned, plus scapolite, a feldspathoid) and 41 are hypothetical.

0.3 The Cages in Zeolite Frameworks

A part of a zeolite framework, which includes a cavity large enough to contain extraframework atoms, may be called a cage or a channel, depending of its finite or (ideally) infinite length. There is no clear-cut difference between the two, because a cage often may be considered as a part of a channel between two constrictions, the free section of which is smaller than the average free section of the channel.

The consideration of the following cages may be useful in describing the natural zeolite structures (Fig. 0.3 A):

a – the 6-ring (usually a ring is not considered as a cage, but it is included here for completeness)

b – the double 6-ring

c – the 8-ring

d – the double 8-ring (δ-cage of Meier 1968)

e – the cancrinite cage (ε-cage of Breck 1974)

f, g – the sodalite cage or truncated octahedron or cube-and-octahedron combination, the octahedral faces of which are regular hexagons (β-cage of Meier 1968)

h – the gmelinite cage (γ-cage of Meier 1968)

i – the chabazite cage

j – the levyne cage

k – the erionite cage

l – the α-cage or truncated cubo-octahedron or cube-octahedron-rhombic-dodecahedron combination in which the cubic faces are octagons and the octahedral faces are regular hexagons.

Fig. 0.3 A. **a** – The 6-ring. **b** – The double 6-ring. **c** – The 8-ring. **d** – The double 8-ring (δ-cage). **e** – The cancrinite cage (ε-cage). **f** – The sodalite cage (β-cage). **g** – Another view of the sodalite cage. **h** – The gmelinite cage (γ-cage). **i** – The chabazite cage. **j** – The levyne cage. **k** – The erionite cage. **l** – The truncated cubo-octahedron (α-cage)

0.4 Cation Sites in Zeolite Frameworks

Following Galli (1975a), the cation sites in zeolites may be classified as follows:

I   The cation is coordinated by framework oxygens only.
II   The cation is coordinated by framework oxygens at two nearly opposite sides, plus some water molecules.
III The cation is bound to framework oxygens on one side only, plus to some water molecules.
IV   The cation is completely surrounded by water molecules.

Some examples for the different coordinations follow:

I   This coordination has been found by Gard and Tait (1972) in offretite, where a K is in the centre of a cancrinite cage, and at bond distance from 6 out of the 12 oxygens which are in the two basal hexagonal rings of the cage; there is not enough room for water in this cage.
II   This is the typical situation in structures with narrow channels, with a cation occupying their centre. Besides fibrous zeolites, this coordination has been found for instance for Sr in brewsterite by Perrotta and Smith (1964).
III This is the most common situation: the cation is attached to the wall of the channel and is covered by water molecules on the other side.
IV   This coordination has been described for instance by Galli (1971) for Ca in stilbite and by Galli (1975b) for Mg in mazzite; this is the only coordination known so far for Mg, probably because this ion is so strongly solvated in solution, that it does not lose its water molecules during the formation of the zeolite crystal.

1 Fibrous Zeolites

1.1 Natrolite, Tetranatrolite, Paranatrolite, Mesolite, Scolecite

Natrolite	Tetranatrolite	Paranatrolite
$Na_{16}(Al_{16}Si_{24}O_{80}) \cdot 16H_2O$	$Na_{16}(Al_{16}Si_{24}O_{80}) \cdot 16H_2O$	$Na_{16}(Al_{16}Si_{24}O_{80}) \cdot 24H_2O$
Fdd2	I$\bar{4}$2d	pseudo-orthorhombic
$a = 18.29$	$a = 13.10$	$a = 19.07$
$b = 18.64$		$b = 19.13$
$c = 6.59\,(\text{Å})$	$c = 6.63\,(\text{Å})$	$c = 6.58\,(\text{Å})$

Mesolite	Scolecite
$Na_{16}Ca_{16}(Al_{48}Si_{72}O_{240}) \cdot 64H_2O$	$Ca_8(Al_{16}Si_{24}O_{80}) \cdot 24H_2O$
Fdd2	F1d1
$a = 18.41$	$a_0 = 18.51$
$b = 56.65$	$b_0 = 18.97$
$c = 6.55\,(\text{Å})$	$c_0 = 6.53\,(\text{Å})$
	$\beta_0 = 90°39'$

or Cc Z = 1/2

	first setting	second setting
	$a' = 6.53$	$a'' = 6.53$
	$b' = 18.97$	$b'' = 18.97$
	$c' = 9.78\,(\text{Å})$	$c'' = 9.85\,(\text{Å})$
	$\beta' = 108°53'$	$\beta'' = 110°00'$
	$\vec{a}' = \vec{c}_0 \quad \vec{b}' = \vec{b}_0$	$\vec{a}'' = \vec{c}_0 \quad \vec{b}'' = \vec{b}_0$
	$\vec{c}' = 1/2\,\vec{a}_0 + 1/2\,\vec{c}_0$	$\vec{c}'' = 1/2\,\vec{a}_0 - 1/2\,\vec{c}_0$

1.1.1 History and Nomenclature

The minerals had been known for a long time, mainly as "fibrous zeolite" or by similar names, when Klaproth (1803) proposed the name natrolite (the etymology is obvious) for a mineral from Högau (= Hegau), Germany. Of the many obsolete synonyms, "mesotype" and "galaktite" were more commonly used and may be found more frequently in old papers. Rinne (1890) introduced "metanatrolite" for dehydrated natrolite. Chen and Chao (1980) proposed the name tetranatrolite for a mineral from Mt. St. Hilaire, Quebec, which had been described eleven years earlier by Krogh Andersen et al. (1969) as "tetragonal natrolite" from Greenland. Chao (1980) introduced also the name paranatrolite, for a mineral from Mt. St. Hilaire, Quebec. Gehlen and Fuchs (1813) introduced, for the calcium-analogue of natrolite, the name "skolezit", from the Greek $\sigma\kappa\acute{\omega}\lambda\eta\xi$ = worm, because it curls up like a worm at the blow-pipe; the current spelling is scolecite in most languages; Fuchs and Gehlen (1816) proposed, for the intermediate member, the name mesolite from the Greek $\mu\acute{\varepsilon}\sigma ov$ = middle and $\lambda\iota\vartheta o\varsigma$ = stone.

1.1.2 Crystallography

The structures of natrolite, mesolite, and scolecite were described by Pauling (1930) but these were proven with experimental data only in 1933 by Taylor et al. for natrolite and scolecite. The structure has topological symmetry $I4_1/amd$ (see Chap. 0.2.1); the actual framework with symmetry $I\bar{4}2d$ is characterized by a rotation of the chains (see Fig. 1.1 A) by an angle ψ which is zero only in the ideal most symmetrical arrangement, but is $\sim 24°$ in natural natrolite, mesolite and scolecite, $\sim 35°$ in dehydrated natrolite (Alberti and Vezzalini 1983b), and $\sim 34°$ in $Rb_2(Ga_2Ge_3O_{10})$, a non-hydrophilic synthetic compound (Jarkow and Klaska 1981); the maximum theoretical ψ-value is $45°$. Tetranatrolite has very probably the deformed structure with $\psi \sim 24°$ and a disordered (Si, Al)-distribution in the tetrahedra (symmetry $I42d$, Fig. 1.1 A d), although no experimental proof thereof has been published yet, due to the difficulty of finding suitable single crystals. Natrolite is the first zeolite for which a definite (Si, Al)-order (see Fig. 1.1 A a) was guessed already in 1933 by Taylor et al. and confirmed later (Meier 1960; Torrie et al. 1964; Peacor 1973). With this order and a deformation with $\psi \sim 24°$, the symmetry coincides with the real symmetry of the crystal Fdd2. Na-cations and water molecules are in the channel parallel to c (see Fig. 1.1 B), as couples around screw-diads, hence at two different levels in the c-projection; the water molecules are near to the port between channels. Even the positions of the hydrogens are known, because Torrie et al. (1964) performed a natrolite refinement by neutron diffraction; the water molecule is linked by bent hydrogen bonds to two framework oxygens, making an $O-O(W)-O$ angle of $134°$; the hydrogens lie in the plane of the three mentioned oxygens, the $H-O-H$ angle being regularly $108°$.

In natrolite dehydrated at $350°$ (Alberti and Vezzalini 1983b), ψ increases to $35°$ (Fig. 1.1 C), and Na migrates to a site nearly in the port between channels, namely not far from the site occupied by water in hydrated natrolite; besides the strong chain rotation around c, the tetrahedra are also rotated around an axis normal to c, so allowing a shortening of c. Peacor (1973) studied the natrolite structure by X-rays at room temperature and also at $88°$, $146°$, and $198°C$, without refining the water occupancy factors, which were

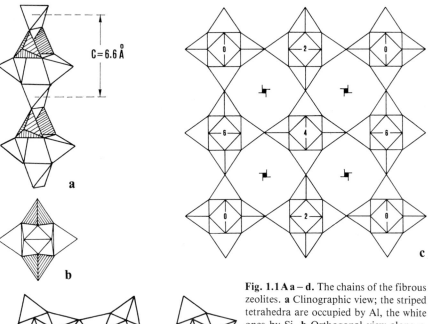

Fig. 1.1A a – d. The chains of the fibrous zeolites. **a** Clinographic view; the striped tetrahedra are occupied by Al, the white ones by Si. **b** Orthogonal view along c, the (Si, Al)-order being shown as in a. **c** The framework of natrolite projected along c in its most symmetric arrangements, with $\psi = 0°$; among the symmetry elements, 4_1 only are shown (this drawing must be compared with Fig. 0.2Cc); numbers give the height of the central tetrahedron as multiples of $c/8$. **d** The framework of natrolite projected along c with a rotation $\psi = 24°$ of the chains; 2_1 only are shown; this structure has been proposed for tetranatrolite, although without an experimental proof; the positions of Na and H_2O are not known, but are probably similar to those of natrolite

fixed at 100%; this value should be nearly true because residuals were low in all cases: there is no relevant change in the structure from room condition up to ~200°C.

Alberti and Vezzalini (1981a) calculated a "disorder index" of 30% from the refinement of a natrolite from Hungary, the corresponding value for the Bergen Hill sample (Peacor 1973) being 13%; this result means that partially disordered natrolites are more frequent than commonly believed. A similar result was obtained by Hesse (1983).

Nothing is known of the structure of paranatrolite, but it certainly shares the same framework with natrolite.

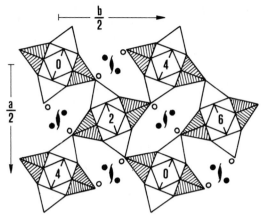

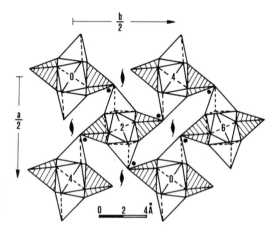

Fig. 1.1B. The structure of natrolite projected along c (the drawing is rotated by 45° in comparison with Fig. 1.1Ad, because so are the crystallographic axes); the symmetry is lowered to Fdd2 by the (Si, Al)-ordering; 2_1 only are shown; *black dots* = Na, *white circles* = H_2O

Fig. 1.1C. The structure of dehydrated natrolite projected along c, with $\psi = 35°$; 2_1 only are shown; *black dots* = Na

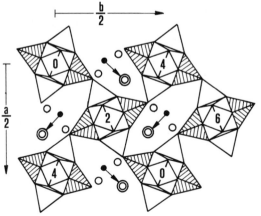

Fig. 1.1D. The structure of scolecite projected along c, symmetry F1d1; *blacks dots* = Ca, *white circles* = H_2O (1 and 2), *double circles* = H_2O(3). The *arrows* show the Ca → H_2O(3) vectors

The structure of scolecite was refined by Adiwidjaja (1972), by Fälth and Hansen (1979) and by Smith et al. (1984). The (Si, Al)-order is the same as in natrolite (Fig. 1.1 A a). The Na-atoms in natrolite are twice the number of Ca in scolecite, so the latter occupy the positions on one side only of the line in the centre of the channels which is a screw-diad in natrolite, but obviously not in scolecite; sites nearly corresponding to the missing Na-atoms are occupied by another water molecule $H_2O(3)$ (Fig. 1.1 D); in short, $Na_2(H_2O)_2$ are substituted by $Ca(H_2O)_3$, and this substitution conditions the symmetry reduction from Fdd2 to F1d1. Within each $Ca(H_2O)_3$ group, the vectors $Ca \rightarrow H_2O(3)$ may have only two orientations, the sum of the two being in the $+a$ direction. Alberti et al. (1982a) put forward the hypothesis that, in Na-rich scolecites, two different domains exist where the sum of the vectors $Ca \rightarrow H_2O(3)$ is in the $+a$ or $-a$ direction, sodium being present at the boundaries of the opposite domains.

Smith et al. (1984) also determined, using X-rays and neutrons, the sites of the hydrogens, which are all linked to framework oxygens, which are nearly in the same plane with the water molecules, as also occurs in natrolite.

Adiwidjaja (1972) also investigated the structure of scolecite heated at 300°C and cooled in dry air. These crystals are still monoclinic, but Fd11 instead of F1d1; they contain two H_2O instead of three and it is just that water molecule which is in the site similar to the Na-site in natrolite, which is absent.

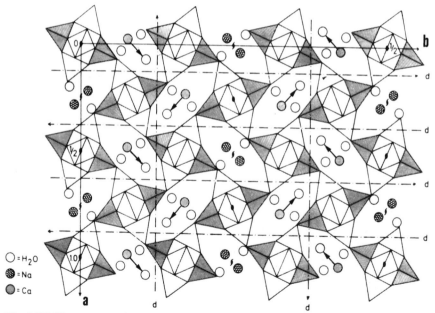

Fig. 1.1E. The structure of mesolite projected along c, symmetry Fdd2 with tripled b; 2 and 2_1 only are shown; $H_2O(3)$ corresponds to the *circle* which appears as the farthest from Ca. The *arrows* show the $Ca \rightarrow H_2O(3)$ vectors

Fig. 1.1F. Natrolite from Altavilla, Vicenza, Italy (×2)

Fig. 1.1G. Natrolite from Altavilla, Vicenza, Italy (×10)

Fig. 1.1H. Natrolite from Auvergne, France (×4)

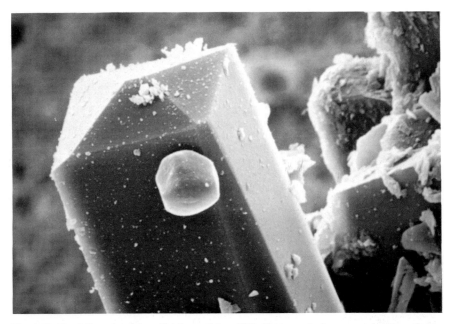

Fig. 1.1I. Partially ordered natrolite from Gulacs Hill, Hungary; its lower part is tetranatrolite (SEM Philips, ×950)

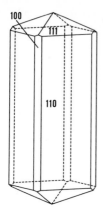

100
111
110

Fig. 1.1J. Common morphology of natrolite

The symmetry change is conditioned by the $Ca - H_2O$ distribution pattern in the different channels (in half channels the three sites are rotated by 180°).

The structure of mesolite was refined by Adiwidjaja (1972) down to $R = 7.2\%$ and by Pechar (1982) to $R = 6.9\%$; the (Si, Al)-order is the same as in natrolite. In the structure (Fig. 1.1 E), all c-channels contained in the same (010)-plane may be all of the natrolite type or all of the scolecite type; there is an alternation of one natrolite plane with $Na_2(H_2O)_2$ with two scolecite planes with $Ca(H_2O)_3$. The vectors $Ca \rightarrow H_2O(3)$ have two orientations in each pair of

Fig. 1.1K. Tetranatrolite from Ilimaussaq, Greenland (SEM Philips, ×5000)

Fig. 1.1M. Scolecite from Brazil (×8)

Fig. 1.1L. Mesolite from Iceland (×8)

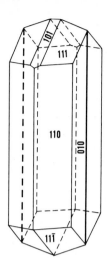

Fig. 1.1N. Common morphology of scolecite; the (100)-twin plane is marked with *continuous or dashed lines*

adjacent planes; if the sum of the $Ca \rightarrow H_2O(3)$ vectors of two adjacent planes is in the $+a$ direction, the corresponding sum is in the $-a$ direction for the subsequent pair of scolecite planes. This structure has b three times larger than natrolite, and the same symmetry Fdd2.

The morphology of natrolite is quite simple (Fig. 1.1 J), with the presence of the dominant forms {110} {111} and smaller {100} and {010}; less frequent are {101} {311} {131} {331} and others; frequent are also radiating fibrous fillings of cavities. The pseudo-tetragonality leads one to suppose that pseudo-merohedral twins may be frequent: they do indeed exist, but are not so easy to find. These twins were described already in 1885 by Stadtlaender, as having the normal to 110 as twin axis, so having a and b interchanged, and (001) as composition plane. Pabst (1971) described an intergrowth "akin to twinning" with a and b interchanged in the two individuals; the composition plane was not given, because the twin was detected by X-ray single crystal photographs; Pabst preferred the line bisecting a and b as twin axis, but the difference from the normal to {110} is half a degree only; he interpreted the twin as a single crystal of disordered tetranatrolite which, after ordering into natrolite, assumed a structure with a continuously linked framework and domains with two different orientations at 90° as far as the (Si, Al)-ordering is concerned; Loewenstein's rule would not hold at the twin boundaries in this model.

Tetranatrolite has the forms {110} and {111} only, and the same is probably true for paranatrolite.

The morphology of mesolite is similar to that of natrolite but simpler, only the forms {110} and {111} having been observed; more commonly the mineral occurs as divergent tufts of prismatic needles. Pseudo-merohedral twins should be possible and also frequent, but have never been described so far.

Scolecite is always (100) twinned, the individuals featuring the forms {110} {111} {$\bar{1}$11} {010}, and a smaller {101} (Fig. 1.1 N).

1.1.3 Chemistry and Synthesis

The crystal chemistry of this genus had been thoroughly investigated in the 1930's by Hey (1932b, 1933, 1936), and then later by Foster (1965), but this review is based on the more recent study by Alberti et al. (1982a). All the zeolites of this genus are always quite near to the stoichiometric formulae of the title page, so that it is sufficient to give one chemical analysis for each species in Table 1.1 I; in all samples, the ratio Si:(Si + Al) is very near to 0.6, and Mg, Sr, Ba are nearly always less than 0.01 and only in some cases 0.02 atoms per 80 oxygens; K is always in the range $0.01 - 0.03$; Fe is normally absent, but may raise to 0.24 atoms per unit cell in some ochreous natrolites; scolecites are purer than natrolites and mesolites. For all these zeolites the water content is near to the stoichiometric values.

In natrolite, the substitution of Na by Ca is less than 0.5 Ca per unit cell, higher values for Ca and K having being reported by Hey (1932b) and Foster (1965); Harada et al. (1967) and Harada and Nakao (1969) reported on samples with 0.8 Mg per unit cell.

In mesolite, the average ratio M:(M + D) (M = monovalent-, D = divalent-cations) is 0.491 with a narrow range $0.455 - 0.517$, smaller than the $0.399 - 0.536$ found by Hey (1933) and than $0.398 - 0.562$ found by Foster (1965); this ratio is 0.406 in a mesolite from Tezuka, Japan (Harada et al. 1968); and even 0.299 in a mineral from Morojose, Japan (Harada et al. 1969a). "High-sodium mesolite", a variety found at Rio Cambone, Sardinia, Italy, at Kladno, Bohemia (see Foster 1965) and at Waniguchi, Japan (Harada et al. 1968) is gonnardite: this has been proven by Alberti et al. (1982a) for the first two occurrences, and is probable for the third.

In scolecite all substitutions are at the lowest value, but Na can be as high as 1.4 per unit cell, as found also by Hey (1936).

The only two known analyses of tetranatrolite (one is in Table 1.1 I) are both within the range of natrolites.

Paranatrolite has again the same composition (Chao 1980), but a higher water content, namely 24 (exactly 23.84) molecules per unit cell.

The first natrolite synthesis, proven by X-ray data, cannot be attributed with certainty. As a matter of fact Koizumi and Kiriyama reported already in 1957 on the presence of some natrolite in the product of hydrothermal treatment at $180° - 250°C$ of the amorphous material obtained by ignition of natrolite. Full evidence for a true synthesis was given by Senderov and Khitarov (1966, 1971), who state: "Gels were the initial material employed for synthesis, but they were seeded to reduce the effect of the low rate of formation of natrolite nuclei": so it is not clear if the synthesis was also possible in the absence of seeds. The composition of the gels was $nNa_2O \cdot Al_2O_3 \cdot 3SiO_2$ with $1 \leqslant n \leqslant 2$, some 0.2N NaCl solution being added; the natrolite was obtained at 120° and 180°C, often with analcime. The synthesis was done by the same authors also starting from mixtures of nepheline and albite and 0.2N NaOH solution at $150° < T < 200°C$, apparently without seeding; 250°C

Table 1.1I. Chemical analyses and formulae on the basis of 80 oxygens and lattice constants in Å

	1	2	3	4
SiO_2	47.60	46.90	47.33	46.63
Al_2O_3	26.40	25.60	24.77	25.68
Fe_2O_3	0.51	0.11	0.01	
CaO	0.10	1.48	9.29	13.89
SrO	0.02		0.02	
BaO	0.09			
Na_2O	15.41	14.00	5.04	0.05
K_2O	0.01	1.12	0.04	
H_2O	9.85	9.59	13.50	13.75
Total	100.00	100.17	100.00	100.00
Si	24.17	24.19	24.69	24.29
Al	15.80	15.57	15.24	15.77
Fe	0.20	0.05		
Ca	0.06	0.82	5.20	7.75
Sr	0.01		0.01	
Ba	0.02			
Na	15.18	14.00	5.10	0.05
K	0.01	0.74	0.03	
H_2O	16.68	16.50	23.50	23.90
$E\%$	+4.3	-5.5	-1.8	+1.4
a	18.296	13.098	18.437	18.509
b	18.631		56.618	18.975
c	6.587	6.635	6.555	6.527
ρ				90°38'
D	2.26	2.28	2.28	2.27

1. Natrolite, Schaffhausen, Germany (N6 of Alberti et al. 1982a)
2. Tetranatrolite, Mt. St. Hilaire, Quebec (Chen and Chao 1980) with 0.06% TiO_2, 1.31% H_2O^-, the value here given as Fe_2O_3 is really FeO
3. Mesolite, Cap d'Or, Nova Scotia (M9 of Alberti et al. (1982a); 1/3 of the unit cell content
4. Scolecite, Poona, India (S4 of Alberti et al. 1982a)

Note: density for numbers 1, 3 and 4 is calculated, for n.2 is measured.

seems to be the highest temperature for natrolite existence in alkaline media. No evidence is given for these synthetic natrolites to be orthorhombic or tetragonal, viz. ordered or disordered.

Höller and Wirsching (1980) reported on the hydrothermal treatment of nepheline with 0.1 N NaOH at 150 °C in an "open system", viz. with frequent change of the reacting solution; analcime was obtained at first, but later on natrolite too appeared. Also in this case, no evidence is available for deciding if this natrolite is disordered (tetragonal) or ordered (orthorhombic). The Ga-analogue of natrolite was synthetized by Ponomareva et al. (1974).

Scolecite was synthetized for the first time by Koizumi and Roy (1960), starting from gel mixtures $CaO \cdot Al_2O_3 \cdot 3 SiO_2$, with 230° < T < 285 °C in the presence of seeds of natural scolecite: the powder pattern of this synthetic scolecite is not detailed enough to decide if the pseudotetragonal phase is really monoclinic or not.

Wirsching (1981) obtained scolecite studying the reaction of basaltic glass with 0.1 and 0.01 N $CaCl_2$ solutions at 200 °C in a simulated "open system", viz. with frequent renewal of the reacting solution; well developed crystals, nearly 5 μm in length were formed without seeds; no powder data are given for this synthetic scolecite.

We are not aware of any mesolite synthesis.

The most obvious comment on these syntheses is that, in spite of the frequent occurrence of natrolite, mesolite and scolecite in nature, these minerals are difficult to synthesize, especially without any natural seed, namely they are very hard to nucleate, as is laumontite.

1.1.4 Optical and Other Physical Properties

The optical properties of natrolite, mesolite and scolecite are well known and show only reduced oscillations around the average values, as predictable from the narrow chemical ranges of these zeolites. The data here given are mainly from Hey (1932a, 1932b, 1936) with due consideration of many other authors. For natrolite (standard deviations in parentheses):

$\alpha = 1.479(1)$ $\beta = 1.481(1)$ $\gamma = 1.491(1)$
$2V = +61°(2°)$ $\alpha \to a$ $\beta \to b$ $\gamma \to c$

For mesolite: $\alpha = 1.505(1)$ $\beta = 1.506(1)$ $\gamma = 1.507(1)$
$2V = +84°(2°)$ $\alpha \to a$ $\beta \to c$ $\gamma \to b$

For scolecite: $\alpha = 1.512(1)$ $\beta = 1.519(1)$ $\gamma = 1.520(1)$
$2V = -35°(1°)$ $\gamma \to b$ $\hat{\alpha c} = 16°(1°)$ in the obtuse angle.

− For tetranatrolite (Chen and Chao 1980): $\omega = 1.480(1)$ $\varepsilon = 1.495(2)$
− For paranatrolite (Chao 1980): indices of refraction slightly higher than those of tetranatrolite, $2V = -10°$ oblique extinction with an angle of 21°.

Natrolite gives a white-yellow luminescence with UV-light (Laurent 1941); it is piezoelectric like mesolite (Ventriglia 1953). Scolecite too is piezoelectric (Bond 1943). IR-spectra of natrolite, tetranatrolite, mesolite and scolecite may be very useful for identification of these zeolites and are given in the Appendix.

1.1.5 Thermal and Other Physico-Chemical Properties

The thermal curves (heating rate $= 20°C/min$) for natrolite are in Fig. 1.1 O. There is only one water site in the unit cell, and only one sharp water loss at 330°C. Peacor (1973 and written communication) studied the structure of natrolite at 88°C, 146°C, 198°C, and found no relevant change in lattice parameters, structure, or water occupancy. Reeuwijk (1972, 1974) published not only similar thermal curves, but also Guinier-Lenné X-ray photographs which show, with a low heating rate (0.3°/min), a phase transition to α-metanatrolite, corresponding to the dehydration, at 285°C; there is a second inversion, to β-metanatrolite at 510°C. α-Metanatrolite has a, b and especially c shortened in comparison with the hydrated natrolite so that cell volume is 20% smaller; its structure, determined by Alberti and Vezzalini (1983 b) (see Chap. 1.1.2), is characterized by a twisting of the chains in two opposite ways in two kinds of domains, each of monoclinic symmetry F2, so as to simulate a global symmetry Fdd2. The single crystals of β-metanatrolite are too badly mosaicized to allow X-ray measurements and the lines of the powder pattern are too broad to allow a reliable indexing, so that nothing can be said with certainty on this phase, although it probably shares the same framework with the original phase; β-metanatrolite becomes amorphous at 775°C. Rehydration is always possible so long as crystallinity is maintained.

The thermal curves of scolecite (Fig. 1.1 P) show two main water losses, the first at 240°C, the second at 420°C. The first broadly corresponds to the loss of $H_2O(3)$, namely the water molecules which occupy one of the two Na-sites of natrolite (see Chap. 1.1.2 and Adiwidjaja 1972); this partially dehydrated scolecite has the same monoclinic symmetry as the original phase and a slightly smaller unit cell (Reeuwijk 1974), but the unique axis is different and the space group is Fd11 instead of F1d1. The second water loss at 420°C broadly corresponds to the remaining two water molecules, which are symmetrically equivalent in natrolite but not in scolecite; over 420°C, the material is amorphous.

The thermal curves of mesolite are more complicated and different for different samples (see Fig. 1.1 Q); there are three and perhaps even four main water losses, which should correspond to the four sites occupied by water in this structure, but no structural study is available to clarify which site is emptied first, although probably the third H_2O near the calcium is lost first, then the H_2O near the sodium, and at the highest temperature the remaining two H_2O near the calcium. The unit cell gets smaller at 200°C and the material

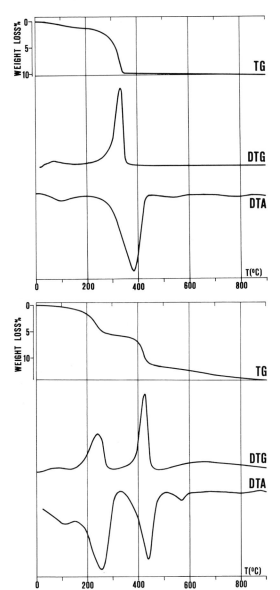

Fig. 1.1O. Thermal curves of natrolite from Tierno, Trento, Italy (N2 of Alberti et al. 1982a); in air, heating rate 20°C/min

Fig. 1.1P. Thermal curves of scolecite from Poona, India (S4 of Alberti et al. 1982a); in air, heating rate 20°C/min

is amorphous at 320°C; these temperature limits are referred to a heating rate as small as 0.3°/min (Reeuwijk 1974).

Data on ion exchange are scanty; in particular, no exchange isotherm has been published. The best data available are still those of Hey (1932b), who gave a full review of the preceeding literature. Hey confirmed that natrolite does not exchange its ions with a Ca-solution, even after 700 h at 100°C, but

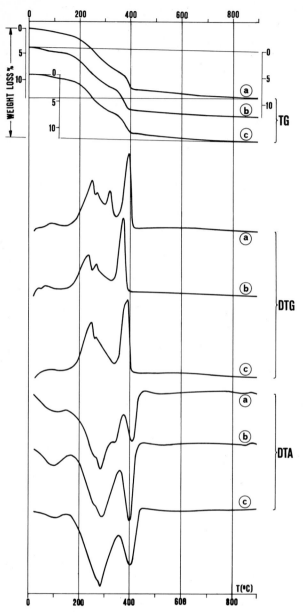

Fig. 1.1Q. Thermal curves of mesolite from (a) Prospect Park, New Jersey (M6 of Alberti at al. 1982a), (b) Cap d'Or, Nova Scotia, Canada (M9 of Alberti et al. 1982a), (c) Harbabkhandi, Iran (M14 of Alberti et al. 1982a); in air, heating rate 20°C/min

he quotes Lemberg who claimed to have obtained scolecite from natrolite via a potassium compound. A minor exchange was obtained with K- and Ag-solutions at room temperature; the majority of sodium was exchanged at $200° - 250°C$ with fused salts of Li, K, Ag, Tl, and NH_4, this last form being anhydrous. Belitskiy and Gabuda (1968) obtained similar results with fused salts of Li, K, Rb, Cs, Ag, Tl; they prepared also the ammonium form in a gaseous medium with an old recipe of Clarke and Steiger (1900), and were able to deduce from NMR measurements on all forms that water diffusion in natrolite involves channels both parallel and perpendicular to the *c*-axis.

Belitskiy (1972) reported on the thermal curves of the same forms and found a relation between the temperature of dehydration and the radius of the replacing cation. Belitskii and Fedorov (1976) investigated the Na − K exchange with fused salts, and found a relevant structural difference which hinders the formation of a continuous isomorphous series: this fact probably explains the impossibility of K − Na-exchange in water solutions from 0° to 100°C.

Hey (1933) obtained similar results with mesolite: nearly no exchange with Na- or Ca-solutions, but high exchange with fused salts (180°C − 270°C) of Na, K, Li, NH_4, Ag, Tl with complete removal of Na and partial exchange of Ca.

Hey (1936) applied the same fusion technique to obtain different forms of scolecite, but nearly no exchange was obtained with NH_4, and a partial substitution with Na, K, Ag, Tl; the Na-form is quite similar to a calcian natrolite, not to mesolite.

Dehydrated natrolite has a contracted structure with highly deformed channels (see Chapt. 1.1.2), so that gas adsorption is not possible in most cases; ammonia is adsorbed in significant quantity, probably with a restoration of the original shape of the channels; the same holds for scolecite (Barrer 1938).

1.1.6 Occurrences and Genesis

A few occurrences have been described where a sedimentary genesis has been attributed to natrolite. Hay (1966) found natrolite as an alteration product of nepheline from the Pleistocene sediments of Olduvai Gorge, Tanzania, which are in contact with sodium carbonate solutions under a hot semiarid climate; the hydrologic situation is similar to a closed system, as typified by the saline lake basins. The Green River formation, including some tuffaceous deposits, is widespread in Wyoming and Colorado; it contains many authigenic minerals, mainly analcime (see Iijima and Hay 1968 and quoted literature), crystallized in a saline environment: Pabst (1971) described fine natrolite crystals, up to 3mm in length, from a drill hole in this formation occurring at Garfield Co., Colorado. Another sedimentary occurrence is in Jurassic sediments of the Vizcaino peninsula, Mexico (Barnes et al. 1984; see later on, the sedimentary occurrence of mesolite).

Hay and Iijima (1968b) proposed "open system" genesis (slow percolation of meteoric water) for the zeolites of the palagonite tuffs of Oahu, Hawaii. At Diamond Head, Punchbowl, Manana Island, Ulupau Head and Salt Lake Tuff, natrolite occurs with thomsonite and gonnardite in the cement of these tuffs.

Deep sea samples may contain natrolite only as hydrothermal deposition in basalts (Bass et al. 1973).

Merritt (1958) found natrolite with analcime in the Permian conglomerate and sandstone of Wichita Mts., Oklahoma. Udluft (1928) and Jérémine (1934) described a fossilized wood in a nephelinite tuff from Mt. Elgon, Uganda, which was found to be mainly natrolite. Utada (1970) does not mention natrolite as a zeolite generated by burial diagenesis, but the zeolite is listed (Utada 1980) as occurring rarely in the analcime zone of the hydrothermally altered alkaline mineral assemblage of Cretaceous or Neogene formations of Japan.

On the whole, natrolite seems quite rare as a sedimentary zeolite but, as a balance, it is very common as a mineral of hydrothermal deposition. Kostov (1969, 1970) found a definite zoning also in the hydrothermal deposition of zeolites in Bulgaria, the fibrous ones, including natrolite, occurring in the highest temperature zone.

Only a brief account of the many occurrences in vugs of massive rocks (with hydrothermal genesis, probably) can be given here. Some classical occurrences, already listed by Hintze (1897) and Dana (1914), are the following:
- in the Tertiary basalts of Northern Ireland (see also Walker 1951, 1959, 1960a);
- in the basalts of Dunbartonshire, Glen Farg, Renfrewshire and other localities of Scotland;
- in the augite syenite of Langesundfjord, Southern Norway;
- in the phonolite from Hohentwiel in Hegau (see also Ramdohr 1978), and from Kaiserstuhl, both in Baden, and in the basalts of Sontra, Hessen, all Germany;
- in the basalts of Usti nad Laben (= Aussig) (see also Hibsch 1915), Zálezly (= Salesl) (see also Hibsch 1917) and Česka Lipa (= Böhmisch Leipa), all Bohemia;
- in the basalt from Puy de Marman, Puy-de-Dôme, France (see also Pongiluppi 1976);
- in the basic volcanic from Tierno, Trento, and Montecchio Maggiore and Altavilla, Vicenza, all Italy;
- in Tertiary basalts of Nova Scotia (see also Walker and Parsons 1922; Aumento 1964; Aumento and Friedlander 1966);
- in a diorite from Bergen Hill, New Jersey; at Copper Falls, Lake Superior;
- in the eleolite-syenite from Magnet Cove, Arkansas.

Some more recent descriptions in basic volcanics are:

Germany: in the Tertiary basalt of Vogelsberg, Hessen (Hentschel and Vollrath 1977) and of Teichelberg, Bavaria (Tennyson 1978), of Rauschermühle, Pfalz (Leyerzapf 1978); at Zeilberg, Oberfranken, Germany, with a partial (Si, Al)-disorder in the framework (Hesse 1983).

Czechoslovakia: in the nepheline-phonolites from the area of Most, Brüx (Hibsch 1929).

Hungary: in the basalts of Balaton Lake (Mauritz 1931, 1938; Koch 1978); in the basalts of Gulacs Hill (Alberti et al. 1983a) as a phase with partial (Si, Al)-disorder in the framework.

U.S.S.R.: in basaltic breccias and tuffs from the Lower Tunguska region (Shkabara and Shturm 1940); in the basalts of Volhynia (Shashkina 1958).

Japan: in the analcime-dolerite from Nemuro, Hokkaido (Suzuki 1938); in the altered basalt from Mazé, Niigata Pref. (Harada et al. 1967; Shimazu and Kawakami 1967).

Taiwan: in an alkali olivine basalt (Juan et al. 1964).

U.S.A.: in the basalts of Table Mt., Colorado (Johnson and Waldschmidt 1925); in the andesite from Valle del Ojo de la Parida, New Mexico (Needham 1938); as needles enclosed in analcime trapezohedra from the basaltic pillow lavas at Coffin Butte, Oregon (Staples 1946); in vesicular basalt of Eckman Creek quarry, Oregon (Groben 1970).

Brazil: in the basalts of southern Brazil (Franco 1952).

Australia: in the Tertiary basalt of Tasmania (Sutherland 1965).

Occurrences in other rock types are:

Norway: in the gneiss of Kragerø (Saebø and Reitan 1959).

England: in the gabbro of Dean quarry, Cornwall (Seager 1969).

Germany: in the Oligocene bituminous shale at the contact with basalt, in the southern Rhinegraben (Heling 1978).

Italy: in the metagabbros of the "Gruppo di Voltri", Liguria (Cortesogno et al. 1975).

Czechoslovakia: in a picrite from Palačov, and in an amphibolite and gneiss from Templstejn (Černý and Povondra 1966).

U.S.S.R.: in the Khibinsky and Lovozersky Mts., Kola Peninsula (Labuntzov 1927); as compact masses with a chalcedonic structure in pegmatite veins in luyavrite and foyaite from Kola Peninsula (Kuzurenko 1950); as a rock forming mineral of dykes in the ultrabasic rocks from Borus Mts., this natrolite being formed by a metasomatic action on pre-existing rocks (Judin 1963).

Japan: in the metagabbros from Yani, Aichi Pref. (Matsubara et al. 1979).

Canada: as giant crystals in the aplite dykes from Johnston asbestos mine, Quebec (Poitevin 1938); in the gneiss from the Sherritt Gordon mine, Manitoba (Brownell 1938); in the pegmatitic dykes in the nepheline syenite at Mt. St. Hilaire, Quebec, with tetranatrolite (Chen and Chao 1980); in the syenite of the Ice River alkaline complex, British Columbia, as fine crystals up to 15 cm in length, well terminated with many forms, with edingtonite (Grice and Gault 1981).

U.S.A.: in weathered nepheline syenite from Wykertown, New Jersey (Milton and Davidson 1950).

The late magmatic deposition of natrolite has also been proposed for some occurrences. There may be a continuous transition from the high hydrothermal genesis up to these "tails" of magmatism. Povarennykh (1954) proposed an "autohydrolitic" phase in the natrolite-analcime deposition at the expense of nepheline in several alkaline complexes from the U.S.S.R. Mitchell and Platt (1982) described the formation of late-stage primary and replacement natrolite in the nepheline syenite of the Coldwell alkaline complex, Ontario.

Tetranatrolite is known only as a mineral of hydrothermal deposition. The type locality may be the first proven occurrence, a vein of albitite in an inclusion of naujaite in the luyavrite in the westermost part of the Tugtup agtakorfia, Ilímaussaq Massif, Greenland, or the occurrence in pegmatite dykes and miarolitic cavities in the nepheline syenite from Mt. St. Hilaire, Quebec, where the samples were collected by Chen and Chao (1980), who proposed the new name. Other occurrences are:

- in ussingite-chkalovite veins traversing urtite in the alkaline intrusion of the Karnasurt Mts., Lovozero Massif, U.S.S.R. (Guseva et al. 1975);
- in the basalt of Gulacs Hill, Hungary (Alberti et al. 1983a);
- in the selbergite of Schellkopf, Eifel, Germany (Hentschel 1983).

The mineral is probably not very rare because many samples could have been misidentified as natrolite. Alberti and Vezzalini (oral communication) suggest that some samples from Aci Trezza, Sicily, and Val di Fassa, Trento, both Italy, and from Portugal are possible tetranatrolites.

Paranatrolite is known not only from the type locality, Mt. St. Hilaire, Quebec (Chao 1980), but also from the selbergite from Schellkopf, Eifel, Germany (Hentschel 1983).

Only one sedimentary occurrence of mesolite has been found thus far: analcime, mesolite and natrolite occur as cement and replacement of plagio-clase grains in Jurassic volcanogenic sedimentary rocks of the Vizcaino penin-sula, Baja California Sur, Mexico; in the underlying Triassic sandstones lau-montite is present; the genesis is due to reaction of pore water with the rock components and may be considered a variant of burial diagenesis (Barnes et al. 1984).

Mesolite, usually as feathery tufts, is a common mineral of vugs of massive rocks, probably of hydrothermal genesis. Some of the occurrences listed by Hintze (1897) and Dana (1914) are the following:

Iceland: in the basalts from Berufjord (see also Kristmannsdóttir and Tómasson 1978; Betz 1981).
Faröes: in the basalts of Naalsö.
Scotland: in the basalts of Skye (see also Sweet 1961); in the basalts from Eigg, Edinburgh, and from Kinron and Hurtfield Moss.
Northern Ireland: in the basalts of Antrim, Portrush, Island Magee (see also Walker 1960a).
Germany: in the basalt of Pflasterkaute, Eisenach, Thuringia.
Canada: in the basalts of Nova Scotia: King's Co., Annapolis Co., Cape Blomidon (see also Walker and Parsons 1922).
U.S.A.: in the basalts of Fritz Island, Pennsylvania; in the basalts of Table Mt., Golden, Colorado (see also Johnson and Waldschmidt 1925).
Brazil: in the augite porphyrite from Serra de Botucatu.

Newer descriptions in basic volcanics are:

Germany: at Oberwiddersheim, Vogelsberg, Hessen, with thomsonite (Hentschel and Vollrath 1977).
Italy: at the Alpe di Siusi, Bolzano (Vezzalini and Alberti 1975).
Czechoslovakia: in the area of Česka Kamenice (= Böhmisch Kamnitz) (Hibsch 1927) with natrolite.
Hungary: at Gulács, near the Balaton Lake (Mauritz 1938); at Nadap (Erdély 1940; see also Koch 1978).
Bulgaria: in the Western Srednogorie (Kostov 1969).
U.S.S.R.: in the Lower Tunguska region (Shkabara and Shturm 1940).
Iran: in North Eastern Azerbaijan (Comin-Chiaramonti et al. 1979).
India: in the Deccan Traps (Sukheswala et al. 1974).
Japan: at Tezuka, Nagano Pref., with scolecite (Harada et al. 1968).
U.S.A.: at Ritter Hot Spring, Grant Co., Oregon (Hewett et al. 1928); at Skookumchuck Dam, Washington (Tschernich 1972).
Brazil: in Southern Brazil (Franco 1952).
Australia: at Kyogle, New South Wales (Hodge-Smith 1929); in the Melbourne area (Vince 1980).

Occurrences in other types of rocks are:

Italy: in a hornblende schist from the Peloritani Mts., Sicily, with analcime (Stella Starrabba 1947); in the metagabbro of the Gruppo di Voltri, Liguria, with natrolite (Cortesogno et al. 1975).
Czechoslovakia: in Upper Turonian metamorphosed marls from Horni Jilove, with chabazite, scolecite, thomsonite (Rychly and Ulrych 1980).
U.S.S.R.: in the Khibinsky and Lovozersky Mts., Kola Peninsula (Labuntzov 1927).
Canada: in the serpentine belt, Eastern Township, Quebec (Faessler and Badollet 1947).

Scolecite is quite a common zeolite, always found in vugs of massive rocks and probably of hydrothermal genesis. Some classical occurrences, already listed by Hintze (1897) and Dana (1914), are:

Iceland: in the basalts of Berufjord (see also Betz 1981); Kristmannsdóttir and Tómasson (1978) found zones with increasing temperature: (1) chabazite, (2) mesolite/scolecite, (3) stilbite, (4) laumontite.
Scotland: in the basalts of Talisker, Skye, and of Staffa, Island of Mull (see also McLintock 1915).
Germany: in the amphibole granite from Pflasterkaute, Eisenach, Thuringia; in the granite of Striegau, Silesia.
Switzerland: in granite/gneisses of Maderaner Valley, Uri, and of Fiesch, Valais.
India: in the Deccan Traps, by Poona (see also Sukheswala et al. 1974).
Greenland: Bøggild (1953) lists 17 occurrences in the districts of Umanak, Godhavn, Julianehaab, East Greenland.

Canada: in a granite dyke in the serpentine region of the Blacklake, Megantic
    Co., Quebec.
U.S.A.: in the basalts of Table Mts., Golden, Colorado.
Brazil: at Serra de Tuberau, Santa Catharina.

The following descriptions in basic volcanics have been published after 1915:

Czechoslovakia: in the phonolite near Ústí nad Lábem (= Aussig), Bohemia
    (Hibsch 1915); in a rock from Horni Jilove, Bohemia as an overgrowth on
    mesolite (Ulrych and Rychly 1980).
Hungary: in the basalts of the Balaton Lake area (Mauritz 1938, 1955; Koch
    1978).
Bulgaria: in the Srednogorian volcanics, in the zone of highest hydro-
    thermalism (Kostov 1969).
Japan: in andesitic to basaltic rocks from Hudakake, Kiyokawa-mura,
    Kanagawa Pref. (Harada et al. 1968).
Taiwan: in the Penghu basalts, with thomsonite, chabazite, phillipsite (Lin
    1979).
Canada: in the traps of Nova Scotia (Walker and Parsons 1922).
U.S.A.: in the basalts of Table Mt., Golden, Colorado (Johnson and Wald-
    schmidt 1925).
Brazil: in the basalts of Southern Brazil (Franco 1952) and of Rio Pelotas, Rio
    Grande do Sul (Franco 1953).
Africa: in the basalts of Alomatà, Eritrea, Ethiopia (Scherillo 1938).
Australia: in the Jurassic dolerites of Tasmania, with chabazite, interpreted by
    Sutherland (1977) as a burial assemblage reflecting stages of unloading of
    the Jurassic land surface.

Occurrences in other rock types are:

Switzerland: in granites and gneisses from Schattig Wichel, Uri (Parker 1922);
    in a biotite oligoclase gneiss in Valle Maggia, Tessin (Grütter 1931); in a
    gneiss from Giubiasco, Tessin (Gschwind and Brandenberger 1932).
Czechoslovakia: in the amphibolite of Červena Hora, Moravia (Novotna
    1926).
Austria: in a gneiss from Stubachtal, Salzburg, with stilbite, heulandite,
    laumontite (Paar 1971).
Italy: in carbonate metamorphic rocks of San Giorgio, Sondrio (Cerio 1980)
    with laumontite, epistilbite, heulandite.
U.S.A.: at the contact of Palaeozoic sediments with a quartz monzonite in the
    Italian Mts., Gunnison Co., Colorado (Cross and Shannon 1927).

1.2 Thomsonite

$Na_4Ca_8(Al_{20}Si_{20}O_{80}) \cdot 24H_2O$
Pcnn
$a = 13.05$
$b = 13.09$
$c = 13.22 (\text{Å})$

1.2.1 History and Nomenclature

Brooke (1820) proposed the name thomsonite for a mineral from Kilpatrick, Dunbartonshire, Scotland, in honour of T. Thomson. One year later, Brewster (1821) proposed the name "comptonite" for a mineral from Vesuvius, Italy, in honour of Lord Compton, chairman of the Geological Society of London; as early as 1825 Monticelli and Covelli stated that the two minerals were identical, but the second name, now fully obsolete, was used for a long time. Other old synonyms are faröelite, mesole, plus some others which had a more restricted use and are listed for instance by Hey (1962). The name thomsonite was proposed by Squires (1821, quoted in Hey 1962) for another different mineral, a calcium silico-carbonate, now discredited, but this second meaning never acquired a wider use. Please note that "kalithomsonite" (Gordon 1924) is ashcroftine (Hey and Bannister 1933).

1.2.2 Crystallography

The framework of thomsonite was described by Taylor et al. (1933) assuming $c = 6.6$ Å, which is half the true value. Meier (1968) proposed an (Si, Al)-order with an alternation of the two atoms in the tetrahedra and so explained the double value of c in comparison with natrolite. Alberti et al. (1981) refined the structure down to an R = 3.5%, and found all cations and water sites; they adopt a cell with a and b interchanged, and hence space group Pcnn. The topological symmetry of thomsonite (see Chap. 0.2.1) Pmma is reduced by rotation of the chains by an angle $\psi = 23°$ (see Chap. 1.1.2 and Fig. 1.2A) to Pman, and by the (Si, Al)-ordering to Pcnn with double c (Alberti and Gottardi 1975). The distribution of (Si, Al) in the tetrahedra is nearly perfect, with a disorder index near to $5-10\%$ (Fig. 1.2B). The axes of the two sets of channels parallel to c coincide neither with diads nor screw diads, there are only inversion centres on them. In the first set of channels, there is a cation site nearly midway between inversion centres, fully occupied by Na and Ca in equal amounts. In the second set of channels there is a Ca-site very near to an inversion centre, so that it may be occupied only at 50%; the whole amount of Sr, which is but small, is located in this position. There are four fully occupied water sites, the first two being in the same channel with (Ca, Na), the second two in the port between channels. Both cation sites are of type II (see Chap. 0.4).

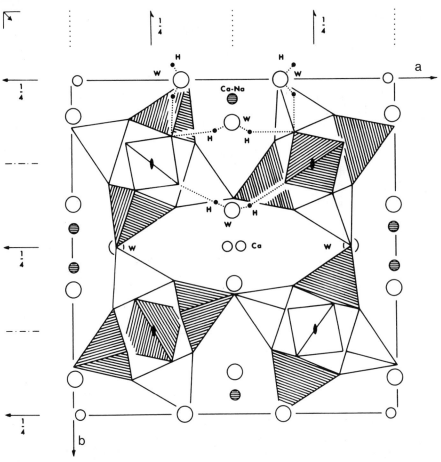

Fig. 1.2A. The structure of thomsonite viewed along c, the chains being rotated by $\psi = 23°$ from the most symmetric arrangement; the central tetrahedra of the two upper chains are at the same height, which is different for the central tetrahedra of the two lower chains (compare with Fig. 0.2Cb); *striped* tetrahedra are occupied by Al, *white* tetrahedra by Si

No single crystal refinement of an, at least partially, disordered thomsonite is available, but Wise and Tschernich (1978b) found some evidence for its existence in the powder pattern of microcrystalline botryoidal thomsonite, which features a weakening of the diffractions with l odd.

Hey (1932a) and Wise and Tschernich (1978b), also recalling old data from the literature, state that thomsonite occurs with three different habits, all with c-elongation: (1) long radiating prisms rich in crystal forms (see Fig. 1.2E). (2) radiating clusters of fibrous crystals, actually blade shaped. (3) fibrous botryoidal spherules. Detailed morphological data are possible in the first case, and simple forms {100} {010} {110} {001} are usually the only ones present, {101} {401} {801} are rare; in the second case, the blades are parallel to {010}.

{110} and {041} twins have been described.

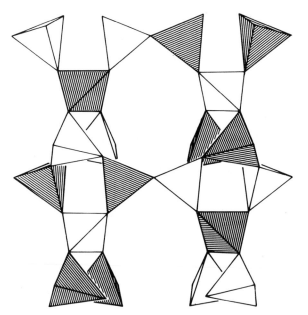

Fig. 1.2B. Two thomsonite chains at the same height with their (Si, Al)-order, shown as in Fig. 1.2A

Fig. 1.2C. Thomsonite from Bohemia (SEM Philips, ×60)

Fig. 1.2 D. Lamellar thomsonite from Skye, Scotland (SEM Philips, ×500)

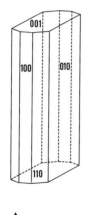

▲
Fig. 1.2 E

Fig. 1.2 F ▶

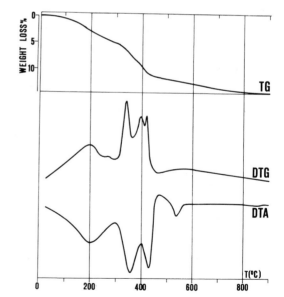

Fig. 1.2 E. Common morphology of thomsonite

Fig. 1.2 F. Thermal curves of thomsonite from Savona, Italy; in air, heating rate 20°C/min (by the authors)

1.2.3 Chemistry and Synthesis

A good report on thomsonite chemistry was given by Hey (1932a); Wise and Tschernich (1978b) gave new evidence for a correlation between crystal habit and chemistry, already found by Hey (1932a).

Fine crystals, showing good prismatic forms, have a composition near to the stoichiometric formula, if any Sr is considered equivalent to Ca: see analyses of Table 1.2I. If the crystal habit corresponds to clusters of fibres or fibrous botryoidal or waxy spheres, then the percent of Si in the tetrahedra may rise from 50% to 55% and the M/D ratio (M = monovalent-, D = divalent-cations) may rise from 0.5 to 0.64 (see Wise and Tschernich 1978b) or even to 2.7 as in sample no. 5 of Table 1.2I. Sr-rich thomsonites are not rare; samples nos. 3 and 4 not being the only examples of this kind. Sample no. 4 is rich in K and Mg, as are many of the samples listed by Hey (1932a). On the whole, the thomsonite structure allows much wider chemical ranges than the natrolite or mesolite or scolecite structure.

Barrer and Denny (1961) obtained from gels $CaO \cdot Al_2O_3 \cdot nSiO_2$ with n = 34 or 41, with hydrothermal treatment at 225° or 245 °C, a phase obviously Na-free but quite similar to thomsonite. Juan and Lo (1969) performed the synthesis from glasses hydrothermally at 200°C − 275 °C when studying the system $Ab - An - H_2O$: thomsonite was obtained when the starting material ranged from $Ab_0 - An_{100}$ to $Ab_{30} - An_{70}$, some garronite and analcime being obtained when Ab > 15%. Wirsching (1981) obtained the zeolite in a simulated "open system" namely by hydrothermal reaction of nepheline with 0.1 or 0.01 N $CaCl_2$ solution, which was frequently renewed; temperature was 150°C, 200°C, 250°C.

1.2.4 Optical and Other Physical Properties

Thomsonite is biaxial positive with $\alpha \to a$ $\beta \to c$ $\gamma \to b$; the few values of refractive indices listed in Table 1.2II represent well the whole range of possible values.

Transparent, colourless or yellowish, streak white. Mohs-hardness: 5. Cleavage: (100) perfect, (010) good.

Although the structure was refined in a centric group, thomsonite is pyroelectric (Wooster in Hey 1932a).

The IR spectrum, given in the Appendix, may be useful for identification.

1.2.5 Thermal and Other Physico-Chemical Properties

The thermal curves (see Fig. 1.2F) and in particular the DTG show four peaks of weight loss under 450°C, plus one shoulder or small peak, in partial accordance with the presence of four water sites in the structure; one more broad peak between 500 and 700°C could be due to the loss of hydroxyls formed by the reaction of the framework with water at lower temperature.

Table 1.2I. Chemical analyses and formulae on the basis of 80 oxygens and lattice constants in Å

	1	2	3	4	5
SiO_2	37.17	36.51	36.75	36.68	40.39
Al_2O_3	31.93	31.98	29.36	29.27	29.39
Fe_2O_3				0.05	
MgO		0.06		1.00	
CaO	13.98	14.69	11.85	5.92	7.09
SrO			2.87	9.72	
BaO			0.03	0.15	
Na_2O	4.00	4.05	3.82	3.02	10.54
K_2O		0.05		2.16	
H_2O	13.35	13.17	15.32	12.19	12.39
Total	100.43	100.51	100.00	100.16	99.80
Si	19.88	19.56	20.48	20.52	21.42
Al	20.12	20.20	19.29	19.31	18.46
Fe				0.01	
Mg		0.05		0.83	
Ca	8.00	8.43	7.08	3.55	4.02
Sr			0.93	3.15	
Ba			0.01	0.03	
Na	4.14	4.21	4.13	3.28	10.84
K		0.03		1.54	
H_2O	23.80	23.53	28.33	22.75	21.94
$E\%$	-0.1	-4.7	-4.2	-3.2	-2.2
a	13.051	13.095	13.047		13.07
b	13.092	13.102	13.089		13.07
c	13.263	13.240	13.218		13.23
D	2.37		2.41	2.44	2.31

1. Glassy prisms from Old Kilpatrick, Dumbarton-shire, Scotland (Hey, 1932a); lattice constants by Nawaz and Malone (1981)
2. Fine crystals from Yanai, Japan (Matsubara et al.,1979)
3. Fine crystals from Death Valley, California (Alberti et al.,1981)
4. Fine crystals from Tyamyr, U.S.S.R. (Yefimov et al.,1966)
5. Fibrous crust from Zalezly (=Salesl), Bohemia (no.5 of Hey, 1932a)

Table 1.2 II. Refractive indices

	1	2	4	5
α	1.529	1.510		1.523
β	1.531	1.512	1.528	1.525
γ	1.542	1.525	1.534	1.537
2 V	+52°	+55°		

Numbers refer to the same samples of Table 1.2 I.

After Reeuwijk (1974), thomsonite looses $8 H_2O$ under 320°C with a slight but definite a-contraction and an even smaller c-expansion; over 320°C there is a strong rearrangement of the structure, and most water is lost under 450°C; at this temperature the lattice is destroyed, but nearly $2 H_2O$ are lost at higher T, as already said.

No data on ion exchange or adsorption are available for thomsonite, as far as we know.

1.2.6 Occurrences and Genesis

Thomsonite from most occurrences is of hydrothermal origin, even that associated with deep sea basalts (Bass et al. 1973). A sedimentary genesis for thomsonite has been proposed for one occurrence only: Hay and Iijima (1968b) found thomsonite, with natrolite and gonnardite, as major constituent of the cement of the palagonite tuffs of Diamond Head, Punchbowl, Manana Island, Ulupau Head and Salt Lake Tuff, all Hawaii; the crystallization happened by slow percolation of meteoric water (hydrologic open system).

Some famous occurrences listed already by Hintze (1897) and Dana (1914) are:

Scotland: in the basalts of Kilpatrick (the type locality), Dunbartonshire, of Kilmacolm and Port Glasgow, Renfrewshire.

Northern Ireland: in the Tertiary basalts with chabazite (see also Walker 1951, 1960a).

Iceland: in the basalts of Berufjord (see also Betz 1981).

Faröes: in the basalt of Skutin, Nolsoy island, as spherical concretions of lamellae, radiating individuals ("faroelite") with stilbite (see also Betz 1981).

Norway: in the altered eleolite from the islands of Låven and Arö in Langesund fjord.

Germany: in the nepheline basalt from Pflasterkaute, Eisenach, Thuringia.

Czechoslovakia: in the basalts near Kadan (= Kaaden), near Usti nad Labem (= Aussig), near Litomerice, all Bohemia (see also Hibsch 1915).

Italy: in the lavas of Somma-Vesuvius (the type locality of "comptonite"); in the porphyrite from the ravine of Bulla (= Puflerloch), South Tyrol (see also Passaglia and Moratelli 1972); in the Monzoni Mts., Trentino, with chabazite (see also Bellinzona 1923).

Canada: in the basalts of Nova Scotia (see also Aumento 1964).

U.S.A.: in the basalts from Duluth and Grand Marais near Lake Superior, Minnesota, with mesolite and scolecite (see also Hanley 1939); in an eleolite from Magnet Cove, Arkansas; in the basalt of Table Mt., Golden, Colorado (see also Johnson and Waldschmidt 1925; Henderson and Glass 1933).

Newer descriptions in basic volcanics are:

Germany: in a basalt from Rossdorf, Darmstadt, as Sr-rich variety (Müller and Deisinger 1971); in the basalt of Oberwiddersheim, Vogelsberg with mesolite and chabazite (Hentschel and Vollrath 1977); in the basalt of Teichelberg, Bavaria, with natrolite (Tennyson 1978).

Italy: in the basalts of Isola dei Ciclopi, Sicily (Di Franco 1932); in the andesite from Montresta, Sardinia (Pongiluppi 1974); in the basalt of Alpe di Siusi (= Seiseralp), South Tyrol (Vezzalini and Alberti 1975).

Hungary: in the basalts North of Lake Balaton (Koch 1978).

Bulgaria: in the "fibrous zeolite zone" of the Srednogorian Mts. (Kostov 1969).

U.S.S.R.: in basaltic breccia and tuff of the Lower Tunguska region (Shkabara and Shturm 1940); in the basalts of Volhynia (Shashkina 1958).

Iran: in the shoshonites from Azerbaijan, as spherical aggregates with mesolite and chabazite (Comin-Chiaramonti et al. 1979).

Japan: in the basalts from Mazé, Niigata Pref., with natrolite, gonnardite (Shimazu and Kawakami 1967; Harada et al. 1967); in the basalt as coating material on the surface of fractures filled with saponite, at Iwano, Saga Pref. (Muchi 1977); in a Pliocene trachybasalt lava from Dōzen, Oki Islands, with levyne and cowlesite (Tiba and Matsubara 1977).

Greenland: in the basalt of Disko Island (Gärtner and Machatschki 1927).

Canada: in the porphyritic trachyte from Yellow Lake near Olalla, British Columbia as stout to elongate prisms with brewsterite and mesolite (Wise and Tschernich 1978b).

U.S.A.: at the contact zone of a quartz monzonite with carbonated sediments in the Italian Mts., Colorado (Cross and Shannon 1927; Hogarth and Griffin 1980); in the basalts from Ritter Hot Spring, Grant Co., Oregon, with heulandite, stilbite, chabazite (Hewett et al. 1928); in the andesite from Ojo de la Parida, New Mexico, with natrolite and analcime (Needham 1938); in the basalt from Drain, Oregon, as bladed crystals with analcime and mesolite (Wise and Tschernich 1978b); in the basalt from Beech Creek, Grant Co., Oregon, as long composite rods with radiating sprays at their ends, or as botyroidal growths, for instance as smooth waxy balls (Wise and Tschernich 1978b); in the basalt from Point Sal, California, as bladed crystals with natrolite and analcime (Wise and Tschernich 1978b); in the basalt of Skookumchuck Dam, Washington, as bladed crystals with mesolite (Tschernich 1972; Wise and Tschernich 1978b).

Occurrences in other rock types are:

Italy: at the contact of serpentinites with prasinites in the Val d'Ayas, Valle d'Aosta, and in the prasinites of Valli di Lanzo, Piemonte, both with scolecite (Gennaro 1929); in a gneiss from Saint Vincent, Valle d'Aosta (Abbona and Franchini Angela 1970); in the metagabbro of the "Gruppo di Voltri" near Tiglieto, Liguria, with natrolite (Cortesogno et al. 1975).

U.S.S.R.: in geodes and crevices of a teschenite from Kurseki Caucasus (Shkabara 1948); in the alkalic rock province from Tyamir, as Sr-rich variety (Yefimov et al. 1966).

Japan: in an altered metagabbro from Yani, Aichi Pref., with natrolite (Matsubara et al. 1979).

Canada: in the aplite dykes at the Caribou chrome pit and in the asbestos pits of Eastern Quebec (Poitevin 1936; Faessler and Badollet 1947).

U.S.A.: in a monzonite dyke from Bearpaw Mts., Montana, with analcime (Pecora and Fisher 1946).

1.3 Edingtonite

$Ba_2(Al_4Si_6O_{20}) \cdot 8H_2O$		
$P\bar{4}2_1m$	or	$P2_12_12$
$a = 9.58$		$a = 9.55$
$c = 6.52\,(\text{Å})$		$b = 9.67$
		$c = 6.52\,(\text{Å})$

1.3.1 History and Nomenclature

This rare barium zeolite was described by Haidinger (1825) and named in honour of Mr. Edington from Glasgow, in whose collection the mineral was found; the type locality is Kilpatrick Hills, Dunbartonshire, Scotland.

1.3.2 Crystallography

The structure of edingtonite was solved by Taylor and Jackson (1933), and refined by Galli (1976), Kvick and Smith (1983) and by Mazzi et al. (1984). The topologic symmetry of this structure (see Chap. 0.2.1) is $P\bar{4}2_1m$, which remains unchanged if the chains are rotated by an angle $\psi = 17°$ (see Chap. 1.1.2 and Fig. 1.3A). There may be a (Si, Al)-order in the tetrahedra, like natrolite, and this fact reduces the symmetry to $P2_12_12$. This order may be nearly perfect, as in the sample from Böhlet Mine, Sweden, with $a - b = 0.12$ Å,

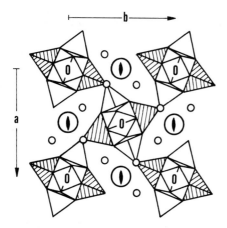

Fig. 1.3 A. The structure of edingtonite viewed along c, the chains being rotated by $\psi = 17°$ (compare with Fig. 0.2 Ca). *Striped* tetrahedra are occupied by Al; *white* tetrahedra by Si; *large circles* are Ba; *small circles* H_2O; the diads only are shown

Fig. 1.3 B. Edingtonite from Ice River, Canada (SEM Philips, ×60)

refined by Galli (1976), or absent as in the disordered samples from Old Kilpatrick and from Ice River with $a = b$, refined by Mazzi et al. (1984). The Ba-atoms are on the diads, and are surrounded by framework oxygens on both sites and also by water molecules (coordination of type II of Chap. 0.4). There are two water sites in the unit cell. Kvick and Smith (1983) refined an orthorhombic edingtonite from New Brunswick with neutron diffraction, obtaining atomic parameters similar to those of Galli (1976) but including also the hy-

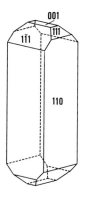

Fig. 1.3C. Common morphology of orthorhombic edingtonite

drogen positions, which are always near the join between a framework oxygen and an oxygen of a water molecule.

There is another mineral with the framework of edingtonite, namely kalborsite $K_6(Al_4Si_6O_{20}) \cdot B(OH)_4Cl$, but it can hardly be considered a zeolite, because it does not contain any water molecules in its channels (Khomyakov et al. 1980; Malinovskii and Belov 1980).

The morphology of the tetragonal edingtonite from Ice River (Fig. 1.3 C) is quite simple, with $\{110\}$ dominant, terminated $\{111\}$ $\{1\bar{1}1\}$ $\{001\}$ (Grice and Gault 1981); the orthorhombic sample from Böhlet shows the forms (Nordenskiöld 1895): $\{110\}$ $\{001\}$ $\{111\}$ $\{1\bar{1}1\}$ $\{121\}$ $\{1\bar{2}1\}$. The morphology of the tetragonal holotype from Kilpatrick (Hintze 1897; Dana 1914) is different with $\{110\}$ $\{112\}$ $\{113\}$ $\{111\}$.

Twins on $\{110\}$ are rare.

1.3.3 Chemistry and Synthesis

The chemical composition of edingtonite is rather near to the stoichiometric formula of the title page, with some substitution of Ba by Ca and Sr (see Table 1.3 I).

Barrer et al. (1974) were able to synthetize a Ba – Li-bearing phase very similar to edingtonite with Ba : Li $\simeq 1 : 2$, and Al : Si $= 2 : 3$; starting materials were metakaolinite and amorphous silica, which were hydrothermally treated at $80 °C$ for one week with much LiOH and a little $Ba(OH)_2$ (Ba : Li $= 1 : 18$). No other author has been able to synthetize a phase which has not only the edingtonite framework but also the same natrolite type (Si, Al)-order and Ba as extraframework cation. Already in 1956 Barrer and Baynham obtained the phase K – F which was later shown to have the same framework with a different kind of (Si, Al)-order; these authors performed the hydrothermal synthesis of K – F at $120°C – 150°C$ from gels $K_2O \cdot Al_2O_3 \cdot nSiO_2$ with $1 \leqslant n \leqslant 3$. Barrer and Marcilly (1970) obtained three K – F-modifications containing the halides KCl, KBr, KI, at $200°C$. The conditions of crystallization of zeolite F at changing K : (Na + K) and Si : Al ratios were established by

Table 1.3I. Chemical analyses and formulae on the basis of 20 oxygens and lattice constants in Å

	1	2	3	4
SiO_2	37.29	35.14	36.58	37.42
Al_2O_3	20.08	20.12	20.20	20.20
MgO			0.15	
CaO	0.10		0.06	
SrO	0.03			0.05
BaO	29.03	31.18	29.65	28.36
Na_2O	0.18		0.28	0.08
K_2O	0.23		0.17	0.52
H_2O	13.04	13.16	12.75	13.36
Total	100.00	99.60	99.91	100.00
Si	6.12	5.95	6.04	6.13
Al	3.88	4.02	3.93	3.90
Mg			0.04	
Ca	0.02		0.01	
Sr				0.01
Ba	1.87	2.07	1.92	1.82
Na	0.06		0.09	0.03
K	0.05		0.04	0.11
H_2O	6.77	7.44	7.02	7.30
E%	+0.1	-2.9	-3.5	+3.1
a	9.584	9.550	9.524	9.581
b		9.665	9.620	
c	6.524	6.523	6.500	6.526
D	2.69	2.78	2.75	2.72

1. Old Kilpatrick, Scotland (Mazzi et al. 1984); H_2O from Heddle (1855); density from Hey (1934).
2. Böhlet Mine, Sweden (Reeuwjik 1974); lattice parameters by Galli (1976).
3. Staré Ransko, Bohemia (Novak 1970) with 0.07% FeO.
4. Ice River, Canada (Mazzi et al. 1984).

Table 1.3 II. Optical data

	1	2	3	4
α	1.542	1.540	1.539	1.535
β	1.550	1.553	1.550	1.542
γ	1.550	1.557	1.553	1.545
2 V	-15°	-53°	-53°	-62°

1. Old Kilpatrick, Scotland (Hey 1934).
2. Böhlet Mine, Sweden (Hey 1934).
3. Staré Ransko, Bohemia (Novak 1970).
4. Ice River, Canada (Grice and Gault 1981).

Bosmans et al. (1973) for T = 90°C only. Similar results were obtained at 80°C by Colella and Aiello (1975). Barrer et al. (1968) showed that the phase Rb − D of Barrer and McCallum (1953) was a variant of K − F. Baerlocher and Barrer (1974) determined the structure of Na-exchanged K − F, and Rb − D from powder data as did Tambuyzer and Bosmans (1976) for K − F itself; all three phases have the edingtonite framework but an (Si, Al)-order given by an alternation of the two atoms in the tetrahedra, as in thomsonite. Evidence for this kind of ordering is given by a value of c (13.2 Å), which is twice that of edingtonite (6.5 Å): in detail, this ordering is evident for K − F and Rb − D, but the Na-exchanged K − F has only two weak peaks which must be indexed with such a c-value.

1.3.4 Optical and Physical Properties

The indices of refraction of edingtonite are not as constant as would be expected from the regularity of the chemical composition: the values are in Table 1.3 II. The mineral is always biaxial negative, even if tetragonal from X-ray measurements; the orientation is $\alpha \rightarrow c$ $\beta \rightarrow b$ $\gamma \rightarrow a$. Transparent colourless; streak white. Cleavage {110} perfect. Mohs-hardness 4.

Edingtonite is piezoelectric (Bond 1943; Ventriglia 1953).

1.3.5 Thermal and Other Physico-Chemical Properties

The thermal curves of edingtonite (Fig. 1.3 D) show four water losses, which are nearly complete at 450°C; from this temperature on, there is a small continuous loss with a low rate. With two independent water sites only, there must be an inversion with symmetry reduction somewhere to allow for the loss in four steps. Reeuwijk (1974) found that the heating causes a small contraction of a and b, and a small expansion of c, before destruction at 400°C; there is a small discontinuity at 100°C which could be related to the inferred symmetry reduction.

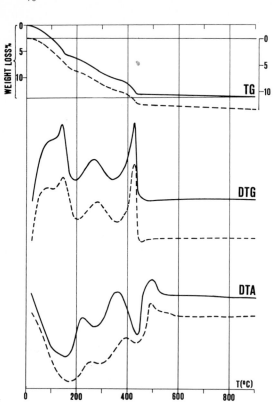

Fig. 1.3 D. Thermal curves of orthorhombic edingtonite from Böhlet Mine, Sweden (*continuous line*), and of tetragonal edingtonite from Ice River, Canada (*dashed line*); in air heating rate 20°C/min (Mazzi et al. 1984)

Some base exchange data on edingtonite are given by Hey (1934) who obtained them by fusion with sodium chlorate, potassium thiocyanate, silver and thallous nitrate at 270°C; the exchange was conspicuous, but never complete.

Base exchange is certainly possible with the synthetic zeolite F, as mentioned in several publications quoted for the synthesis.

No data of gas adsorption are available for edingtonite.

1.3.6 Occurrences and Genesis

Only five occurrence of edingtonite are known: all are probably of hydrothermal deposition.

Two are tetragonal:

Scotland: in the basalt of Kilpatrick Hills (Old and West), Dunbartonshire (Hintze 1897; Dana 1914).

Canada: in the syenite of the alkaline complex from Ice River, British Columbia, with natrolite (Grice and Gault 1981).

Three are orthorhombic:

Sweden: Böhlet Mine, Westergotland (Hintze 1897; Dana 1914).

Czechoslovakia: in an altered basic rock from Staré Ransko, Eastern Bohemia (Novak 1970).

Canada: in the Brunswick no. 12 mine near Bathurst, New Brunswick (Kvick and Smith 1983).

1.4 Gonnardite

$Na_5Ca_2(Al_9Si_{11}O_{40}) \cdot 12 H_2O$
tetragonal or orthorhombic
$a = 13.35$
$b = 13.35$
$c = 6.65 (\text{Å})$

1.4.1 History and Nomenclature

The name was proposed by Lacroix (1896) for a zeolite occurring as spherules of fibrous materials filling the cavities in a doleritic basalt from Gignat, Puy-de-Dôme, France, in honour of F. Gonnard. Hey (1932a) showed that the outer part of these spherules was thomsonite, whereas the inner part was really a different mineral.

1.4.2 Crystallography

Bannister (in Hey 1932a) obtained a fibre photograph from pure gonnardite and measured tetragonal constants (see title) similar to those of thomsonite, but with c halved, and to those of tetranatrolite; Hey (1932a) decided on an orthorhombic symmetry because of the biaxial optics. Gonnardite is considered here to be of unkown structure, but it certainly belongs to the fibrous zeolites with chains of tetrahedra along c; Alberti and Gottardi (1975) emphasized that gonnardite has the unit cell dimensions and perhaps also the symmetry of a "disordered" thomsonite, but, at least, the dimensions are in common with tetranatrolite. Amirov et al. (1972) claimed to have resolved the structure by single crystal X-ray diffraction, but these single crystals were never definitely characterized as gonnardite, which occurs always as aggregates of microcrystalline fibres. Moreover, a team of four scientists (Amirov et al. 1978) including two of the former group, published a structure of thomsonite grossly wrong because of an assumed c-value of 6.6 Å (instead of 13.2 Å) and a consequently wrongly deduced (Si, Al)-disorder. As a matter of fact, the atomic parameters found for their "gonnardite" and "thomsonite"

Fig. 1.4A. Gonnardite from Höwenegg, Hegau, Germany
(SEM Philips, ×500)

are nearly the same, so that probably both studies were made incorrectly on thomsonite crystals.

So, apart from the very probable presence of tetrahedral chains similar to natrolite in its framework, nothing else has been proven by reliable experimental evidence on this structure, but two facts are worth mentioning in this context: (1) as far as the powder pattern (see Appendix) is concerned, gonnardite is more similar to tetranatrolite than to any other fibrous zeolite; (2) the IR spectrum of gonnardite (see the Appendix) is also very similar to that of tetranatrolite. On the whole, the problem of the gonnardite symmetry and structure is still open.

The morphology is characterized by fibrous aggregates.

1.4.3 Chemistry and Synthesis

The chemistry of gonnardite (see Table 1.4I) seems to be quite variable when looking at the available data (besides those here reported, only three other analyses have been published), but the true state of the matter could be different because most gonnardite samples are an intimate intergrowth of the mineral plus thomsonite as in the holotype, or plus tetranatrolite. The only sample which seems to be rather pure is from Klöch, Styria; the formula given in the title page is an informed guess only. Anyway, the ratio $Si:(Si+Al)$ is always near to 57%, with small oscillations; the ratio $Na:(Na+Ca)$ is quite variable, from 0.6 to 0.9; potassium is very low or absent, and so are Mg, Sr, Ba.

Aoki (1978), when studying the hydrothermal stability of chabazite, found gonnardite and Ca-analcime (= disordered wairakite) as decomposition products at intermediate conditions (240°C, 0.2 Kbar): this is the only report on a synthesis of gonnardite.

1.4.4 Optical and Other Physical Properties

In view of the absence of single crystals of suitable dimensions, optical data are uncertain. Table 1.3 II lists some data, but they are incomplete and must be considered with caution. The changing sign of elongation could be explained as due to the coincidence of β with c, which has been proposed for some samples.

Gonnardite is colourless white or yellowish, with white streak. The IR spectrum, given in the Appendix, is similar to that of tetranatrolite.

1.4.5 Thermal and Other Physico-Chemical Properties

Figure 1.4B shows the thermal curves of gonnardite. There is one main water loss at 330°C plus some very small and ill defined ones at 50°, 130°, 200° and 420°C. Reeuwijk (1974) showed that the lattice undergoes a marked contraction along a and b between 20° and 50°C; the lattice is destroyed just above 300°C.

No data are available for ion exchange or gas adsorption in gonnardite.

1.4.6 Occurrence and Genesis

Gonnardite is known mainly from vugs of basalts and andesites, but Hay and Iijima (1968b) found gonnardite with thomsonite and natrolite in the cement of the palagonite tuffs from Oahu, Hawaii. Typical occurrences in vugs of massive rocks (mostly basalts) are:

France: Chaux de Bergonne, Gignat, Puy-de-Dôme with thomsonite, phillipsite, chabazite (Lacroix 1896; Pongiluppi 1976).

Germany: Weiberg, Rhineland, with phillipsite (Meixner et al. 1956); Höwenegg, Hegau (Hentschel, written communication).

Czechoslovakia: Vinaricka Hill, Kladno, Bohemia, with thomsonite: this mineral was identified as mesolite by Antonín (1942) but it is actually gonnardite (Alberti et al. 1982a).

Table 1.4I. Chemical compositions and formulae on the basis of 40 oxygens

	1	2	3	4	5	6	7
SiO_2	41.85	42.80	43.20	42.75	40.03	42.71	43.86
Al_2O_3	27.02	28.15	27.90	27.36	27.88	27.43	26.92
Fe_2O_3					0.85	0.14	
MgO					0.22		
CaO	9.29	4.26	3.61	7.77	6.03	5.28	7.13
Na_2O	7.25	12.65	13.16	8.15	10.05	11.12	7.67
K_2O		0.13		0.15	0.40		0.02
H_2O	14.36	11.85	11.74	13.44	14.22	13.91	13.00
Total	99.77	99.84	99.61	99.62	100.29	100.59	98.60
Si	11.28	11.24	11.34	11.39	10.89	11.35	11.67
Al	8.58	8.72	8.64	8.59	8.94	8.59	8.44
Fe					0.17	0.03	
Mg					0.09		
Ca	2.68	1.20	1.02	2.22	1.76	1.50	2.03
Na	3.79	6.44	6.70	4.21	5.30	5.73	3.96
K		0.04		0.05	0.14		0.01
H_2O	12.91	10.38	10.28	11.94	12.90	12.33	11.53
E%	-6.2	-1.9	-1.1	-1.2	-0.2	-1.3	+5.1
D	2.26	2.28	2.27	2.27			

1. Chaux de Bergonne, Gignat, France (Hey, 1932a)
2. Kloch, Styria, Austria (Meixner et al., 1956)
3. Aci Trezza, Sicily (do)
4. Aci Castello, Sicily (do)
5. "High-sodium mesolite", Rio Cambone, Sardinia (Deriu, 1954)
      plus TiO_2 0.02%, FeO 0.59%
6. Mazé, Niigata Pref., Japan (Harada et al., 1967)
7. Halap Hill, Hungary (Alberti et al., 1983a)

Table 1.4II. Optical data

	1	2	3	4	5	6
α		1.498	1.497	1.506	1.504	1.500
β					1.505	
γ		1.502	1.499	1.508	1.507	1.504
	$\alpha \rightarrow \underline{c}$	$\beta \rightarrow \underline{c}$	elong.	elong.	$\beta \rightarrow \underline{c}$	$\alpha \rightarrow \underline{c}$
	2V less than -50°		+	-		

Numbers refer to the same samples of Table 1.4 I.

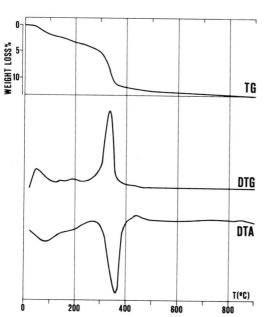

Fig. 1.4B. Thermal curves of gonnardite from Klöch, Styria, Austria; in air, heating rate 20°C/min (by the authors)

Austria: Klöch, Styria, with phillipsite, chabazite: this sample is quite pure (Meixner et al. 1956).

Hungary: Halap Hill, with tetranatrolite (Alberti et al. 1983a).

Italy: Capo di Bove, Rome (host-rock = leucite tephrite) (Meixner et al. 1956); Aci Trezza and Aci Castello, Sicily (Meixner et al. 1956): this gonnardite is intergrown with tetranatrolite (Alberti, private communication); Rio Cambone, Sardinia, identified as mesolite by Deriu (1954) but Alberti et al. (1982a) showed it to be gonnardite; Monastir, Sardinia with analcime (Pongiluppi 1974).

India: in a basic xenolith from the syenite of the carbonatite complex of Korati, Tiruppattur, Tamil Nadu, with phillipsite and chabazite (Ramasamy 1981).

Japan: Mazé, Niigata Pref., with the deposition sequence: thomsonite and/or gonnardite, natrolite, analcime (Harada et al. 1967; Shimazu and Kawakami 1967); Waniguchi, Niigata Pref., with thomsonite; this zeolite was identified as mesolite by Harada et al. (1968), but Alberti et al. (1982a) stated that it is gonnardite, probably.

Australia: Tasmania, with chabazite, phillipsite, natrolite, stilbite (Sutherland 1965).

2 Zeolites with Singly Connected 4-Ring Chains

2.1 Analcime, Wairakite, Viséite, Hsianghualite

Analcime	Wairakite
$Na_{16}(Al_{16}Si_{32}O_{96}) \cdot 16\,H_2O$	$Ca_8(Al_{16}Si_{32}O_{96}) \cdot 16\,H_2O$
Ibca	I2/a
$a = 13.73$	$a = 13.69$
$b = 13.71$	$b = 13.64$
$c = 13.74\,(\text{Å})$	$c = 13.56\,(\text{Å})$
or	$\beta = 90°30'$
$I4_1/acd$	
$a = 13.72$	
$c = 13.69\,(\text{Å})$	
or	
$a = 13.72$	
$c = 13.73\,(\text{Å})$	
or	
Ia3d	
$a = 13.73\,(\text{Å})$	

Viséite	Hsianghualite
$Na_2Ca_{10}(Al_{20}Si_6P_{10}O_{60}(OH)_{36}) \cdot 16\,H_2O$	$Li_{16}Ca_{24}(Be_{24}Si_{24}O_{96}) \cdot F_{16}$
cubic	$I2_13$
$a = 13.65\,(\text{Å})$	$a = 12.88\,(\text{Å})$

2.1.1 History and Nomenclature

Analcime has been known since antiquity, because it is common and easy to recognize; Hintze (1897) writes that Dolomieu was the first scientist to report in 1784 on this mineral from the Isola dei Ciclopi, Sicily, but it is difficult to accept that nobody described it prior to this date. Haüy (1801) proposed the name "analcime" from the Greek ἄναλκις = forceless because of its poor ability in acquiring frictional electricity. The spelling "analcite" is widely used in English-speaking countries, but should be dropped, because there is no good reason to abandon the first proposed name.

Wairakite was found by Steiner (1955) at Wairakei, New Zealand, and named for this locality.

Viséite was first described by Mélon (1943), as a new mineral found at Visé, Belgium, hence the name.

Hsianghualite (also spelled hsinghualite) was described by Wen-Hui et al. (1958) as a new mineral from Hunan, China; the name is from the Chinese words meaning "fragrant flower".

Kehoeite (Headden 1893; McConnell and Foreman 1974) also has the analcime framework, but it will not be considered here, being a phosphate.

2.1.2 Crystallography

The structure of analcime was determined in cubic space group Ia3d by Taylor (1930) and refined in the same space group by Calleri and Ferraris (1964), Knowles et al. (1965), Ferraris et al. (1972). These authors did not find any Si/Al order. The sodium atoms were placed at the centers (S) of the cage of Fig. 0.2Ga with an occupancy factor of 66%, as the multiplicity of this site is 24, but there are only 16Na in the unit cell; the 16 water molecules fully occupy the site coded W in Fig. 0.2Ga, with multiplicity 16. Please remember (see Fig. 0.2Gb) that each tetrahedral node is part of three cages and so each W-site, whereas each S-site pertains to one cage only.

It was long known, not only on the basis of optical observations, but also from X-ray photographs, that analcime shows deviations from cubic symmetry (e.g. Coombs 1955). Mazzi and Galli (1978) showed that at least two different symmetries, $I4_1/acd$ and Ibca, may be present in different non-cubic analcimes, for the tetragonal ones both $a > c$ and $a < c$ being possible. This may be explained in the following way: (please reconsider here the three cages of Fig. 0.2Gb):

1. If the crystal is cubic all three cages are equivalent and so are all tetrahedral nodes and all three S-sites (this is nearly true for sample ANA-5 of Mazzi and Galli).
2. If the crystal is tetragonal, two cages are equivalent and the third is not, so the Na-occupancy in the first two S-sites may be higher and the Al-fraction correspondingly higher in nearby tetrahedral nodes, with both Na-occupancy and Al-fraction lower in the third cage (this is true for samples ANA-1, ANA-2, ANA-3 of Mazzi and Galli, with $a > c$); the two equivalent cages may also have a lower Na-occupancy and Al-fraction than the third non-equivalent cage (sample ANA-6 of Mazzi and Galli, with $a < c$).
3. If the crystal is orthorhombic, the three cages have three different Na-occupancies and correspondingly proportionally different Al-fractions in nearby tetrahedral nodes (this is true for samples ANA-4 and ANA-7 of Mazzi and Galli).

◀ **Fig. 2.1A.** Analcime
from Alpe di Siusi,
Italy (×4)

Fig. 2.1B. Analcime
from San Giorgio,
Laverda, Italy (SEM
Philips, ×250)
▼

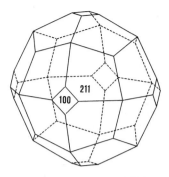

◀ **Fig. 2.1C.** Common morphology of analcime

Fig. 2.1D. Sedimentary analcime from Ischia, Italy (SEM Philips, ×950)
▼

   Hazen and Finger (1979) confirmed the variability of analcime, measuring 15 samples whose symmetry was either cubic, or tetragonal, or orthorhombic, or even monoclinic or triclinic.

   In wairakite (Takéuchi et al. 1979), Ca is placed in only one out of the three cages, the other two being empty, or more precisely, sparsely occupied by Na; correspondingly, Al fully occupies the two tetrahedral sites near Ca, the other Al-fractions being very near to zero. The symmetry of the thus described wairakite would be $I4_1/acd$: it is really $I2$, because such distortion allows more proper $Ca - O$ distances. Nakajima (1983) described a wairakite which probably has a disordered (Si, Al)-distribution, because of its "tetragonal" powder pattern, which is non-cubic but without the line splittings typical of the monoclinic system.

Fig. 2.1E. Sedimentary analcime from Ischia, Italy (SEM Philips, ×1900)

Fig. 2.1F. Wairakite from Wairakei, New Zealand (SEM Philips, ×120)

The structure of viséite is not known, but probably it shares the same framework with analcime in view of several crystallographic analogies.

Hsianghualite (Section of etc. 1973) has a perfect alternation of the Be and Si in the tetrahedra of the framework; the 24 Ca-atoms occupy all the 24 S-sites and the 16 F-atoms all the 16 W-sites (see Fig. 0.2 G); the 16 Li-atoms are in tetrahedral cavities, each coordinating one F and three O of the framework.

The morphology of analcime (Fig. 2.1 A to E) is quite simple: it nearly always features the (pseudo)cubic trapezohedron or icositetrahedron or leucitohedron {211}; other forms, like {100}, {110}, {210}, {112}, have been observed, but are rare. Pseudo-merohedral twins are easily developed in all non-cubic analcimes: two sets of twin-lamellae at 90° are frequent (contact plane {101}?). The morphology is also the same (Fig. 2.1 D and E) in sedimentary samples (see also Mumpton and Ormsby 1976 and 1978).

Wairakite shows occasionally pseudo-octahedral or pseudo-icositetrahedral faces. Pseudo-merohedral twins are always present as in analcime.

Viséite was found as small wart-like masses.

Hsianghualite gives granular masses, but also trisoctahedra or dodecahedra (?), these last ones being more in agreement with the symmetry if either pentagonal or rhombic dodecahedra.

2.1.3 Chemistry and Synthesis

The chemistry of analcime is rather simple; normal hydrothermal analcimes have a composition quite near to the stoichiometric formula of the title page; also sedimentary analcimes have only sodium as extra-framework cation, but the ratio Si/Al may range with continuity from 2 to 3: analysis no. 2 in Table 2.1 I is only one example of this "acid" (i.e. silica-rich) analcime, but Iijima and Hay (1968) examined 69 analcimes from the Green River Formation, Wyoming, finding $1.95 < \text{Si/Al} < 2.81$, every value in between being represented, with two frequency maxima at 2.15 and 2.75; as a matter of fact, they did not perform chemical analyses for all these samples, but they deduced the Si/Al ratios from the unit cell parameters using a diagram of Saha (1959). Similar results were obtained also by Coombs and Whetten (1967).

Wairakite (no. 6 in Table 2.1 I) may be quite near to the stoichiometric formula. A continuous isomorphous series exists between analcime and wairakite, as sufficiently proven by analyses nos. 3, 4, and 5 of Table 2.1 I. Aoki and Minato (1980) gave detailed data on the lattice constant variations at the wairakite side of this series. Similar data on the whole series were plotted by Harada et al. (1972) and by Harada and Sudo (1976); these last authors put forward the hypothesis that the β-angle is not exactly 90° at the analcime end either, the difference being only 10' or so, at least in some cases (they write of a "monoclinic sodium analogue of wairakite").

Analysis no. 7 was obtained from a sample of the only known occurrence of hsianghualite.

Table 2.1I. Chemical analyses and formulae on the basis of 96 oxygens and lattice constants in A

	1	2	3	4	5	6	7
SiO_2	54.26	58.62	57.41	56.27	57.07	55.90	36.15
Al_2O_3	23.26	19.37	20.11	21.32	20.64	23.00	0.50
Fe_2O_3		0.95			0.48		0.14
BeO							16.04
MgO		0.23	0.70	0.43	0.04		0.17
CaO	0.04	0.15	1.42	6.17	8.30	11.70	34.89
Li_2O							5.72
Na_2O	13.78	11.00	10.79	6.57	4.14	1.06	0.08
K_2O	0.06	0.38	0.07	0.17	0.14	0.16	0.04
H_2O	8.60	8.22	9.04	8.94	9.43	8.50	1.28
F							7.54
Total	100.00	99.02	99.54	99.87	100.24	100.37	99.42
Si	31.95	34.32	33.64	32.87	33.33	32.30	23.23
Al	16.14	13.37	14.01	14.79	14.40	15.67	0.37
Fe		0.42			0.20	0.02	0.07
Be							24.77
Mg		0.20	0.62	0.38	0.04		0.16
Ca	0.03	0.09	0.90	3.89	5.26	7.24	24.03
Li							14.80
Na	15.73	12.49	12.36	7.50	4.75	1.19	0.10
K	0.05	0.28	0.05	0.13	0.11	0.12	0.03
H_2O	16.89	16.05	17.50	17.44	18.18	16.38	2.74
F							15.32
E%	+2.0	+3.2	-9.3	-8.5	-5.6	-1.0	
D	2.27		2.25	2.25	2.26	2.26	2.98
a	13.729		13.70	13.66	13.66	13.69	12.897
b				13.66	13.66	13.68	
c	13.709			13.65	13.58	13.56	
β				90°20'	90°20'	90°30'	

1. Tetragonal analcime, Val Duron, Trento, Italy (specimen ANA-3 of Mazzi and Galli, 1978; see also Monese and Sacerdoti, 1970); D is calculated.

2. Sedimentary silica-rich analcime, Big Sandy Formation, Arizona (Sheppard and Gude, 1973) with 0.06% TiO_2, 0.02% SrO, 0.02% BaO.

3. Calcian analcime, Tanzawa Mts., Japan (specimen T-3 of Seki and Oki, 1969).

4. Sodian wairakite, Tanzawa Mts., Japan (sample T-2, do).

5. Sodian wairakite, Tanzawa Mts., Japan (sample T-1, do).

6. Wairakite, Wairakei, New Zealand (Steiner, 1955) with 0,017% Rb_2O, 0.05% SrO.

7. Hsianghualite (Beus, 1960), total less 3.13% O for F.

As chemical composition for viséite, the average values of Mélon (1943) can be assumed, but the $H_2O\%$ value of McConnell (1952) has to be preferred, thus obtaining the composition: P_2O_5 20.49%, SiO_2 9.93%, Al_2O_3 26.15%, CaO 14.79%, Na_2O 1.77%, H_2O 26.11%, F 0.67%, total 99.91%. Assuming a $D_{calc} = D_{obs} = 2.2$, the following chemical formula is calculated and framed in the analcime-scheme

$$Na_{1.94}Ca_{8.96}(Al_{17.43}Si_{5.62}P_{9.81}[H_4]_{12.08}O_{96}) \cdot 25.09\,H_2O$$

F has been disregarded in this formula calculation; the substitution $SiO_4 \leftrightarrow (OH)_4$, common in various silicates, for instance in hydrogrossular, has been assumed as present in viséite.

After Breck (1974, p. 252), the first synthesis of analcime was done by de Schulten in 1880, but the first synthesis proven by X-ray data is due to Barrer and White (1952) who obtained an easy hydrothermal crystallization from gels $Na_2O \cdot Al_2O_3 \cdot nSiO_2$ with $3 \leqslant n \leqslant 5$, at a pH ~ 8 or with 100% NaOH excess with $1 \leqslant n \leqslant 5$, in both cases with $200 < T < 250°C$.

The literature on synthesis and equilibrium diagrams of analcime is so wide that only the main results will be reviewed here.

Ueda and Koizumi (1979) were able to synthesize crystals of slightly silicic analcime a few μm in diameter at a temperature as low as 100°C, starting from sodium aluminosilicate gels containing an extremely small quantity of alumina.

Guyer et al. (1957) reported on the possibility of obtaining phases with the analcime structure with the Si/Al ratio ranging from 1.5 (typical of natrolite) to 2 (the stoichiometric value) to 5 (a high ratio typical of an acid zeolite like mordenite). Saha (1959) presented a good correlation between the refractive index and unit cell edge of both synthetic and natural analcime with the Si/Al ratio, at least for the range $1.5 - 3.0$. Yoder and Weir (1960) discovered that, at 25°C, in the pressure range from 0.001 to nearly 10 Kbars, analcime features a compressibility which increases with pressure: such abnormal behaviour usually preceeds a phase transition, and as a matter of fact one of the investigated samples (from Golden, Colorado) presented a transition at 8.4 Kbars to a slightly denser phase with a higher birefringence, and hence possibly a lower symmetry. Hazen and Finger (1979) confirmed the results of Yoder and Weir (1960) finding that two samples (Isola dei Ciclopi, Sicily, and Golden, Colorado) undergo two symmetry transitions at 4 and 12 Kbars and two volume discontinuities at 8 and 18 Kbars. At 1 bar, the sample from Sicily is tetragonal with $a < c$ and the one from Golden is orthorhombic: at the first transition, both loose orthogonality and become monoclinic centered with pseudo-cubic [110] as unique axis, and triclinic at the second transition; the first volume discontinuity is 1.5%, the second one 0.2%. Hazen and Finger consider these transitions as a particular kind (polyhedral tilting) of second order transformation.

Saha (1961) reported on the stability field of analcime solid solutions in the $NaAlSiO_4 - NaAlSi_3O_8$ range, with $P(H_2O) \lesssim 3$ Kbars and $T < 800°C$, and

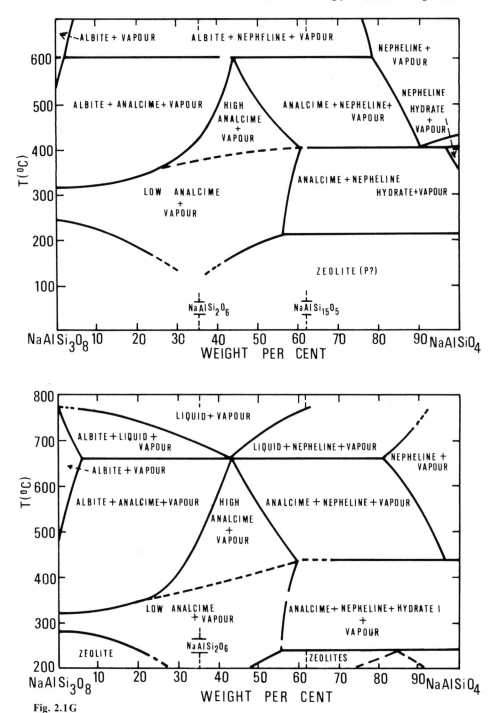

Fig. 2.1G

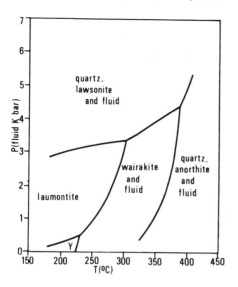

Fig. 2.1H. P(H$_2$O)-T stability diagram of laumontite-wairakite after Liou (1971b) and Zeng and Liou (1982); yugawaralite is stable in the field marked *Y*

gave improved diagrams of refractive indexes and lattice parameters versus Si/Al ratio. Kim and Burley (1971, 1980) published two complete isobaric diagrams NaAlSiO$_4$ – NaAlSi$_3$O$_8$ at P(H$_2$O) = 2 and 5.15 Kbars (both are reproduced in Fig. 2.1 G). These diagrams confirm the known possibility of analcime solid solutions over a wide range of Si/Al, and contain a definite, quite large field of a "high" analcime, with a unit cell edge smaller than normal or "low" analcime. Kim and Burley give evidence for excluding the possibility that the low-high transition may be due to dehydration and favour the idea of some structural change; they suggest a possible relation of their "high" analcime with the high pressure phase first described by Yoder and Weir (1960), both transitions being characterized by a unit cell volume reduction.

From all these reports (Guyer et al. 1957; Saha 1961; Kim and Burley 1971, 1980), it is obvious that all equilibrium diagrams involving analcime must be considered with caution, first because all equilibria are sluggish at temperatures lower than 200 °C, and second, because the analcime involved in reactions may have different compositions. The results of Liou (1971c) are difficult to reconcile with the diagrams of Fig. 2.1 H. In fact he gives a P(H$_2$O)/T diagram for the decomposition of a (slightly silicic) analcime into albite + nepheline + fluid which, at P(H$_2$O) = 2 Kbars should occur at 570 °C, but in these conditions we are in the field albite + analcime + fluid of Kim and

---

Fig. 2.1G. Isobaric (*top* P(H$_2$O) = 2 Kbar; *bottom* P(H$_2$O) = 5.15 Kbar) binary diagram of the system NaAlSi$_3$O$_8$ – NaAlSiO$_4$, after Kim and Burley (1980); the stability fields of both high- and low-analcime are wide, and extend up to the albite composition; zeolites are probably species P (gismondine-like)

Burley; albite was probably associated with analcime under 570°C also in Liou's runs. Thus, the diagrams for the reactions

analcime + quartz + fluid = albite + fluid

given by Liou (1971c) and Thompson (1971b) cannot be related in a simple way to the diagrams of Kim and Burley (1980); in fact, according to the first two authors mentioned, the reaction boundary is nearly constant at ~200°C from 2 to 5 Kbars, and according to Kim and Burley, with an albite (= analcime + quartz) composition and both at $P(H_2O) = 2$ and 5 Kbars, a zeolitic phase (probably P, see Chap. 3.1.3) is stable up to 250°C, but a phase with analcime structure and albite composition is stable in a temperature interval around 300°C, which is more in agreement with the previous results of Coombs et al. (1959). We think that metastability and variability of composition of analcime could explain these discrepancies. Hamilton and Ryabchikov (1976) calculated P/T curves for the two reactions:

analcime + albite = jadeite + fluid

analcime = jadeite + nepheline + fluid .

These results cannot be reviewed here, but evidence for the difficulty of the problem is given simply by the second reaction, which would be meaningless if components were stoichiometric

$$NaAlSi_2O_6 \cdot H_2O = NaAlSi_2O_6 + NaAlSiO_4 + H_2O$$

but the real and correct reaction is

$$Na_{1.07}Al_{1.07}Si_{1.93}O_6 \cdot H_2O = 0.815\, NaAlSi_2O_6 + 0.278\, Na_{0.92}Al_{0.92}Si_{1.08}O_4$$
$$+ H_2O \, .$$

Roux and Hamilton (1976) emphasized the impossibility of analcime coexisting with melt in the albite-orthoclase-nepheline-kalsilite-water system, if not in the interval $600 < T < 640°C$, $5 < P(H_2O) < 13$ Kbars. Arima and Edgar (1980) gave evidence for the very slow rate of analcime decomposition into albite + analcime + fluid, especially when $P(total) > P(H_2O)$, thus explaining preceeding results which were wrong due to the failure to detect albite.

The synthesis of analcime too has been obtained many times starting from natural materials; more recent publications also describe reactions similar to natural processes. Franco and Aiello (1968) obtained analcime treating hydrothermally halloysite plus enough NaOH so as to have $Na_2O : Al_2O_3 = 2$ with $200 \leqslant T \leqslant 300°C$. Höller (1970) succeeded in recrystallizing hydrothermally several minerals and rocks to analcime: nepheline, nephelinite, sodalite, nepheline-basanite, basaltic glass, the sodifying agent being NaOH from 0.01 to 4N; he found a high crystallization rate at high T and pH, but also no crystallization at all if pH was too high at high T, in this case only silica phases being the product. Boles (1971) used heulandite and clinoptilolite as starting materials, and added NaOH 0.1 M or Na_2CO_3 0.1 to 0.01 M: in three weeks at 100°C analcime was formed, in conditions not far from those prevailing in a saline-alkaline lacustrine environment. Similar to natural conditions is also the analcimization of rhyolitic tuff and green tuff at 200°C in 50 – 260 h, with 12% Na_2SiO_3 added (Abe et al. 1973). Abe and Aoki (1976) later used a clinoptilolite tuff which was treated hydrothermally with Na_2CO_3 or $NaHCO_3$ 0.25 to 0.5 M at 100°C – 300°C for 240 h ($8 \leqslant pH \leqslant 11.2$); analcime was formed also at pH = 8 in the temperature interval 150°C – 220°C, but if pH was 10 or higher, the temperature could be as low as 100°C. Kirov et al. (1979) used an autoclave with a temperature gradient from 100°C to 260°C and studied the reaction of perlite and basaltic glass with NaOH (+ KOH in some cases) from 0.2 to 1 M; analcime was obtained from both materials mainly at high T (>150°C), together with mordenite and clinoptilolite from perlite, with zeolite P and tobermorite from basaltic glass. Data on other alterations of natural materials to analcime may be found in Kirov et al. (1975) and Kirov and Pechigargov (1977).

The researches by Wirsching (1975) and by Höller and Wirsching (1978) are quite interesting because the treatment with mineralizing liquids was performed both in closed systems (autoclaves) and in simulated open systems (autoclaves with frequent renewal of the fluid), thus approaching two typical natural situations. Starting materials were three glasses: basaltic, phonolitic, rhyolitic. In closed system, at T = 180°C or 200°C or 250°C, with H_2O only or with NaOH 0.01 N, analcime was easily obtained both from the basaltic and the phonolitic glass at the two higher temperatures, but not at 180°C, where chabazite and phillipsite appeared in the runs with NaOH 0.01 N; almost no analcime was obtained from the rhyolitic glass. In the simulated open system, at 250°C, analcime was obtained from the phonolitic glass both with H_2O and NaOH 0.01 N, but only with NaOH from the rhyolitic glass. On the whole, the chemistry of the starting glass is very important in closed systems, but much less in open systems.

Boussaroque and Maury (1972) gave evidence for the impossibility of simultaneous synthesis (and hence simultaneous genesis) of analcime and dawsonite $NaAl(CO_3)(OH)_2$, frequently associated in saline lake deposits (see Chap. 2.1.6); the easiness of formation of dawsonite from a sodic solution of aluminium sulphate at about 100°C, and the easy transformation of dawsonite into analcime at 200°C in the presence of colloidal silica explain the dawsonite-analcime paragenesis.

The first certain synthesis of wairakite, proven by X-ray, was performed by Ames and Sand (1958) from gels of composition $CaO \cdot Al_2O_3 \cdot 5SiO_2$ at $330 < T < 450°C$, P ~ 1 Kbar; they were unable to obtain wairakite from gels of the stoichiometric composition $CaO \cdot Al_2O_3 \cdot 4SiO_2$, nor did they control the chemical composition of the obtained phase. Coombs et al. (1959) made the synthesis starting from oxide mixes or glass (of unknown composition) or from several calcium minerals, and were able to state that wairakite is stable from about 370°C to 450°C and up to 4 Kbars: the wairakite field is limited at lower T by epistilbite, at higher T by anorthite, and at higher P by prehnite. Koizumi and Roy (1960), operating from gels, obtained the phase from 300 °C to 500 °C at P ~ 1 Kbar, also from the stoichiometric composition, but more easily with SiO_2 excess. Barrer and Denny (1961) had similar results with gels $CaO \cdot Al_2O_3 \cdot nSiO_2$ with $3 \leqslant n \leqslant 9$ and $250 < T < 450$ °C: the powder pattern of the phases obtained in the lower T range (250 °C – 350 °C) could be indexed as tetragonal (as the "disordered" wairakite of Nakajima 1983: see Chap. 2.1.2), but phases of the higher T range (350 °C – 450 °C) gave a cubic powder pattern. Results varied using gels $3 CaO \cdot Al_2O_3 \cdot nSiO_2$, because wairakite-like phases were present only if $5 \leqslant n \leqslant 9$, the temperature influence on symmetry being the same.

Harada et al. (1966) obtained wairakite starting from other calcium zeolites (stilbite and heulandite).

Seki (1968) emphasized that the powder pattern of all these synthetic phases could be considered cubic, or tetragonal (with the 400/004 splitting), but never monoclinic as in the natural wairakites (with the $332/\overline{3}32$, $422/\overline{4}22$ and similar splittings). Liou (1970) was able to synthesize true monoclinic wairakites starting from stoichiometric oxide mixes, or laumontite or anorthite + quartz; metastable disordered (cubic or tetragonal) wairakite was obtained first and easily, but it took from one to two months of continuous hydrothermal treatment at 350 °C and 2 Kbars = P (fluid), to achieve a material with a well defined monoclinic powder pattern, and hence with (Si, Al)-order in the framework.

Liou (1970, 1971 b) and Zeng and Liou (1982) studied several equilibrium diagrams involving wairakite: Fig. 2.1 H gives a somewhat simplified summary of all these results. Wairakite has a definite stability field, as has yuga-waralite in a low P low T region. Epistilbite has probably only one field where metastable growth is possible. On the lower T side, wairakite goes to laumontite, which has the same chemical formula but with more zeolitic water; on the higher T side, wairakite disappears in favour of anorthite plus quartz; over the high P limit, a less hydrated silicate like lawsonite appears, but the transformation into prehnite is also possible. Somewhat different limits of stability were found by Juan and Lo (1971).

Wirsching (1981) synthetized a definitely non-cubic wairakite, due to the 400/004 splitting, but without any proof of monoclinicity, using a simulated "open system", namely studying the hydrothermal reaction of rhyolitic glass, basaltic glass, nepheline, oligoclase with 0.1 to 0.01 N $CaCl_2$ solutions at T = 200 and 250 °C; the solution was repeatedly renewed during the experiment.

There is no report on syntheses of viséite and hsianghualite.

2.1.4 Optical and Other Physical Properties

The mean refractive indices of analcime, of some intermediate members of the analcime-wairakite series and of viséite are reported in Table 2.1 II. Saha

Table 2.1 II. Optical properties

	1	2	3	4	5	7
n_{mean}	1.487	1.483	1.489	1.496	1.499	1.613
		1.488				

Numbers refer to the same samples of Table 2.1 I.

(1959) reported for synthetic analcimes with increasing Si/Al ratios the following refractive indices:

$Na_{1.2}(Al_{1.2}Si_{1.8}O_6) \cdot H_2O$ 1.494

$Na_1(Al_1Si_2O_6) \cdot H_2O$ 1.488

$Na_{0.75}(Al_{0.75}Si_{2.25}O_6) \cdot H_2O$ 1.482

Many analcimes show some birefringence, of around 0.001 with optical sign negative and 2V small.

Analcime is colourless, white, grey, greenish or reddish, with vitreous luster (red analcime contains haematite, green analcime contains chlorite, after Aurisicchio et al. 1975); Mohs-hardness = 5 – 5.5, cleavage {100} poor.

IR and Raman-spectra (Velde and Besson 1981) are negligibly influenced by pressure; this is completely true for T – O vibrations, some band splitting being detected at least in two O – OH vibrations; there are three sites of hydrogen bonding in analcime.

For wairakite: $\alpha = 1.498$ $\gamma = 1.502$ $2V_\gamma$ from 70 to 105° with $\alpha \to b$ and, nearly, $\beta \to a$ and $\gamma \to c$.

Colourless to white, vitreous lustre; Mohs-hardness = 5.5 – 6, cleavage poor and uncertain.

Viséite is isotropic with n = 1.530, white or bluish or yellowish; Mohs-hardness = 3 – 4.

Hsianghualite is isotropic with n = 1.6132, colourless, transparent to translucent; Mohs-hardness = 6.5, no cleavage.

2.1.5 Thermal and Other Physico-Chemical Properties

Thermal curves of analcime are reported in Fig. 2.1 I. Both the DTG and DTA show only one maximum at 350°C, corresponding to the loss of water, which is in only one symmetry independent site in the cubic structure, the deviation from cubicity being too low to be felt in such a thermal analysis. Rehydration is possible and easy. The first small water loss at 100°C is probably due to surface adsorption.

The water loss of wairakite (see Fig. 2.1 J) happens in two steps at 360°C and 500°C, correspondingly to the distribution of water into two symmetrically independent sites (Takéuchi et al. 1979). Rehydration is easy.

Thermal curves for viséite and hsianghualite are not known.

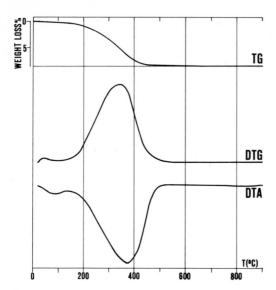

Fig. 2.1I. Thermal curves of analcime from Val Duron, Italy; in air, heating rate 20°C/min (by the authors)

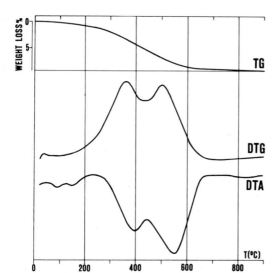

Fig. 2.1J. Thermal curves of wairakite from Wairakei, New Zealand; in air, heating rate 20°C/min (by the authors)

  The ion exchange properties of phases with this framework are quite interesting and have been the subject of many publications. To explain the results, it is necessary to remember that, strictly speaking, there is no structural K-equivalent of analcime, because in leucite $K(AlSi_2O_6)$ the potassium atoms occupy the W(ater)-sites, the S-sites being empty. So Barrer (1950) and Barrer and Hinds (1953) were able to obtain leucite-like waterless phases by ion-exchange of analcime with solutions containing K, NH_4^+, Tl, Rb, at $T = 200°C$ or higher, and gave also evidence for the coexistence of two phases when the exchange was incomplete; at T ~ 100 °C, they obtained full exchange only with Tl. The Ca- and Ag-exchanged phases were also obtained, these have a hydrated analcime structure. Barrer and Hinds (1953) also give an exchange isotherm for Na, K in the analcime-leucite system, total molarity = 3.5, T = 110 °C; the system shows a large hysteresis and the isotherm is such that the Na-concentration in the solid phase remains nearly 90%, if the K/Na ratio in the solution is not over 3.

Balgord and Roy (1971) tested the exchange properties of three analcimes with Al/Si = 0.66, 0.50, 0.33; the results with the stoichiometric (0.50) phase are as follows: complete exchange with K, NH_4^+, Rb, Tl, was possible at 100 °C (at 250 °C with Tl), and gave phases with leucite structure; an extended immiscibility gap was proven in the K – Na series. A continuous series with the analcime-type structure was obtained by Li- and Ca-exchange at 250 °C; Na could be replaced only in part with Mg and Sr (and also with Ni and Co). The exchange properties of the other two analcimes were good with Ca, but more difficult with K. The reverse exchange of leucite with Na-solutions is quite important for geologists, and was studied by Sersale (1962), who found that at T = 235 °C and P = 200 Kg/cm^2 the ion exchange Na→K proceeded at a substantial rate, given a proper Na-concentration in the contact solution, and by Gupta and Fyfe (1975), who found that leucite is ion-exchanged to analcime in a few days if the contact solution is a sodian one but also if it is a solution similar to sea water, at T = 150 °C; they give sufficient evidence for claiming that also at 25 °C and with a ratio K/Na = 1/3, the exchange will proceed at a fast rate, at least from the geological point of view: these results are a confirmation of the above mentioned ones by Barrer and Hinds (1953).

As for the adsorption of gases, Breck (1974) quotes Milton (U.S. Pat. 2,996,358) who found that, of several analcime specimens, only one was able to adsorb argon, nitrogen, methane and ethane, this adsorbing specimen being calcium-bearing, in contrast to non-adsorbing specimens, which were calcium-free.

2.1.6 Occurrences and Genesis

Analcime features a wide spectrum of genetic types, from sedimentary, both continental and marine, to metamorphic of low grade, to magmatic, and also hydrothermal, a type of genesis responsible for all fine exhibition specimens.

Surface genesis has been attributed to analcime of the bauxite beds of North Onega, USSR, by Kal'berg and Levando (1962), who related the zeolite crystallization to a high pH in the surface crust; a successive reduced alkalinity permitted the crystallization of kaolinite and boehmite but, later on, Nasedkina (1980) reinterpreted this analcime as low hydrothermal alteration of the volcanic rock which is under the bauxite bed. The necessary high pH may be the result of the concentration of sodium bicarbonate-carbonate by evaporation: this is the case of saline, alkaline soils of the San Joaquim

Valley, California (Baldar and Whittig 1968) and Ruzizi, Burundi (Frankart and Herbillon 1970), where analcime occurs with clay minerals.

Analcime is a typical mineralogical component of the tuff-bearing saline alkaline lake deposits; on the basis of a wide review of many occurrences of this kind (Teels Marsh, Nevada; Owens, China, Searles Lakes and Kramer, California; Lake Natron, Peninj Beds, Olduvai Gorge, Tanzania; Green River Formation, Wyoming; Lockatong Formation, New Jersey), Hay (1966) showed that the crystallization of analcime is related more to the chemical condition (alkalinity, salinity) of the brines than to the chemistry of the glasses and other components of the sediments, thus excluding a direct formation of analcime from volcanic glasses, he found good evidence for the possibility of such a growth, at least in one occurrence, Lake Natron. These saline alkaline lake deposits ("closed systems" of Surdam and Sheppard 1978) usually feature a series of concentric zones at different distances from the lake: typically the innermost feldspar zone is surrounded by the analcime zone, which in turn is followed by the zone "other zeolites", and by a fresh glass zone with some clay minerals (montmorillonite). Typical examples of this lateral zonation are Lake Tekopa, California (Sheppard and Gude 1968), the Big Sandy Formation, near Wikieup, Arizona (Sheppard and Gude 1973), the Green River Formation, Wyoming (Surdam and Parker 1972). All these analcimes are silica-rich, as stated in Chap. 2.1.3. The pioneering research of Ross (1928, 1941) lead to the discovery of the sedimentary analcime in Yavapai Co. and near Wikieup, Arizona, which both certainly pertain to this kind of genesis. Taylor and Surdam (1981) re-described as closed system the occurrence of analcime at Teels Marsh, Nevada, where interstitial brines convert phillipsite to analcime.

The formation of a sediment that is quite rich in analcime in the Tuzla salt basin, Bosnia, Yugoslavia, is probably similar to a closed system genesis; these Miocene sediments contain neither any other zeolite, nor authigenic feldspar, and do not show lateral zoning either; the evaporitic character is the only feature in common with the saline lake deposits: if a comparison is possible, we would say that the analcime zone only was developed (Gottardi and Obradovic 1978). Another "closed system" with analcime only, is "La Limagne de Clermont-Ferrand", France, where Giot and Jacob (1972) found from 5% to 10% of the zeolite spread in the sediments of the former lake.

Analcime is also known to occur as alteration product of tuffaceous deposits in "open systems", namely by slow water percolation: this could be the case of the Quaternary tufo verde (green tuff) Epomeo of the island of Ischia, Italy, rich in analcime and phillipsite (Gottardi and Obradovic 1978). In the Vieja Group, Texas, a thick layer of Oligocene tuff is altered in the same way, and a typical vertical zoning is developed over the water table with an upper montmorillonite zone and an intermediate clinoptilolite zone overlaying a lower analcime zone: a growing concentration of the percolating solution may be hypothesized to explain the zoning (Walton 1975). Hay and Iijima (1968a, b) described an occurrence of analcime which crystallized by slow percolation of meteoritic water through palagonite tuffs of Koko Craters, Oahu, Hawaii; the palagonite in turn is the product of alteration of sidero-melane; chabazite, phillipsite, montmorillonite, opal, calcite, aragonite, gypsum are also present, the analcime averaging 5 to 10% in porous pala-gonite tuffs.

The most common zeolites in marine sediments are clinoptilolite and phil-lipsite, but analcime is not rare. The first occurrence of this type was dis-covered already in 1891 by Murray and Renard; an up to date review may be found in Kastner and Stonecipher (1978; see also Bass et al. 1973). Analcime is always found associated with basaltic glass or components, but its direct derivation from these materials has been proved in only one case (see below). Sedimentary materials associated with analcime include smectite, chlorite, celadonite and the zeolites phillipsite or clinoptilolite (perhaps more frequent-ly this last one). The abundance of analcime increases with age of the sedi-ments, but not with burial depth. On the whole the hypothesis that most anal-cime grew at the expense of less stable, pre-existing zeolites is probably true but not definitely ascertained. A case of proven derivation from pre-existing phillipsite was described at Cape Breton Island, Nova Scotia, by von Bitter and Plint-Geberl (1980), who found analcime pseudomorphs after phillipsite in marine carbonates of Mississipian age. A case of analcime formation directly from volcanic glass was described by Müller (1961) for recent sub-marine sediments in the Gulf of Naples, Italy; analcime crystallized from glass is associated with analcimized leucite. Sameshima (1978) found significant percentages of analcime in the matrix of volcanoclastic beds in the Miocene marine sedimentary sequence of the Waitemata group, New Zealand. Djourova (1976) found a green layer 10 km long, 2 km wide, and 5 m to 190 m thick in the Oligocene pyroclastic marine horizons of the Northeastern Rhodopes; this layer is rich in analcime (over 50% in some places), with quartz, potassium feldspar and chlorite, and is sometimes overlain by a clin-optilolite layer.

The most typical sedimentary occurrences of analcime are certainly those which can be attributed to burial diagenesis or to burial metamorphism of very low grade, a clear separation of the two being always questionable. Thick sequences of volcanoclastic sediments are the most suitable environments for the crystallization of several zeolites, and also for analcime. A vertical zoning, extending also for several km, may be developed, analcime being present from the surface to the middle zones. Coombs et al. (1959), when introducing the zeolite facies, considered a heulandite-analcime stage as the lowest grade of the facies, in other words as a transition to diagenesis. As a matter of fact, these authors found analcime from surface down to considerable depth in several volcano-sedimentary formations in New Zealand. The three stages of the zeolite metamorphic facies after Coombs et al. (1959) are:
1. heulandite-analcime stage
2. laumontite stage (2 – 3 albite-chlorite stage)
3. prehnite-pumpellyite stage

Utada (1965, 1970, 1971), on the basis of many researches on the Tertiary sedimentary sequences of Japan, extended the zeolite facies concept to the diagenesis side and considered the different zones with authigenic minerals including both diagenesis and very low metamorphism, as caused by burial depth and/or diffused low hydrothermalism; he proposed the following scheme:

Zone I   Altered glass zone (with glass, both fresh or altered, opal, montmorillonite and, rarely, also chabazite and phillipsite).
Zone II   Clinoptilolite-mordenite zone (with these two zeolites, plus opal, montmorillonite, celadonite, corrensite, carbonates).
Zone III   Analcime-heulandite zone (with these two zeolites, plus feldspar, opal, quartz, montmorillonite, mixed-layer clay minerals, chlorite).
Zone IV   Laumontite zone (with laumontite, feldspar, montmorillonite, chlorite, sericite, quartz).
Zone V   Albite zone (with albite, adularia, quartz, chlorite, sericite, calcite).

Clearly enough, the first two zones correspond to an increasing diagenesis, and the last three broadly match the three metamorphic stages of the zeolite facies. In case of diagenesis and/or low metamorphism, the succession is a vertical one and related to burial depth; in case of low hydrothermalism, it is inversely related to the distance from the thermal source (Utada 1980). Analcime is hence considered as typical of zone III at the boundary between diagenesis and metamorphism; in the same zone wairakite may substitute for analcime under proper chemical conditions. The crystallization of analcime is highly favoured in this zone, and highly improbable or impossible in subsequent zones, probably possible in all preceeding zones but only in a favourable chemical environment: evidence for this possibility is given by the proven genesis of analcime at surface or at shallow depth.

Many other occurrences of analcime can be framed in such a scheme: so the three zones of the diagenetic sequence of Pahute Mesa, Nevada (Moncure et al. 1981) closely correspond to the first three of Utada, with analcime abundant as pseudomorphs of clinoptilolite and as a pore cement in the third zone extending from 1.2 to 3 km of depth. Moiola (1970) described a sedimentary tuffaceous Neogene sequence nearly 3 km thick, in the Esmeralda Formation, Nevada; there is a typical zoning with analcime and potassium feldspar in the lowest 180 m, clinoptilolite in the large intermediate zone, and phillipsite in a thin zone on the top.

Analcime has been found in sediments whose mineralogical compositions have been related neither to a definite burial depth nor to a definite diagenesis or metamorphism. So the first known description of a sedimentary zeolite, analcime in Lower Cretaceous ferriferous clay of Duingen near Alfeld/Leine, Germany (Seebach 1862) is due to an undetermined grade of diagenesis. A typical diagenetic origin may be assumed for the analcime which builds up the cement of many sandstones, like the Lower Devonian red sandstone in the northern Caucasus, U.S.S.R. (Rengarten 1950) and like the Buntsandstein of Triassic age in northern Germany (Füchtbauer 1967); the pore-filling water was found to be saline and alkaline. Abundant analcime was found also in the cement of red and white fine-grained feldspathic sandstones called the Stormberg Red Beds, Upper Permo-Triassic Karroo System, South Africa (Fuller and Moir 1967); this analcime too is certainly diagenetic, but cannot be attributed to any particular zone.

Echle (1975) described an occurrence of analcime of intermediate type in West Anatolia, Turkey; a thick (~ 500 m) deposit of saline alkaline lacustrine tuffaceous sediments is overlain by an evaporitic layer; abundant ($\sim 10\%$) analcime is present at all depths and its genesis is related to the presence of a pore-filling brine which originated either from the upper evaporite, or by reaction of pore-filling water with the volcanic glass of the tuffitic layer. Ataman and Beseme (1972) reported on an Oligocene analcime-bearing layer (nearly 100 m thick), included in a thick (up to 1500 m) sequence of continental sediments in North West Anatolia, Turkey; this layer has been deposited in a saline lacustrine environment and also contains trachytic ash: so the origin of analcime could have occurred at the surface as "closed system", but a diagenesis at some depth cannot be excluded, at least as a continuation of the first.

A similar sedimentary analcime corresponding to an uncertain diagenesis occurs in the Upper Flowerpot Shale, northwestern Oklahoma (Wu 1970); the high alkalinity of the evaporite-bearing Flowerpot Shale beds certainly favoured the crystallization of analcime. A similar occurrence in Argentina was described by Teruggi (1964). Uncertain is also the degree of diagenesis of the layers (5 cm thick) between sandstone and siltstone in the Lower Carboniferous of Tuva, U.S.S.R. (Bur'yanova 1954).

Analcimolites are sedimentary rocks so named for their high content of analcime. They all probably underwent a diagenesis or metamorphism whose intensity is difficult to quantify, because of their very simple paragenesis or because of the absence of any zoning in the formation containing them. Joulia et al. (1958) described a Cretaceous rock from the central Sahara, which is several tens of metre thick and is built up only by analcime or nearly so; no pyroclastica are known to occur in the area. Vanderstappen and Verbeek (1959) found similar analcimolites as thick layers in a Mesozoic formation of Zaire and excluded a genesis from volcanic components, but Vernet (1961) found some evidence (the association with montmorillonite) favouring the hypothesis of volcanic parent material. Morelli and Novelli (1967) found analcimolite (nearly 80% of the mineral) in a Cretaceous rock of Libya, with minor illite and quartz; in this case also no volcanic parent material is present in the rock, nor is any known in the area; the authors favour the hypothesis of a genesis by reaction of alumino-silicates with lacustrine brines.

The "Pietra verde" is a Ladinian rock which occurs in the Dolomites, Italy (Callegari 1964) and also in Montenegro (Obradovic and Stojanovic 1972) and in Croatia (Obradovic 1977); its actual composition always includes analcime, albite, quartz and celadonite, the original rock being a tuff (a tuffite in some cases) which was subjected to a very low metamorphism (zone III of Utada). Similar is the occurrence at Currabula, New South Wales, Australia (Wilkinson and Whetten 1964) where analcime is associated not only with albite, but also with heulandite and montmorillonite; the original rock was a tuff, later on subjected to a very low metamorphism in the zeolite facies.

A low hydrothermal genesis is usually assumed for all zeolites which occur in fissures, veinlets and geodes in different rock types, more frequently in basic lavas; the occurrences are always interpreted as a low $(100°C - 200°C)$ hydrothermal deposition with little or no interaction between fluid and wall rock. Obviously there is no definite boundary between this deposition from warm fluids and diagenesis: if the newly formed crystals are quite small and the interaction not negligible, the genesis will be somewhere in between the two schematic situations. As already said, Utada (1980) presents reliable evidence for considering the diagenesis zoning as applicable to low hydrothermalism: this can be true only in the intermediate situation mentioned. Ten years earlier, Kostov (1970) analyzed all the zeolite occurrences of typical hydrothermal deposition in the volcano-sedimentary sequences of the Srednogorian Zone and the Rhodopes, Bulgaria; he found a definite zoning very similar to the sequences with zeolites of diagenetic origin; the different zones correspond to different geographical sites, depending on the nearest source of hydrothermalism: so high-alumina zeolites, including analcime and fibrous zeolites, are present where the temperature of the fluids was highest, namely where post-intrusive hydrothermal mineralization is now observable. Kristmannsdóttir and Tómasson (1978) also found a zoning in an area of Iceland with widespread hydrothermal deposition of zeolites; they found that analcime was not confined to any one of the four zones: (1) chabazite, (2) fibrous zeolite, (3) stilbite, (4) laumontite.

The occurrences of analcime in vugs of massive rocks (probably of hydrothermal genesis) are so common that only selected localities, where well developed fine crystals occur, will be listed here. Classical occurrences, reported by Hintze (1897) and Dana (1914) are the following:

Italy: in the basalts of the Isola dei Ciclopi, Sicily (see also Di Franco 1926); in the porphyrites of the Alpe di Siusi, South Tyrol and of Val di Fassa, Trento.

Germany: in the silver mines of Andreasberg, Harz.

Czechoslovakia: in the basalts from Usti nad Labem (= Aussig), Bohemia.

Scotland: in the basalts of Kilpatrick, Dunbartonshire, of Salisbury Crags and Ratho Quarry, Edinburgh, of Campsie Hills, Stirlingshire.

Northern Ireland: in the Tertiary basalts of Antrim (see also Walker 1960a).

Canada: in the basalts of Five Islands, Cape d'Or and other localities of Nova Scotia (see also Walker and Parsons 1922; Aumento and Friedlander 1966; Gardiner 1966); at Mt.St. Hilaire, Rouville Co., Quebec.

U.S.A.: in the diabase of Bergen Hill, Hudson Co., New Jersey (see also Manchester 1919); at Burger's Quarry, West Paterson, Passaic Co., New Jersey.

More recent descriptions are the following:

England: in the gabbro of Dean quarry, St. Keverne, Cornwall, as fine crystals up to 4.5 cm in diameter (Seager 1969, 1978).

Iceland: 19 occurrences are listed by Betz (1981).

Faröes: 20 occurrences are listed by Betz (1981)

Hungary: in the andesite of Dunabogdany (Reichert and Erdély 1935), also in the Nadap and Balaton Lake area (Koch 1978).

France: in the Tertiary basalts of the Central Massif (Pongiluppi 1976).

Germany: in the nosean-sanidinite from Rocheskyll, in the lower Rhine volcanic area (Kalb 1939) and in the basalts of Vogelsberg (Hentschel and Vollrath 1977).

Italy: in the Triassic submarine basic lava of Val Duron, Trento (Monese and Sacerdoti 1970).

U.S.S.R.: in veins of the volcanic tuffs of the Crimea, as water-clear crystals.

U.S.A.: in the andesite near Challis, Custer Co., Idaho (Ross and Shannon 1924), in the Tertiary basalt of Ritter Hot Spring, Grant Co., Oregon (Hewett et al. 1928); in basaltic pillow lavas at Coburg quarry, Oregon, as crystals perched on natrolite fibers, and at Coffin Butte, Oregon, as trapezohedra enclosing needles of natrolite (Staples 1946).

Brazil: in the basalts of southern Brazil (Franco 1952).

Australia: in the Tertiary basalt of Kyogle, New South Wales (Hodge-Smith 1929).

New Zealand: in the Tertiary basalt near Pahau river, as fine crystals up to 1 cm (Mason 1946a).

The possibility of a magmatic genesis for analcime has long puzzled petrologists, and the problem is still open. Roux and Hamilton (1976) showed that analcime can coexist with melt only if $600 < T < 640°C$ and $5 < P < 13$ Kbars; this means that true primary analcime phenocrysts can be present in a (sub)volcanic rock only if it has been rapidly quenched to surface conditions from the above mentioned temperature and pressure intervals, which in turn favour the formation of hydroxyl-bearing ferromagnesian minerals (mica, amphibole). The other possible origin, or secondary origin, for phenocrysts of analcime present in magmatic rocks with groundmass, is the sodification of leucite (see Chap. 2.1.5); as leucite has a Si/Al-disordered framework, when sodified it should originate (Galli et al. 1978) a disordered, hence cubic, analcime, that can be recognized because optically isotropic and without the 200-peak in the powder pattern. Fornaseri and Penta (1960) found that Rb and Cs are low (10 ppm) in hydrothermal analcimes, but high (1000 ppm) in analcimized leucites and also high in leucite hydrothermally sodified to analcime in the laboratory; so if one finds analcime, low in Rb and Cs, as phenocrysts in a magmatic rock, the primary origin would be proven or at least probable.

Cundari and Graziani (1964) described in detail the process of analcimization of leucite of the lavas occurring near the Vico lake, central Italy, where both fresh tephrites with leucite as well as the same rock with leucite replaced by analcime can be found; this analcime is definitely secondary; it is isotropic (Galli, pers. commun.) as predictable.

Another well known rock rich in analcime is the analcime-phyric volcanic (blairmorite) of the Crowsnest Formation, Alberta, Canada (Pearce 1970); as this analcime is optically anisotropic, and gives a powder pattern with the 200-peak (Aurisicchio et al. 1975), as its Rb and Cs content are very low (Barbieri et al. 1977), one should conclude that this analcime is primary, but the presence in the rock of anhydrous ferriferrous minerals (garnet, clinopyroxene) only, is not in agreement with this hypothesis.

The choice primary-secondary is much more uncertain in all other occurrences of porphyritic analcime in volcanic rocks:

- in the Upper Cretaceous analcime-olivine-basalts of Georgia, U.S.S.R. (Dzotzenidse 1948);
- in the analcimite of Teic Dam, Azerbaijan, Iran (Comin-Chiaramonti 1977);
- in the Kulait lava and in leucite-basanite, The Lecks, North Berwick, in leucite-melabasanite of Garvald, Haddington, in analcime-basanite of Kidlaw, all Scotland (Bennet 1945, who favours a secondary genesis);
- in mafic analcime-phonolites of Highwood Mts., Montana and in analcime tinguaite of the Dunedin volcanic complex, New Zealand (Wilkinson 1968); the coexistence of clinopyroxenes instead of amphiboles favours the secondary genesis, for which also the author describes convincing petrological evidence;
- in the analcime-tinguaite of the Square Top Intrusion, Nundle, New South Wales (Wilkinson 1965);
- in the phonolites (blairmorites) of Lupata Gorge, Moçambique (Woolley and Symes 1976);
- in an analcime-picrite at Magnet Heights, Sekukuniland, South Africa (Strauss and Truter 1950).

Girod (1965) lists 29 occurrences of analcime-basanites from all over the world; some of these contain hydroxyl-bearing minerals, hence their analcime is probably primary.

The presence of analcime in intrusive magmatic rocks is well known, and it raises again the choice between a primary crystallization from the magma in the restricted conditions given by Roux and Hamilton (1976), and a secondary origin by hydration of albite and/or nepheline as a result of a retrograde increase of the vapour pressure during cooling.

Intrusive rocks with analcime are mostly sills or small magmatic bodies. Special names for analcime-bearing rocks are: Teschenite = theralite with analcime = analcime-gabbro of Streckeisen (1976); for further information see Wilkinson (1958) who favours a primary origin for the analcime of the teschenite from New South Wales, and Phillips (1968), who favours a secondary origin for the analcime of the teschenite from the Lugar Sills, Scotland.

Glenmuirite = essexite with analcime; essexite is a nepheline-monzodiorite/monzogabbro after Streckeisen (1976); rocks of this kind have been studied

for instance by Lonsdale (1940).
Lugarite = ijolite with analcime; ijolite is a foidolite after Streckeisen (1976); for further information see Tyrell (1923).

Barker and Long (1969) favour a secondary origin for the analcime of a feldspathoidal syenite from New Jersey.

Analcime has also been reported to occur in pegmatites, for instance by Stewart (1941) in a pegmatitic patch in borolanite (foyaite), and by Roques (1947) in a nepheline-syenite pegmatite; in both cases these are probably late hydrothermal depositions in a pre-existing pegmatite.

Wairakite may be formed under conditions of diagenesis, low metamorphism and low- to medium-hydrothermalism.

Wairakite, 0.1 to 1 μm in size, builds up more than 50% of the finest fraction of a clay from Rosamund, California, (Kiss and Page 1969); the presence in a clay and the fine grain both give evidence for an origin under surface conditions, or nearly so.

Wise (1959) described for the first time a wairakite of low metamorphic origin in the tuffs and tuff-breccias at Mt. Ranier, National Park, Washington. Later many metamorphic occurrences were found in Japan: by Seki (1966), as thin veins of epidote, chlorite, quartz, wairakite and prehnite in weakly metamorphosed andesitic tuff-breccia of the Yugami distr., Fukui Pref.; by Seki and Onuki (1971), as an assemblage wairakite, albite, epidote, chlorite and quartz, typical of the zeolite facies, in Miocene volcanics at Kawaji damsite, Tochigi Pref. Seki (1973) reviewed all the metamorphic occurrences of wairakite in Japan. The presence of wairakite in a spilitic rock of the Virgin Islands (Donnelly 1962) may also be attributed to a low metamorphic origin. Similar seems the genesis of the "disordered" wairakite found by Nakajima (1983) in a volcaniclastic formation from Hikihara, Hyogo Pref., Japan.

Wairakite-analcime solid solutions were found in low-grade metamorphic rocks of the Tanzawa Mts., central Japan (Seki and Oki 1969); the association with laumontite and mixed-layer chlorite (corrensite?) gives evidence for the metamorphic grade.

The type locality of wairakite (Steiner 1955) is Wairakei, North Island, New Zealand; the mineral occurs in tuffaceous sandstones and breccias, vitric tuffs and ignimbrite, originally of plagioclase-rhyolitic composition; wairakite was formed during the hydrothermal alteration of these rocks at T = 200°C to 250°C and $P(H_2O)$ = 55 to 265 atms. There are many other occurrences of this type. So Steiner (1958) described the second find of the mineral in a sandstone of the geothermal field of The Geysers, California. Seki et al. (1969a) found wairakite with yugawaralite, laumontite and mordenite in the Onikobe geothermal field, the host rock being Pliocene-Pleistocene volcanic: wairakite occurs in the deepest and hottest (~175 °C) zone. Bargar et al. (1981) found wairakite with many other calcic hydrous silicates in the hot spring and geyser area of the Yellowstone National Park, Wyoming.

Seki (1973) reviewed many Japanese occurrences in geothermal areas, both active now or in the past.

Cortelezzi (1973) described the first typical hydrothermal occurrence of wairakite in the vesicles of a basalt from Cerro China Muerta, Argentina, with epistilbite and levyne.

Kristmannsdóttir and Tómasson (1978) found several zones in a volcanic area of Iceland with widespread hydrothermal depositions: wairakite is present only in the hottest ($\sim 250\,°C$) zone, with analcime.

Viséite occurs with delvauxite at Visé, Belgium (Mélon 1943).

Hsianghualite occurs with fluorite, liberite, taaffeite, nigerite and white mica in the contact zone between granite and Devonian limestone at Hunan, China (Section etc. 1973).

2.2 Laumontite, Leonhardite

Laumontite	Leonhardite
$Ca_4(Al_8Si_{16}O_{48}) \cdot 16\,H_2O$	$Ca_4(Al_8Si_{16}O_{48}) \cdot 14\,H_2O$
C2/m	C2/m
$a = 15.04$	$a = 14.75$
$b = 13.17$	$b = 13.07$
$c = 7.71\,(\text{Å})$	$c = 7.60\,(\text{Å})$
$\beta = 113°12'$	$\beta = 111°54'$

2.2.1 History and Nomenclature

The zeolite described as "zéolithe éfflorescente" by Haüy (1801) was laumontite, but the mineral was considered an independent species by A. G. Werner (in Karsten 1808) who named it "lomonite" in honour of Gillet de Laumont who collected the sample studied by Haüy; Haüy (1809) changed the spelling to "laumonite"; finally, the name laumontite was proposed by Leonhard (1821) and is now currently used. Obsolete synonyms are sloanite, caporcianite and schneiderite. The name leonhardite, proposed by Blum and Delffs (1843) for a partially dehydrated laumontite is still used, though it should be considered only a variety. The transition laumontite to leonhardite is normally reversible; it is sufficient to soak a powder of leonhardite to change it back to laumontite; this is impossible in a variety called "primary leonhardite" by Fersman (1909).

2.2.2 Crystallography

The structure was determined at nearly the same time by Bartl and Fischer (1967) ($R = 12.6\%$) who refined in space group Cm, and by Amirov et al.

Fig. 2.2A. Laumontite from Poona, India (×4)

Fig. 2.2B. Laumontite from Osilo, Sardinia, Italy (SEM Philips, ×240)

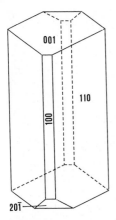

◀ **Fig. 2.2 C.** Common morphology of laumontite

Fig. 2.2 D. Sedimentary laumontite from Bečiči, Yugoslavia (SEM Philips, ×1900)
▼

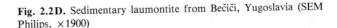

(1967) (R = 16.3%) who preferred space group C2. Schramm and Fischer (1970) gave a new refinement (R = 9.4%) in space group Cm, but Bartl (1970) obtained R = 6% with symmetry C2/m starting from neutron diffraction data. The higher symmetry was confirmed by Amirov et al. (1971) with X-ray data. It is necessary to point out that Bartl and Fischer (1967), Schramm and Fischer (1970) and Bartl (1970) use a reference system different from that given in the title page, and used by most authors ever since. The transformation matrix from the old normal axes to the new ones is 001/100/010.

The Si/Al distribution is perfectly ordered in the framework, the Al occupying the black nodes in Fig. 0.2 H.

Fig. 2.2E. Sedimentary laumontite from Bečiči, Yugoslavia (SEM Philips, ×1900)

There is one fourfold Ca-site on a mirror plane, in the larger channel parallel to c; the coordination polyhedron of this cation is nearly a trigonal prism with two vertices occupied by water molecules and four by framework oxygens, which are part of the Al-tetrahedron. The Ca-site is of type II, because there are oxygens on both sides of the cation, which hence should not change its position in the dehydrated zeolite, whose structure has not yet been studied. The water molecules are placed in two eightfold sites, which are not fully occupied.

All these structural studies were performed in air at room temperature; in these conditions, laumontite is not in its highest degree of hydration and corresponds to the variety called leonhardite. In view of the absence of any structural study of the fully hydrated zeolite, one can suppose that in those conditions the two eightfold water sites, partially occupied in leonhardite, are fully occupied: this hypothesis needs confirmation.

The most common habit of the crystals is given by the simple combination of $\{\overline{2}01\}$ $\{110\}$, sometimes with $\{001\}$ and $\{100\}$ (Fig. 2.2C). Good crystals are not uncommon, but they effloresce, loosing some water rapidly after opening the geodes they grew in; the result may be simply the original transparent crystals becoming earthy whitish, yellowish or reddish, but sometimes they turn to powder; the original beauty of the samples may be maintained by keeping them in water. Sedimentary or low-metamorphic laumontites probably feature the same morphology.

2.2.3 Chemistry and Synthesis

The chemistry of laumontite is normally very near to the stoichiometric formula given in the title page. The eight chemical analyses given in Table 2.2I show a nearly constant Al/Si ratio. Percentages of sodium and potassium may be relevant; in sample no. 8 ("primary leonhardite" of Fersman 1909) the ratio $(Na+K)/Ca$ is higher than 1, and this chemical character is certainly related to the impossibility of getting from this sample the fully hydrated phase.

Table 2.2I. Chemical analyses and formulae on the basis of 48 oxygens

	1	2	3	4	5	6	7	8
SiO_2	51.41	52.04	50.63	52.60	51.86	51.90	54.10	51.00
Al_2O_3	22.20	21.46	22.07	20.55	22.44	21.30	20.44	22.48
Fe_2O_3	0.05	0.12	0.73		0.07		1.70	0.21
MgO			0.40		0.04		0.45	
CaO	12.01	11.41	10.72	9.70	10.47	9.75	8.65	6.96
SrO				0.25		0.25		
Na_2O	0.07	0.20	1.08	0.10	0.74	1.35	2.60	2.58
K_2O	0.12	0.66	0.45	1.75	0.97	0.80	0.55	4.36
H_2O	14.24	13.80	14.10	15.05	13.26	14.65	11.45	12.34
Total	100.11	99.69	100.23	100.00	100.06	100.00	100.10	99.99
Si	15.91	16.12	15.73	16.47	15.93	16.19	16.33	15.82
Al	8.10	7.84	8.08	7.59	8.13	7.83	7.27	8.22
Fe	0.01	0.03	0.17		0.05		0.39	0.05
Mg			0.19		0.02		0.20	
Ca	3.98	3.79	3.57	3.26	3.45	3.26	2.80	2.31
Sr				0.05		0.05		
Na	0.04	0.12	0.65	0.06	0.44	0.82	1.52	1.55
K	0.05	0.26	0.18	0.70	0.38	0.32	0.21	1.73
H_2O	14.70	14.26	14.61	15.72	13.59	15.24	11.53	13.81
E%	+0.7	-1.1	-1.0	+3.1	+5.5	+1.2	-1.0	+4.6

1. Huelgoat, Brittany, France (no.1 of Pipping, 1966).
2. Hungary (no.2 of Coombs, 1952).
3. Otama, Southland, New Zealand, with TiO_2 0.05% (no.1 of Coombs, 1952).
4. Blairmore group, Alberta, Canada (average of samples OG, Miller and Ghent, 1973).
5. Kuhmoinen, Finland, with FeO 0.14%, TiO_2 0.02%, P_2O_5 0.05% (no.6 of Pipping, 1966).
6. Blairmore group, Alberta (average of samples MC of Miller and Ghent, 1973).
7. Devon Well, New Plymouth, New Zealand, with TiO_2 0.11%, MnO 0.05% (no.12 of Coombs, 1952).
8. "Primary leonhardite", Kurtsy, Crimea, with MnO 0.03%, P_2O_5 0.03% (no.7 of Pipping, 1966).

Table 2.2 II. Lattice constants of laumontites-leonhardites

		1	5
	a	15.041	15.053
Fully hydrated	b	13.180	13.160
to 16 H_2O	c	7.710	7.721
	β	113°8'	113°26'
Partially	a	14.770	14.782
dehydrated	b	13.056	13.076
to 14 H_2O	c	7.595	7.602
(leonhardite)	β	112°48'	112°43'
	D	2.27	2.27

Numbers refer to the same samples of Table 2.2 I.

Table 2.2 II reports on some lattice constants of laumontite, both in fully hydrated and in partially dehydrated modification; it is a pity that no lattice constants on "primary leonhardite" are available.

Laumontite has never been synthetized, although the growth of crystals on natural seeds is possible, and several equilibria involving laumontite have been studied experimentally and confirmed by thermodynamical calculation. Briefly, the stability of laumontite (see Fig. 2.1 H) is limited by the following reactions (the arrow shows the direction the reaction proceeds with increasing T or P):

At 170°C – 180°C (Liou 1971a)

$$CaAl_2Si_7O_{18} \cdot 7\,H_2O \rightarrow CaAl_2Si_4O_{12} \cdot 4\,H_2O + 3\,SiO_2 + 3\,H_2O$$
stilbite   laumontite   quartz

This low temperature limit is hence valid only in the presence of excess silica.

At 250°C – 280°C (Thompson 1970; Liou 1971b)

$$CaAl_2Si_4O_{12} \cdot 4\,H_2O \rightarrow CaAl_2Si_4O_{12} \cdot 2\,H_2O + 2\,H_2O$$
laumontite   wairakite

At 3 Kbar $P(H_2O)$ (Thompson 1970; Liou 1971b)

$$CaAl_2Si_4O_{12} \cdot 4\,H_2O \rightarrow CaAl_2Si_2O_7(OH)_2 \cdot H_2O + 2\,SiO_2 + 2\,H_2O$$
laumontite   lawsonite   quartz

In the low P, low T part of the diagram, yugawaralite is stable (Zeng and Liou 1982). The presence of CO_2 hinders the formation of laumontite, also if $P(CO_2):P(H_2O + CO_2)$ is 0.01 or even lower (Thompson 1971a). Other information on laumontite stability is given by Juan and Lo (1971), Senderov (1974), Ivanov and Gurevich (1975). Senderov (1980) estimated the Gibbs energy for laumontite and wairakite.

Table 2.2 III. Optical data for laumontite-leonhardite

	1	2	3	5	8
α	1.505	1.507(1.514)	1.505(1.510)	1.504	1.501
β	1.513	1.516(1.522)	1.514(1.518)	1.510	1.508
γ	1.518	1.518(1.524)	1.517(1.522)	1.513	1.512
2 V	−42°	−26° (−33°)	−44° (−39°)	−82°	−33°
$\overset{\wedge}{c\,\gamma}$	32°	32° (8°)	33° (10°)	32°	30°

Numbers refere to the same samples of Table 2.2 I; data in
parentheses refers to the fully hydrated laumontite, other data
to the air stabilized phase or leonhardite.

2.2.4 Optical and Other Physical Properties

Laumontite is optically biaxial negative, with $\beta \rightarrow b$ and $\overset{\wedge}{c\gamma} \simeq 30°$ in the acute angle β. Optical data for five out of the eight samples of which chemical analyses are given in Table 2.2 I, are shown in Table 2.2 III. The data are nearly constant; the refractive indices for fully hydrated laumontite (available for two samples only) are a little higher than for the normal air stabilized phase. Laumontite is fluorescent in UV-light (Laurent 1941).

Mohs-hardness = 4. Cleavage {010} and {110} perfect, {100} imperfect. Laumontite is piezoelectric (Ventriglia 1953) in spite of its centric symmetry.

2.2.5 Thermal and Other Physico-Chemical Properties

TG and DTA curves (starting from 100°C) were given by Pipping (1966) for all his seven samples; Fig. 2.2 F shows our thermal curves for a sample from India. All these data refer to air dried samples. Water is lost in three steps: the first, with a peak at 100°C, corresponds to $3\,H_2O$, the second, with a peak at 240°C to $5\,H_2O$, the third with a peak at 400°C, to $5\,H_2O$. The behaviour of the "primary leonhardite" is anomalous. It is difficult to reconcile this three-step dehydration with the presence of only two eightfold water sites in the structure.

Data on the sorption of several gases (NH_3, CO_2, SO_2, C_2H_2, O_2, H_2) on laumontite were published by Sameshima and Hemmi (1934).

2.2.6 Occurrences and Genesis

Genesis in weathering, diagenetic, metamorphic (zeolite facies) and hydro-thermal environments has been proposed. The boundaries between these kinds of origin may be ill-defined, because diagenesis shades into meta-morphism with increasing P/T; the crystallization of larger crystals in veins and geodes, hence from a fluid without a major interaction with the wall rock, is certainly to be considered as hydrothermal, but the distinction is not so

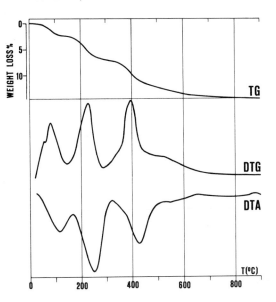

Fig. 2.2F. Thermal curves of laumontite from Nasik, India; in air, heating rate 20°C/min (by the authors)

sharp if veins and fissures are small: the smaller they are the less the difference from a metamorphic and/or diagenetic origin.

Metamorphic formation (and the concept of metamorphic facies) implies a complete chemical equilibrium of the rock forming minerals, or at least a clear tendency towards this equilbrium; but in the low P/T conditions of the zeolite facies, equilibrium is certainly attained with difficulty, and very low metamorphic formation is similar to diagenesis. Also a sedimentary origin by surface weathering has been claimed and supported with good evidence by some authors. Hence, from a genetic point of view, laumontite is unique among zeolites.

Capdecomme (1952) described an occurrence of laumontite with lesser smectite in a granitic sand at Pla des Aveillans, France, and claimed a surface weathering genesis by interaction of percolating water from dissolved snow with the plagioclase; the sands on the sunny slopes, where the snow lasts for a shorter period, are without laumontite. An origin by weathering was also considered for the laumontite of some rocks of North Carolina by Furbish (1965). In a vermiculite-bearing glacial clay from Sundsvall, Sweden, there is laumontite which crystallized by reaction of vermiculite with Ca-ions dissolved in the pore-water (McNamara 1966).

Sands and Drever (1978) identified laumontite in a DSDP-core, at Site 323 (Bellinghausen Abyssal Plain), ranging in age from Cretaceous to Upper Miocene; the highest temperature experienced by the deepest sample was found to be 60°C; a high Ca-concentration in the interstitial water of these sediments is related to the crystallization of laumontite instead of the more common clinoptilolite; this is the only known laumontite of marine diagenesis.

McCulloh et al. (1981) found that laumontite crystals (from 1 μm to 4 mm in size) are deposited in cracks and fractures in transported blocks and stones of gneiss, granite, arkosic sandstone and mudstone mineralized by the thermal waters of Sespe Hot Springs ($43 < T < 89\,°C$), California. There is here a clear evidence for a genesis at a temperature quite near to that of weathering phenomena.

A typical diagenesis may be attributed to many laumontites which built up the cement of sandstones, mostly rich in plagioclases. The cement of the Miocene feldspathic sandstone of San Joaquim, California, is mostly laumontite (Kaley and Hanson 1955). The cement of the Cretaceous sandstone of the Lena coal district, U.S.S.R., is composed of laumontite, chlorite, hydromica and calcite (Zaporozhtseva 1958). In the Mesozoic coal measures of Tabargataj area, U.S.S.R., the cement of sandstones and conglomerates contains laumontite (or heulandite) (Bur'yanova and Bogdanov 1967). Laumontite with barian-strontian heulandite occurs with other authigenic minerals as pore-filling cement in the non-marine sandstones of the Lower Cretaceous Blairmore Group, Canada (Miller and Ghent 1973). Microscopic ($20 - 150$ μm) authigenic laumontite with analcime fills small cavities in the limonitic-dolomitic cement of a red sandstone in the basin of the river Malki, U.S.S.R. (Rengarten 1950).

Burial diagenesis is attributed by Obradovic (1979) to the laumontite occurring in several volcanoclastic series (tuffs, tuffites and tuffaceous sediments) of Yugoslavia, where the variety leonhardite occurs frequently as crystals smaller than 50 μm in the mass of the rocks and in microveins.

A metamorphic genesis was attributed to the laumontite of the Taveyanne formation, Switzerland (Vuagnat 1952; Martini and Vuagnat 1965, 1970; Coombs et al. 1976); these volcanic greywackes with intercalated shales feature a wide "laumontite zone" with transitions to pumpellyite-actinolite and pumpellyite-prehnite zones of higher grade, and to a (restricted) heulandite zone of lower grade. Lippmann and Rothfuss (1980) emphasized the presence of corrensite in paragenesis with laumontite in most parts of the Taveyanne formation and claimed the incompatibility of metamorphic genesis for laumontite and the sedimentary-only genesis proper for a mixed-layer mineral like corrensite. The boundary diagenesis-low metamorphism, which is usually placed at 200°C (see for instance Winkler 1974) is more or less arbitrary, because *"natura non facit saltus"*; Lippmann and Rothfuss (1980) admit that the corrensite genesis was recognized as possible up to 200°C; so, on the whole, a genesis in a temperature range around 200°C is possible and the designation "very low metamorphic" can still be used. The Petrignacola reworked tuffite, Italy (Elter et al. 1964) has a similar content of laumontite and a similar genesis.

A metamorphic genesis (zeolite facies) was firstly attributed to laumontite by Coombs (1952) for an aggregate of interlocking grains 0.5 mm $- 0.3$ mm across, grown from the glass shards of an andesitic tuff in the lower part of the Triassic succession in the Taringatura district, New Zealand; some of these zeolitized beds contain up to 80% of laumontite.

A Neogene volcaniclastic formation with metamorphic zeolites is present in the Tanzawa Mts., Japan (Yoshitani 1965) with an upper mordenite zone, a middle heulandite zone and a lower laumontite zone.

Metasomatism was invoked by Eskola (1960) to explain the transformation of the feldspars of an Archean migmatite of Finland into laumontite; we think that this laumontite could be also considered of low metamorphic origin. Hydrothermal metamorphism in the zeolite facies is the genesis of the laumontite which is the transformation product of a porphyritic granodiorite of Piriul cu Raci, Romania (Kräutner 1966). Murata and Whiteley (1973) described a zonal transformation of the tuffaceous Miocene Briones Sandstone of the Coastal Ranges, California, with three stages of diagenetic history: (1) Alteration of pyroclastica to clinoptilolite + montmorillonite; (2) Crystallization of heulandite; (3) Widespread development of laumontite where the sandstone appears to have been downfolded and faulted to greater depths than elsewhere. Laumontite occurs both as pervasive cement of the sandstone and as filling of fractures. This genesis is no different from some others attributed to burial metamorphism in the zeolite facies.

Hydrothermal genesis is attributed to those crystals filling veins and fractures with no obvious reaction of the mineralizing fluid with the wallrock.

Hydrothermal occurrences of laumontite are so common that already Hintze (1897) and Dana (1914) gave a long list, and only some of these are reported here:

Scotland: in the basalts of Long Craig, Dumbarton Muir, Old Kilpatrick, all Dunbartonshire.

Germany: in the diabase of Niederscheld, Nassau; in vugs of a syenite from Plauenschen Grunde, Saxony.

Czechoslovakia: at Jamolice, Moravia.

Austria: in Zillertal, Tyrol.

Italy: in the granite of Baveno, Lago Maggiore (see also Pagliani 1948).

India: in the Deccan Traps (see also Sukheswala et al. 1974).

Canada: in the traps of Nova Scotia (see also Walker and Parsons 1922; Aumento 1964).

U.S.A.: in the basalt of Table Mt., Colorado.

More recent descriptions in basic lavas are:

Northern Ireland: in the basalts of Antrim (Walker 1960a).

Czechoslovakia: in the andesite of Sklene Teplice (Čaikova and Haramlova 1959).

Hungary: in the amphibole andesite of Nadap (Reichert 1924) with chabazite.

Italy: in the spilitic lava of Toggiano, Modena (Gallitelli 1928).

U.S.A.: in the serpentine of Grants Pass, Oregon (McClellan 1926); in the basaltic lava of Lanakai Hills, Hawaii (Dunham 1933) with nontronite.

Brazil: in the basalts of southern Brazil (Franco 1952).

Japan: in the thick volcaniclastic formation at Onikobe (Seki and Okumura 1968) with yugawaralite.

Kostov (1969) emphasized a zoning of zeolites in a series of basic volcanics, similar to a typical zoning in a low metamorphic sequence, even if the zeolites are clearly of hydrothermal deposition, namely crystallized in amygdales and veins. This interpretation is largely referred to the Sredno-gorian Zone and the Rhodopes, where (Kostov 1970) volcanics ranging from andesitic to rhyolitic, including both lavas, tuffs and tuffites, contain fibrous zeolites in the eastern parts, laumontite and stilbite in the central part, mordenite, chabazite, heulandite in the western part. This zoning from high to low temperature zones correlates also with sites of sulphide mineralization at different temperatures.

A hydrothermal genesis is also obvious for many laumontites which occur in fissures of feldspar-bearing intrusive rocks, like granites, monzonites, gabbros, as in the following:

Norway: in a gneiss at More (Saether 1949).
Switzerland: in an amphibolite of Val Bavona (Fagnani 1948) with prehnite.
Hungary: in the granite of Zsidovar (Takáts 1936) with stilbite.
Poland: in the granitoids of Strzelin (Stepisiewiecz 1978).
New Zealand: in a dioritic rock at Bluff (Mason 1946b).

In the granitic pegmatites of the Russian Karelia near Sinjuja Pala, U.S.S.R., Fersman and Labuntzov (1925) found laumontite with quartz, feldspar, mica, sulphides, allanite, uraninite; this may be interpreted as a late hydrothermal crystallization on the true pegmatitic minerals (see for comparison dachiardite). Similar is the occurrence in the pegmatitic dykes, including allanite and sulphides, near Prosetin and Horka, Skutec, Czechoslovakia (Ulrich 1930).

Other host-rocks for hydrothermal laumontites are:

Norway: in the hornfels not far from the gabbro of Honningvåg (Saebø et al. 1959).
Austria: in the calc-silicate-marble of Hinterhaus, Spitz (Sedlacek 1949).
U.S.S.R.: in the epidote-magnetite-garnet skarn of Dashkesan (Sumin 1955).

2.3 Yugawaralite

$Ca_2Al_4Si_{12}O_{32} \cdot 8 H_2O$
Pc
$a = 6.73$
$b = 13.95$
$c = 10.03$ (Å)
$\beta = 111°30'$

Table 2.3I. Atomic coordinates $x/a, y/b, z/c$ (X10'000) of the main atoms of the yugawaralite structure recalculated with the origin in a pseudo-center of symmetry; atoms in the same box are related by the pseudo-centrosymmetry

Kerr and Williams (1969)		Leimer and Slaughter (1969)	
Si(1)	9683 8989 1669	9705 8732 1338	Si(8)
Al(2)	0224 1054 8377	0198 1209 8673	Si(4)
Si(3)	0361 8753 8792	0286 8959 8303	Si(3)
Si(5)	9860 1233 1457	9754 1014 1584	Si(7)
Al(1)	6277 7571 1837	6554 7253 1227	Si(1)
Si(6)	3675 2481 8068	3416 2860 8759	Si(6)
Si(2)	3380 7871 3760	3730 7418 3159	Si(2)
Si(4)	6542 2260 6328	6361 2496 6959	Si(5)
Ca	6716 9657 6090	6753 0322 6001	Ca(2)
		3216 9626 3868	Ca(1)

2.3.1 History and Nomenclature

The name was proposed by Sakurai and Hayashi (1952) for the locality of the first find, Yugawara Hot Springs, Japan.

2.3.2 Crystallography

The structure has been determined and refined by Kerr and Williams (1967, 1969) (= KW) and refined also by Leimer and Slaughter (1969) (= LS). First of all, both refinements give parameters which are difficult to use because, although the structure is pseudocentro-symmetric C2/m (topologic symmetry of the framework), the authors failed to choose one of the pseudocentres of inversion as origin; this can be done shifting the origin to (0.3723;0.25;0.8163) for KW data, and to (0.3426;0.25;0.3696) for LS data; this change of coordinates also allows an easier comparison of the two sets of atomic parameters (see Table 2.3I). Please note that the equivalent sites are (x, y, z) and $(x, 1/2 - y, 1/2 + z)$ in space group Pc with this new origin, which is not chosen in agreement with the "International Tables".

The KW refinement, carried out on a crystal from Iceland (R = 6.5%) shows a perfect (Si, Al) ordering which reduces symmetry from C2/m to Pc as shown in Fig. 2.3A, which is very similar to Fig. 0.21b. The single cation site Ca is in the main channel, which runs parallel to c; the Ca-ions are attached to the wall of the channel, so that they coordinate framework oxygens on one side and water molecules on the other side (coordination of type III of Chap. 0.4).

The LS refinement carried out on a crystal from Yugawara (R = 14%) shows a partial (Si, Al) ordering which could be compatible also with symmetry P2$_1$/c; as a matter of fact, space groups Pc and P2$_1$/c may be distinguished on the basis of the 0k0 diffractions with k = odd, present in the former and absent in the latter; but LS do not list intensities for 0k0 diffractions. LS found two different cation sites Ca(1) and Ca(2), the latter being equivalent to the single cation site of KW. On the whole, the difference of the structure of the crystal from Yugawara in comparison with that from Iceland, is not convincing and needs confirmation.

The crystals are tabular {010} with a six-sided contour determined mainly by the two pinacoids {100} and {001}, and by the prisms {120} {011} {$\bar{1}$11}; a less developed form is {110}. Striation on {010} parallel to c was observed in the large crystals from Bombay.

2.3.3 Chemistry and Synthesis

The unit cell content is very near to the stoichiometric formula $Ca_2Al_4Si_{12}O_{32} \cdot 8H_2O$, as shown in Table 2.3II; only one minor cation, sodium, is always present with calcium as an extraframework cation.

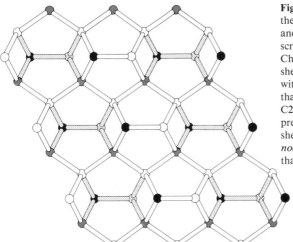

Fig. 2.3 A. The (Si, Al)-order in the yugawaralite sheet after Kerr and Williams (1969); in the description of the framework in Chapter 0.2.2, all nodes and all sheets are assumed to be equal with mirror planes in-between, so that the topologic symmetry is C2/m; in real yugawaralite, Al is present in the *black nodes* in one sheet and in the *dark grey, striped nodes* in the succeeding sheet, so that the symmetry is Pc

Fig. 2.3 B. Yugawaralite from Sardinia, Italy (×20)

Fig. 2.3C. Yugawaralite from Sardinia, Italy (SEM Philips, ×60), with a small icositetrahedron of analcime overgrown on the {010} face

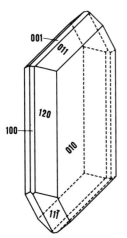

Fig. 2.3D. Common morphology of yugawaralite

Barrer and Marshall (1964a, 1965) synthesized hydrothermally a phase Sr – Q with a powder pattern similar to yugawaralite, from gels $SrO \cdot Al_2O_3 \cdot nSiO_2$ ($5 \leqslant n \leqslant 9$) at 300 °C. Hawkins (1967a) made a similar synthesis with $5 \leqslant n \leqslant 7$ at 350 °C.

Zeng and Liou (1982) studied the yugawaralite-wairakite equilibrium, and found that the first is stable under 230 °C and ca. 0.5 Kbar $P(H_2O)$ (see Fig. 2.1H).

Table 2.3 II. Chemical analyses and formulae on the basis of 32 oxygens and lattice constants in Å

	1	2	3	4
SiO_2	59.29	60.01	58.47	61.74
Al_2O_3	17.43	16.14	17.38	16.67
Fe_2O_3	0.31	0.30	0.03	0.11
MgO	0.11		0.01	0.08
CaO	9.90	10.15	8.51	8.29
SrO			0.21	
Na_2O	0.26	0.39	0.06	0.11
K_2O		0.15	0.09	0.17
H_2O	12.95	12.39	12.12	12.77
Total	100.25	99.53	96.88	99.94
Si	11.80	11.99	11.91	12.18
Al	4.09	3.80	4.17	3.88
Fe	0.05	0.05	0.01	0.02
Mg	0.03			0.02
Ca	2.11	2.17	1.86	1.75
Sr			0.03	
Na	0.10	0.15	0.02	0.04
K		0.04	0.02	0.04
H_2O	8.60	8.25	8.24	8.40
E%	-5.75	-15.19	+9.44	+7.05
a	6.729		6.725	6.724
b	14.008		13.997	14.003
c	10.050		10.043	10.054
β	111°11'		111°11'	111°11'
D	2.22	2.19	2.22	2.22

1. Yugawara (Seki and Haramura 1966); lattice constants by Eberlein et al.(1971).
2. Shimoda (Sameshima 1969).
3. Alaska (Eberlein et al. 1971).
4. Sardinia (Pongiluppi 1977).

2.3.4 Optical and Other Physical Properties

A good and complete set of optical data is given by Eberlein et al. (1971) for specimens from Alaska and Yugawara; the sample from the first locality is colourless, transparent, optically negative with

$$\alpha = 1.492 \quad \beta = 1.498 \quad \gamma = 1.502$$

$2V = 71°$ $\gamma \rightarrow b$ $c\hat{\beta} = 12°$ in the obtuse angle.

The optical data for the Yugawara specimen are similar, as are those from Sardinia (Pongiluppi 1977) and from Bombay (Wise 1978).

The other physical properties after Eberlein et al. (1971) are: cleavage {011} imperfect, {041} and {001} distinct; fracture conchoidal; Mohs-hardness 5 + ; piezoelectric and pyroelectric; non magnetic; streak white; lustre vitreous, nearly iridescent on {010}.

2.3.5 Thermal and Other Physico-Chemical Properties

KW found four water sites in the structure, each with a 75% occupancy and a multiplicity 2, whereas the total number of H_2O molecules per unit cell is nearly 8. The water loss was determined by Eberlein et al. (1971) by two methods: by weighing in a covered platinum crucible after suitable heating, and by TG. After the first method, $2H_2O$ are lost at 100°C, another 2 at 350°C, most of the remaining 4 at 435 °C and a small quantity is given up at 900°C – 1000°C. The inspection of the thermograms (Fig. 2.3E) reveals a loss of $2H_2O$ at 70°C – 100°C, another $2H_2O$ at 200°C – 300°C, and the remaining 4 at 450°C, in good agreement with the first method. The crystals re-hydrate in 5 h after heating at 106°C, in 13 h after heating at 350°C, and in 3 days after a half-an-hour heating at 435 °C (a longer heating destroys the zeolite with formation of another phase). When 935 °C is reached, the mass melts in a glass which becomes transparent at 970°C; with continued heating at 1050°C, crystals of anorthite and β-cristobalite develop in the glass.

The cation exchange capacity was measured by Starkey (in Eberlein et al. 1971), who found a value of 0.045 meq/g, which is nearly 1% of the maxi-

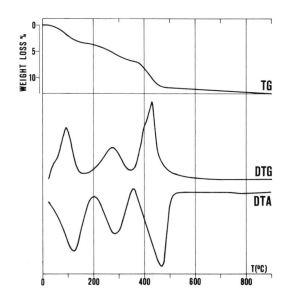

Fig. 2.3E. Thermal curves of yuga-waralite from Sardinia, Italy; in air, heating rate 20°C/min (by the authors)

mum possible value (3.34 meq/g). If the exchange capacity is really so low, yugawaralite is not a zeolite; the low value could be due to the exchanging solution, 1 N NH_4Cl; further reasearch on these equilibria is needed, because the structure of yugawaralite with channels contoured by 8-tetrahedra rings, is a certain premise for good exchange properties. Hawkins (1967b) measured a much higher exchange capacity with Ca on his synthetic strontium yugawaralite, from 0.14 to 1.0 meq/g, which is still only 30% of the theoretical maximum value; he found also a strong selectivity for Sr in the same synthetic yugawaralite, the separation factor (see Chap. 5.4.5) being 2.0 – 2.5.

2.3.6 Occurrences and Genesis

All but one of the occurrences described so far have been found in vugs of massive rocks, and hence have probably a hydrothermal genesis. They are:

Iceland: in a propylite from Heinabergsjokull (Barrer and Marshall 1965; Jakobsson 1977, as reported by Betz 1981).

Italy: at Osilo, Sardinia (Pongiluppi 1977); in cavities and fractures of an altered trachyandesite with laumontite, heulandite, stilbite, chabazite, epistilbite, mordenite, barite.

India: in the cavities of a basalt (?) from Khandivaly quarry, Bombay (Wise 1978), with quartz, calcite, gyrolite, hydroxyapophyllite, laumontite, okenite, prehnite.

Canada: in British Columbia (Wise and Tschernich, in preparation).

U.S.A.: at Chena Hot Springs, Alaska (Eberlein et al. 1971); associated with quartz, laumontite, stellerite and stilbite, in a siliceous xenolith 50×20 m near the edge of a small porphyritic quartz-monzonite pluton.

Japan: Yugawara Hot Springs (the type locality), Kanagawa Pref. (Sakurai and Hayashi 1952); in veinlets or crystals in cavities of an altered andesite tuff; at Onikobe, Miyagi Pref., (Seki and Okumura 1968) associated with laumontite in a borehole in Pliocene-Pleistocene volcanic rocks; at Shimoda, Shizuoka Pref., (Sameshima 1969).

Seki et al. (1969b) described in detail the low metamorphic facies in the Tanzawa mountains, Central Japan. Five zones are distinguished: after zone I (zeolite facies), the zone II (pumpellyite-prehnite greenstone facies) contains two calcium zeolites (wairakite and yugawaralite) in stable association with prehnite, pumpellyite, chlorite, albite and quartz in several metamorphosed basaltic and andesitic rocks. Yugawaralite is also present in cavities and druses with quartz, chlorite, vermiculite, laumontite, wairakite, calcite.

2.4 Roggianite

$Ca_{16}(Al_{16}Si_{32}O_{88}(OH)_{16})(OH)_{16} \cdot 26 H_2O$
I4/mcm
$a = 18.33$
$c = 9.16 (Å)$

2.4.1 History and Nomenclature

Passaglia (1969) described the new mineral from Alpe Rosso, Val Vigezzo, Novara, Italy and proposed the name in honour of Aldo G. Roggiani, teacher of Natural Sciences in the "Liceo" of Domodossola and an enthusiastic mineralogist.

2.4.2 Crystallography

Galli (1980) resolved the crystal structure of roggianite (see Chap. 0.2.2) which is a silicate with an imperfect framework with some OH-vertices which are not shared by two tetrahedra, plus some other OH which are not part of the framework. Al and Si are perfectly ordered, with Al in all tetrahedra with an unshared vertex. The Ca cations lie in the cavities where the framework is interrupted, and the same holds for the 16 extraframework OH. Water molecules are all in the large channels parallel to c, limited by 12-tetrahedra rings.

The morphology of roggianite (Figs. 2.4 A, B, and C) is given by very elongated prisms, with {100} {010} {110} and no termination; aggregates of thin fibres, less than 1 μm in diameter, are usual.

2.4.3 Chemistry and Synthesis

The best analysis of roggianite was obtained by Vezzalini and Mattioli (1979) on a sample from Pizzo Marcio, 1 km from the type locality, which was purer than the first find. This microprobe analysis is as follows: SiO_2 44.11, Al_2O_3 18.54, Fe_2O_3 0.05, CaO 19.97, SrO 0.04, BaO 0.09, Na_2O 0.01, K_2O 0.02, H_2O (by diff.) 17.17, Total 100.00 (all %), which corresponds to the unit cell content

$$Na_{0.02}K_{0.02}Ca_{15.57}Sr_{0.01}Ba_{0.03}(Al_{15.88}Fe_{0.02}Si_{32.10}O_{88})(OH)_{31.30} \cdot 25.99 H_2O$$

which is very near to the stoichiometric formula of the title page.

$D_{calc} = 2.36$.

No synthesis of this phase has yet been done.

Fig. 2.4A. Roggianite from Pizzo Marcio, Italy (SEM Philips, ×120)

Fig. 2.4B. Roggianite from Pizzo Marcio, Italy (SEM Philips, ×3750)

Fig. 2.4C. Roggianite from Pizzo Marcio, Italy (SEM Philips, ×7500)

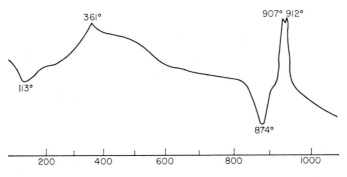

Fig. 2.4D. DTA-curve of roggianite from Alpe Rosso, Italy (Passaglia 1969); in air, heating rate 10°C/min

2.4.4 Optical and Other Physical Properties

Roggianite is uniaxial positive with $\omega = 1.527$ and $\varepsilon = 1.535$. Transparent colourless, streak white.

2.4.5 Thermal and Other Physical Properties

Only the DTA curve is available (see Fig. 2.4D) for the holotype, which was impure. However, this curve is sufficient to recognize a first zeolitic water loss at 113 °C, and a hydroxyl loss at 874 °C; there is a narrow exothermic peak at 910°C certainly due to re-crystallization, and a very broad one around 350°C – 550°C, due to the burning of the organic substance present as an impurity.

2.4.6 Occurrence and Genesis

The type locality is Alpe Rosso, Val Vigezzo, Novara, Italy, where the mineral occurs in an albitite dyke; thomsonite may be associated (Passaglia 1969). Vezzalini and Mattioli (1979) described it from another albitite dyke from Pizzo Marcio, at 1 km from the first locality; this sample has larger crystals, up to 50 μm in diameter. Both are probably late hydrothermal depositions in the albitite.

3 Zeolites with Doubly Connected 4-Ring Chains

3.1 Gismondine, Garronite, Amicite, Gobbinsite

Gismondine	Garronite
$Ca_4(Al_8Si_8O_{32}) \cdot 16H_2O$	$NaCa_{2.5}(Al_6Si_{10}O_{32}) \cdot 13H_2O$
$P2_1/c$	$I4_1/amd(?)$
$a = 10.02$	$a = 9.85$
$b = 10.62$	$c = 10.32\,(\text{Å})$
$c = 9.84\,(\text{Å})$	
$\beta = 92°25'$	

Amicite	Gobbinsite
$Na_4K_4(Al_8Si_8O_{32}) \cdot 10H_2O$	$Na_5(Al_5Si_{11}O_{32}) \cdot 11H_2O$
$I2$	$Pnm2_1$ or $Pn2_1m$
$a = 10.23$	$a = 9.80$
$b = 10.42$	$b = 10.15$
$c = 9.88\,(\text{Å})$	$c = 10.10\,(\text{Å})$
$\beta = 88°19'$	

3.1.1 History and Nomenclature

Gismondi (1817) was the first to describe the mineral from Capo di Bove, Rome, Italy, and named it "zeagonite", but immediately thereafter Leonhard (1817) proposed the name gismondine for the same mineral, and the first name has been abandoned since then. The spelling "gismondite" has also been used, but should be dropped. More than a century later, Walker (1962a) found a new mineral similar to gismondine at Garron Plateau, Antrim, Northern Ireland, and called it garronite. Alberti et al. (1979) found the third mineral with the same framework in Hegau, Germany, and named it amicite for G.B. Amici, the inventor of the Amici-lens. Nawaz and Malone (1982) found gobbinsite in the Gobbins, Island Magee, Northern Ireland. Some obsolete synonyms are given by Dana (1914).

3.1.2 Crystallography

Following Gottardi (1979) and Alberti and Vezzalini (1979), the maximum symmetry compatible with this framework (topological symmetry) is $I4_1/amd$; if the tetrahedra are alternately occupied by Si and Al, the symmetry is reduced to Fddd (with a and b at 45° from the corresponding axes in the former space group); if the framework is slightly deformed, as it really is in gismondine and amicite, its symmetry is $I2/c$, a space group with a general eightfold position. As in both minerals there are four extraframework cations of each kind which cannot enter special positions, it is necessary to lower the symmetry to subgroups whose general multiplicity is four: $P2_1/c$, I2 and Ic. The first is the symmetry of gismondine, the second of amicite. Garronite and gobbinsite probably have a disordered (Si, Al) distribution, and hence higher symmetries; $I4_1/amd$ is the most probable symmetry of garronite; Nawaz (1983) states the space group of gobbinsite to be $Pnm2_1$ or $Pn2_1m$.

The crystal structure of gismondine was determined by Fischer (1963) (see Chap. 0.3) and refined by Fischer and Schramm (1971), without giving the chemical analysis. There is a perfect Si/Al order, given by an alternation of the two atoms at the centers of the tetrahedra. The Ca are placed in a fourfold position, and bound to two framework oxygens on one side only, plus some water molecules (type III of Chap. 0.4). Water is distributed on six sites.

Amicite (Alberti and Vezzalini 1979) also has a perfect (Si, Al) order; the four Na occupy two of the sites where Ca is located in gismondine, plus another two; the four K occupy sites which are filled by water in gismondine. Both Na and K coordinate framework oxygens and water molecules (type III of Chap. 0.4).

Garronite was found to give single crystal data consistent with space group $I4_1/amd$ by Gottardi and Alberti (1974) and probably is the disordered equivalent of gismondine with a higher Si/Al ratio (1.6 instead of 1.0), or, more likely, grew as a disordered phase and later on underwent a partial ordering, assuming a structure with four differently oriented domains.

Gobbinsite is probably the sodium equivalent of garronite, due to the similarity of its powder pattern with that of the synthetic zeolite Na – Pt; Na – P1, a similar phase, was shown by Baerlocher and Meier (1972) to have the gismondine framework.

The morphology of gismondine (Fig. 3.1 C) has long puzzled mineralogists, until a paper by Nawaz (1980b) clarified the matter. The crystals exhibit a "pseudo-octahedron" or "pseudotetragonal bipyramid" with an angle around the horizontal edge of ~88°. Gismondine being metrically pseudocubic, this form could be interpreted as a {101}, which would correspond to {101} {10$\bar{1}$} angle of ~87.4°, but Nawaz studied a gismondine from Osa, Rome, Italy, and found good evidence for the following scheme: the crystals are twinned on the normal to (100) with composition planes (100) and (001); in a (010) section, the square contour is cut by two diagonals in four sectors, the opposite ones being iso-oriented. The pseudo-octahedron is hence

Fig. 3.1A. Gismondine on ettringite from Schellkopf, Brenk, Eifel, Germany (×40) (by courtesy of G. Hesser, G. Hentschel and of "Lapis")

Fig. 3.1B. Gismondine from St. Agrève, France (×10)

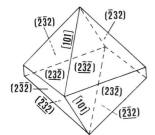

Fig. 3.1C. Clinographic view of a gismondine twin after Nawaz (1980); indices of faces and edges of the two individuals may be distinguished by the *presence or absence of underlining*

Fig. 3.1D. Garronite from Goble, Oregon (×5)

due to four faces of the monoclinic prism $\{\bar{2}32\}$ of one individual plus the four faces of its twin with the same indices. Please note here that the pseudotetragonal axis in monoclinic gismondine coincides with *b*.

Only compact radiating aggregates are known for garronite.

Amicite crystals simulate a tetragonal bipyramid with the forms $\{011\}$ and $\{110\}$.

Gobbinsite crystals are fibrous, elongated parallel to the *c*-axis; single crystal photographs show the presence of intergrowths akin to twinning on (101).

3.1.3 Chemistry and Synthesis

The chemical composition of gismondine (see Table 3.1I) may be quite close to the stoichiometric formula of the title page, as in sample no. 5, or may

Fig. 3.1E. Amicite from Höwenegg, Germany (×5); the grey nucleus is calcium carbonate, the white outer aggregate is amicite, merlinoite being present at the contact with the carbonate

Fig. 3.1F. Amicite from Höwenegg, Germany (SEM Philips, ×120)

Table 3.1I. Chemical analyses and formulae on the basis of 32 oxygens and lattice constants in Å

	1	2	3	4	5	6	7	8	9
SiO_2	36.26	34.50	40.38	35.36	34.64	32.40	43.21	36.38	49.21
Al_2O_3	27.22	27.06	25.00	26.87	27.79	27.95	24.20	29.46	23.64
CaO	14.51	14.83	11.76	14.80	15.37	14.81	10.64	0.22	1.58
SrO	0.24							0.03	0.36
Na_2O	0.24	0.37	2.02	0.48		0.37	2.94	8.22	9.85
K_2O	0.77	0.04	0.18	0.09			0.54	12.96	0.66
H_2O	20.76	23.20	21.05	22.40	22.20	24.09	18.62	12.80	13.55
Total	100.00	100.00	100.41	100.00	100.00	99.72	100.17	100.07	100.00
Si	8.47	8.30	9.25	8.41	8.22	7.94	9.62	8.24	10.25
Al	7.49	7.67	6.75	7.53	7.77	8.07	6.35	7.86	5.80
Ca	3.63	3.82	2.89	3.77	3.91	3.89	2.54	0.05	0.35
Sr	0.03								0.04
Na	0.11	0.17	0.90	0.22		0.17	1.26	3.61	3.98
K	0.23	0.01	0.05	0.03			0.15	3.75	0.18
H_2O	16.17	18.60	16.08	17.76	17.57	19.68	13.82	9.67	9.41
E%	-2.2	-2.0	+0.3	-3.3	-0.5	+1.5	-2.2	+5.4	-4.1
a	10.016	10.027		10.030	10.020	10.027	9.85	10.226	10.145
b	10.626	10.622		10.610	10.637	10.594		10.422	
c	9.824	9.829		9.840	9.832	9.843	10.32	9.884	9.788
β	92°26'	92°28'		92°22'	92°34'	92°33'		88°19'	
D	2.28		2.20			2.22	2.16	2.06	2.15

1. Gismondine, Capo di Bove, Italy (Vezzalini and Oberti 1984); H_2O from Caglioti (1927).
2. Gismondine, Zálezly (=Salesl), Bohemia (Vezzalini and Oberti 1984).
3. Gismondine, Hoffell, Iceland (G.37 of Walker 1962b) plus 0.02% Fe_2O_3.
4. Gismondine, Ballyclare, Ireland (Vezzalini and Oberti 1984).
5. Gismondine, Montalto di Castro, Italy (Passaglia and Rinaldi, written communication).
6. Gismondine, Liguria, Italy (Cortesogno et al. 1975) plus 0.10% MgO, 0.03 Mg in the formula.
7. Garronite, Glenariff Valley, Northern Ireland (Walker 1962a) plus 0.02% Fe_2O_3; lattice constants by Taylor and Roy (1964).
8. Amicite, Heqau, Germany (Alberti et al. 1979).
9. Gobbinsite, Gobbins, Northern Ireland (Nawaz and Malone 1982) with 0.04% Fe_2O_3, 1.00% MgO, 0.12% BaO, which correspond to Fe 0.01, Mg 0.31, Ba 0.01 in the formula.

contain relevant quantities of alkalies, up to 0.90 Na atoms (2.02% Na_2O) as in sample no. 3, up to 0.23 K atoms (0.77% K_2O) like sample no. 1 which is from the type locality. Ba is always absent (less than 0.01%); a small amount of Sr may be present. The ratio Si : (Si + Al) is normally very near to 50%, but may be a little higher (58%) as in sample no. 3. Walker (1962b) emphasized the existence of "low-potash" gismondines in contrast to most gismondines known up to that year, which seemed to be "high-potash": in fact the old analyses showed 2.86% K_2O for the Capo di Bove sample (Caglioti 1927), or 2.35% K_2O for the Zálezly (= Salesl) sample (Sachs 1904, quoted in Doelter 1921), but Vezzalini and Oberti (1984) showed that the high potash content was apparent only, and was really due to phillipsite inclusions, the real potash percentage being sometimes only a little higher than in the "low-potash" samples of Walker (1962b), and sometimes even lower.

Garronite (no. 7) is a little more silica-rich (60% of the tetrahedra occupied by Si) than gismondines and this is well in accordance with its disordered (Si, Al)-distribution: remember the influence of the Si : Al ratio on the (Si, Al)-order in the anorthite side of the plagioclases! Garronite is also higher in soda.

Amicite and gobbinsite are near to their stoichiometric formulae, with some Ca for Na substitution.

There is now general agreement that all synthetic zeolites coded mostly P, but also (Linde) B, or CASH II, share the same framework with gismondine, since Baerlocher and Meier (1972) attributed this kind of structure to "cubic" Na – P, which hence is only metrically cubic, but actually has $I4_1/amd$ as maximum symmetry; its real symmetry is I4 or lower. Barrer et al. (1959) gave the first full report on the synthesis of Na – P, namely of three phases, a cubic, a tetragonal and an orthorhombic one, with chemical formula $Na_2O \cdot Al_2O_3 \cdot nSiO_2$ with $3.3 < n < 5.3$, from gels having $1 < n < 12$ and 200% – 300% alkali excess. A previous paper by Breck et al. (1956) contained minor information on a similar synthesis. Taylor and Roy (1964) gave a comprehensive report on phases obtained by ion exchange from Na – P. Baerlocher and Meier (1970) synthesized a TMA – P phase (TMA = tetramethylammonium); Colella and Aiello (1975) showed how the synthesis is possible also when the ratio Na/(Na + K) ranges from 1 to 0.8 and how to drive the synthesis to P phases with different $a : c$ by changing the alkalinity. On the whole, all these synthetic phases probably correspond more or less to the minerals with disordered (Si, Al) distribution, namely Na – P to gobbinsite, and Ca – P obtained from Na – P by Ca exchange, to garronite. It is also possible that the different varieties of Na – P correspond to different degrees of low (Si, Al) order and to different Si/Al ratios.

3.1.4 Optical and Other Physical Properties

Gismondine is biaxial negative (Table 3.1 II) with 2V ≃ 86° and the following orientation (Nawaz 1980b): $\beta \rightarrow b \ \hat{c}\gamma = 42°$ in the obtuse angle, α being approximately normal to [101], the edge formed by the faces $(\bar{2}32) (\bar{2}\bar{3}2)$. Gar-

Table 3.1 II. Refractive indices

	2	3	4	6
α	1.531	1.515	1.532	1.532
β	1.540		1.539	1.540
γ	1.548	1.523	1.543	1.544

2. Gismondine, Zálezly (=Salesl), Bohemia (Hibsch 1917).
3. Gismondine, Hoffell, Iceland (G.37 of Walker 1962b).
4. Gismondine, Ballyclare, Northern Ireland (average of three values given for G.1 by Walker 1962b).
6. Gismondine, Liguria, Italy (Cortesogno et al. 1975).

ronite is uniaxial with positive or negative sign (Walker 1962a, gives $\omega = 1.504$ $\varepsilon = 1.506$ in one case, and $\omega = 1.512$ $\varepsilon = 1.510$ in another one for the holotype). Amicite (Khomyakov et al. 1982) is biaxial negative with $2V = -82°$ $\alpha = 1.485$ $\beta = 1.490$ $\gamma = 1.494$ and $\alpha \rightarrow b \; \hat{c\gamma} \; 12°$. Gobbinsite is uniaxial negative with $\omega = 1.494$ $\varepsilon = 1.489$ (Nawaz and Malone 1982).

For gismondine: colourless to whitish or greyish; transparent to cloudy; vitreous lustre; white streak; Mohs-hardness 4.5; cleavage $\{\bar{2}32\}$ easy.
For garronite: colourless, transparent; two cleavages at 90°.
For amicite: colourless, transparent, white streak, no cleavages, Mohs-hardness 4.5.
For gobbinsite: chalky white colour, vitreous lustre, no cleavage.

Gismondine is piezoelectric (Ventriglia 1953) in spite of its centricity.

3.1.5 Thermal and Other Physico-Chemical Properties

The thermal behaviour of gismondine is quite complex (see Fig. 3.1 G). The DTG and DTA show five main dehydration steps before 300°C plus some minor ones between 300°C and 600°C. This is in agreement with Fischer and Schramm (1971), who listed six different water sites in their refinement. These curves are similar to those of Reeuwijk (1971), who was also able to show by high temperature X-ray photographs that gismondine undergoes some reversible lattice contractions and rehydrates on cooling if heated up to 300°C; from this temperature on, the formation of anorthite is initiated.

Garronite (see Fig. 3.1 H) shows only three water loss peaks in the DTG at 60°C, 170°C, and 320°C, hence less peaks than gismondine, well in accordance with its higher symmetry.

Amicite (see Fig. 3.1 I), like garronite, has only three water losses at 60°C, 140°C and 320°C.

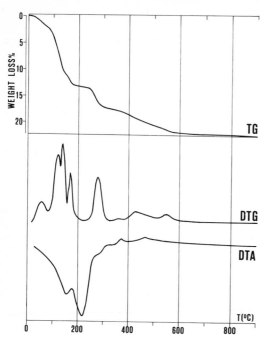

Fig. 3.1G. Thermal curves of gismondine from Montalto di Castro, Italy; in air, heating rate 10°C/min for TG (and DTG), 20°C/min for DTA (by the authors)

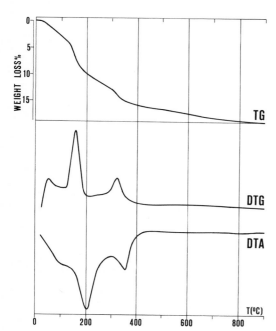

Fig. 3.1H. Thermal curves of garronite from Goble, Oregon; in air, heating rate 20°C/min (by the authors)

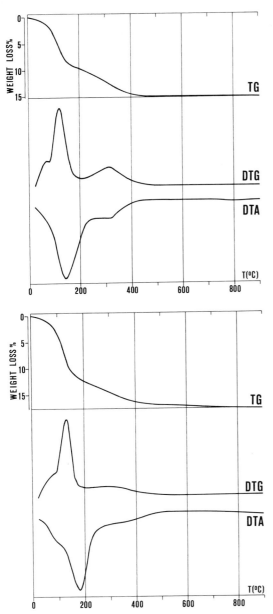

Fig. 3.1I. Thermal curves of amicite from Höwenegg, Germany; in air, heating rate 20°C/min (by the authors)

Fig. 3.1J. Thermal curves of gobbinsite from Gobbins, Northern Ireland; in air, heating rate 20°C/min (by the authors)

Gobbinsite (see Fig. 3.1 J) has also three water losses with one shoulder in the DTG at ca. 80°C, a peak at 150°C and a continuous loss between 200°C and 400°C with almost no peak in the DTG.

The paucity of natural crystals prohibited any study of ion exchange of gismondine (and even more so for garronite, amicite, gobbinsite), but comprehensive studies are available for the synthetic phase Na – P, with the same framework. Taylor and Roy (1964) prepared pure exchanged phases with Li, K, Ca, Ba, measured their lattice constants emphasizing the non-rigid character of the framework, and revealed the existence of two-phase regions in the Na – Ca, Na – K, Na – Ba systems; the temperature of decomposition in hydrothermal conditions is ~200°C for Na, Li, Mg, Ba phases, ~300°C for K and Ca phases; the temperature of decomposition is much lower (150°C – 250°C) in dry conditions. Barrer and Munday (1971) studied exchange isotherms for the Li, Na, K, Rb, Cs, Sr, Ba phases, and found a selectivity series

$$Li < Na < K \sim Rb \sim Cs \quad \text{and} \quad Na < Sr < Ba \ .$$

3.1.6 Occurrences and Genesis

Gismondine is a rather rare zeolite; its sedimentary genesis has been considered only once, when Iijima and Harada (1969) detected it in the cement of a palagonitized alkali-basalt and of a melilite-nephelinite tuff from Oahu, Hawaii, with several other zeolites, phillipsite, chabazite, gonnardite, natrolite, analcime, and faujasite; this latter mineral and gismondine being less common.

Hydrothermal origin is attributed to all other occurrences in vugs of massive rocks: some already described by Hintze (1897) and Dana (1914) will be repeated here.

The type locality is Capo di Bove, in the southeastern suburbs of Rome, Italy, where gismondine occurs in the vugs of a leucitite, with phillipsite; similar are the nearby occurrences at Vallerano. Mostacciano and Osa.

Many occurrences in Germany are listed by Hintze (1897): at Hohenberg, Buehne, Westfalia, large crystals in a nepheline-basalt (this specimen was used for the crystal structure determination); at Frauenberg, Fulda, Hessen, with phillipsite in a basalt; at Gedern and Burkards, Vogelsberg, Hessen, in basalts; at Shiffenberge and Steinbach, Giessen, Hessen, in basalts; at Pflasterkaute, Eisenach, Thuringia, in a nepheline-basalt; at Schlauroth, Goerlitz, Silesia, with natrolite (and thomsonite ?).

At Alexander Dam, Kauai, Hawaii islands, Dunham (1933) found gismondine in an ankaratrite with stilbite and other silicates. Mauritz (1938) described the mineral with stilbite, natrolite, phillipsite in the cavities of a basalt at Halap, Balaton lake, Hungary. Whitehouse (1938), in a review of the minerals of Queensland, Australia, mentioned gismondine both from basalts and granites, in the latter occurring with laumontite and other calcium minerals. Hibsch (1939) reports on the presence of fine crystals with no other zeolites, in the cavities of a leucite-tephrite at Schickenberges, Bohemia; this is the second occurrence in Bohemia, after that of Zálezly (= Salesl), Schieferberge (Hibsch 1917). Hentschel (1983) reported on gismondine with phillipsite and ettringite in the selbergite from Schellkopf, Brenk, Eifel, Germany.

Walker (1962b) greatly contributed to the mineralogy of gismondine, listing 19 localities of Co. Antrim, Northern Ireland, and 21 localities in east Iceland; a small map is also published. In Northern Ireland the host rock is a normal Tertiary olivine-basalt, the associated zeolites being mostly chabazite and thomsonite, less frequently levyne. In Iceland the rock is usually, but not always, a Tertiary olivine-basalt and the associated zeolites are again chabazite and thomsonite, but phillipsite is also quite common. He noticed that all his samples were much lower in potassium than most gismondine previously described.

Cortesogno et al. (1975) described in detail the occurrence of crystals with thomsonite, chabazite, phillipsite in the vugs of a metagabbro of the "Gruppo di Voltri" formation, Liguria, Italy.

Nambu (1970) lists Nagahama, Shimane Pref., as a Japanese occurrence.

Galli and Passaglia (pers. commun.) found gismondine in the already known occurrence at the Ferme Chabane, St. Agrève, but also at Aubenas, both in the Ardèche, France; the host rocks are alkali-basalts, phillipsite and thomsonite being associated.

Garronite has been found in a very restricted number of localities but is certainly not so rare. Walker (1962a), in his description of garronite as a new mineral, listed 10 localities in Co. Antrim, Northern Ireland, and 22 in eastern Iceland. Type locality is the Glenariff Valley, in the Garron Plateau, Co. Antrim, where it occurs with phillipsite, chabazite, thomsonite, analcime, levyne, natrolite, in amygdales in an olivine-basalt; the other occurrences are similar but paragenesis sometimes includes gismondine, gmelinite and mesolite.

The second find of garronite is due to Feoktistov et al. (1969), who found it in the amygdales of a diabase porphyry dyke in Southern Siberia. Walenta (1974) found it in the vugs of a melilite-nephelinite at Höwenegg, Hegau, Germany, with phillipsite, harmotome, mountainite; in the schist ejecta of the Quaternary volcanic district in Eifel, Germany, Hentschel (1978) found garronite as white octahedra or cubo-octahedra 0.1 mm in diameter. Gignat, Puy-de-Dôme, France, is the type locality for gonnardite: Pongiluppi (1976) found there garronite as milky-white microcrystalline layers with phillipsite, chabazite, thomsonite.

Amicite was found for the first time by Alberti et al. (1979) at Höwenegg, Hegau, Germany, in a melilite-nephelinite with merlinoite and calcite. The second find is due to Khomyakov et al. (1982), who described amicite in natrolite veinlets cutting ijolite-urtite pegmatites and apatite-nepheline rocks at Kuhisvumchorr, Kibinis Mt., Kola peninsula, U.S.S.R.

Gobbinsite (Nawaz and Malone 1982) occurs as aggregates of fibrous crystals either alone or associated with gmelinite in a basalt at Hills Port, south of Gobbins, Co. Antrim, Northern Ireland.

3.2 Phillipsite, Harmotome

Phillipsite	Harmotome
$K_2(Ca_{0.5}, Na)_4(Al_6Si_{10}O_{32}) \cdot 12 H_2O$	$Ba_2(Ca_{0.5}, Na)(Al_5Si_{11}O_{32}) \cdot 12 H_2O$

$$P2_1/m$$

$a = 9.88$	$a = 9.88$
$b = 14.30$	$b = 14.14$
$c = 8.67 (\text{Å})$	$c = 8.69 (\text{Å})$
$\beta = 124°12'$	$\beta = 124°49'$

or in pseudo-orthorhombic setting
$B2_1/m$ (double cell, $Z = 2$)

$a = 9.88$	$a = 9.88$
$b = 14.30$	$b = 14.14$
$c_0 = 14.34 (\text{Å})$	$c_0 = 14.27 (\text{Å})$
$\beta_0 = 89°28'$	$\beta_0 = 90°10'$

with $\vec{c}_0 = \vec{a} + 2\vec{c}$

3.2.1 History and Nomenclature

The first description of harmotome is lost in the mists of time, but the name was introduced by Haüy (1801), from the Greek $\alpha\varrho\mu\acute{o}\varsigma$ = joint and $\tau o\mu\acute{\eta}$ = cut, because of its tendency to split along "joints" (contact planes of the twins, probably). No other synonym has any importance to-day. Phillipsite was considered the same as harmotome at first, then the name was introduced by Lévy (1825), in honour of W. Phillips, English mineralogist, for the samples from Aci Castello, Sicily, Italy. Des Cloizeaux (1847) proposed the name christianite for minerals which later on were shown to be phillipsite; the name was for the Danish king Christian VIII; the synonym was used for a long time, but now is forgotten. The name wellsite, proposed by Pratt and Foote (1897) after prof. H. L. Wells of New Haven, Connecticut, is still used for a barian variety of phillipsite.

3.2.2 Crystallography

The structures of harmotome and phillipsite were resolved nearly simultaneously by Sadanaga et al. (1961) and by Steinfink (1962) who found the same framework. In both cases a symmetry lower than the topological symmetry of the framework, Bmmb, was adopted: $P2_1$ for harmotome, and B2mb for phillipsite. Rinaldi et al. (1974) refined both structures showing that the real symmetry is always $P2_1/m$ and locating exactly all extraframework cations.

The framework (see Chap. 0.2.3) shows no Si, Al order. There are two cation sites outside the framework: the first is twofold and fully occupied by Ba or K, and causes the lowering of symmetry from the maximum possible

Fig. 3.2A. Harmotome from Finland (×4)

Fig. 3.2B. Harmotome from Andreasberg, Germany (SEM Philips, ×15)

Fig. 3.2C. Harmotome from Andreasberg, Germany (SEM Philips, ×30)

Fig. 3.2D. Harmotome from Drio le Pale, Italy (SEM Philips, ×120)

Fig. 3.2E. Harmotome from Drio le Pale, Italy (SEM Philips, ×60) with a less common morphology (see Fig. 3.2Hd)

Fig. 3.2F. Phillipsite from Éspalion, France (SEM Philips, ×60)

Fig. 3.2G. Phillipsite from St. Jean le Centenaire, France ($\times 40$)

Bmmb to $P2_1/m$ (Gottardi 1979): in fact O(1) and O(3) would be equivalent in the higher symmetry, but they are not in the lower one, which allows the two O(3) to be nearer (~ 2.90 Å) to cations, whereas the two O(1) are further away (~ 3.40 Å), and this fact allows a more proper set of oxygen-cation distances. These cations have a coordination of type II (Chap. 0.4). The second cation site is fourfold and partially occupied: $\sim 15\%$ by Na, Ca in harmotome, $\sim 40\%$ by Ca, Na in phillipsite. These cations have a coordination of type III, that is, they are at bond distance with framework oxygens at one side only; the site is in a niche of the double crankshaft.

Stacking faults on (100) and (010) by a slip of $a/2$ or $b/2$ are certainly possible in all frameworks with the double crankshaft (Sato and Gottardi 1982) and hence also in phillipsite; if extensively present, the $a/2$ faults may transform a phillipsite into a crystal with both phillipsite- and merlinoite-domains; these faults are more likely to occur in sedimentary phillipsite, whose powder patterns are sometimes without the peak at 6.4 Å, typical for phillipsite and absent in merlinoite.

The morphology of harmotome and phillipsite is characterized by short prismatic pseudo-orthorhombic crystals with a-elongation; these are always twins or even fourlings, or twins of higher rank. The monoclinic setting is used here, as in Hintze (1897) and in Dana (1914), but they adopted a doubled c. The simplest crystals are twins (see Fig. 3.2Ha) of four individuals I, II, III, IV with the forms {110} {010} {001} and rarely also {100}; I and III are iso-

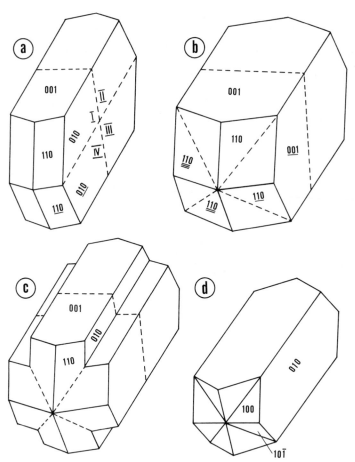

Fig. 3.2 H a – d. Common morphology of harmotome and phillipsite. **a** Fourling (compare with Fig. 3.2 G), I and III (and so II and IV) are iso-oriented, twin- and composition-plane between I and II (and so between III and IV) is (2̄01), which is nearly normal to *a*, twin- and composition-plane between I and IV (and so between II and III) is (001). **b** Prismatic eightling (compare with Fig. 3.2 D) where a fourling of Fig. 3.2 Ha is combined on {011} with another fourling. **c** Cruciform eightling (compare with Fig. 3.2 B and C) similar to the preceding one, but with a different face development. **d** Less common kind of fourling (compare with Fig. 3.2 G), also with {011} as twin- and composition-plane, occurring at St. Jean le Centenaire, France, at Drio le Pale, Italy, and at Asbach, Germany (Hintze 1897)

oriented; for I and II (and so for III and IV) twin- and composition-plane is {201̄} which is nearly normal to *a*; for I and IV (and so for II and III) twin- and contact-plane is {001}. These fourlings are very often further combined (see Fig. 3.2 H b and c) on {011} to eightlings; these may in turn twin on {110}. The most complex twins are more frequent in phillipsite.

The non-existence of untwinned crystals of phillipsite and harmotome means that the twinning exists already in the crystallites at the time of nucleation.

Fig. 3.21. Sedimentary phillipsite from Ischia, Italy (SEM Philips, ×3800)

3.2.3 Chemistry and Synthesis

Selected analyses of these zeolites are collected in Table 3.21. From the reviews on the chemistry of phillipsite by Galli and Loschi Ghittoni (1972), of deep-sea phillipsite by Sheppard et al. (1970) and by Stonecipher (1978), of wellsites and harmotomes by Černý et al. (1977), the following general picture may be drawn. Hydrothermal phillipsite displays a wide range of chemical composition, perhaps the widest among zeolites, so that the formula in the title page is a scheme which may be quite far from individual real cases: the percentage of tetrahedra occupied by Si may be as low as 54% but also quite high, 74%; the most frequent number of K per unit cell is certainly 2, but the lowest value is 0.7 and highest 3.5; the corresponding range for Na is 0.1 – 4.2 and for Ca 0 – 2.1; Sr is always very low, Ba may be zero, or high (see wellsite later on in this chapter). Deep-sea phillipsites feature a more constant chemistry, with the following unit cell content: Si very near to 11.5 (72% of the tetrahedra centered by Si), K nearly 2, Na from 0.5 to 2.3, Ca from 0 to 1.3, Sr absent, Ba very low, Mg higher (up to 0.25) than in other phillipsites. Data on other sedimentary phillipsites, which crystallized from volcanic rhyolitic glass in lacustrine environments, are scanty and never coupled with the corresponding unit cell data, but on the whole it may be stated that they are high in Si (12 atoms per unit cell, or 75% of the tetrahedra with Si, and even more) and in Na + K, and low in alkaline earths, hence similar to deep-sea phillipsites, although still more acid. Sedimentary phillipsites from basic volcanic glasses are more similar to normal hydrothermal samples, namely higher in Ca than those derived from acid glasses, and not so high in Si (Passaglia, pers. comm.).

Wellsite, the barian zeolites on the phillipsite-harmotome join, may have nearly every intermediate Ba-content (see Table 3.21). The number of Ba per unit cell may be 0.4 (no. 5), 0.7 (no. 7), 0.8 (no. 6), 1.15 (no. 8), 1.54 (Black 1969), up to nearly 2 as in most harmotomes. Remarkable is the analysis no. 6 ("sodic harmotome" of Sheppard and Gude 1971) for its high sodium content, which cannot but be considered very strange in the frame of the chemistry of the whole group. Wellsites have a rather constant Si content, nearly 11 per unit cell or 70% of the tetrahedra.

Harmotomes are also rather constant in composition with 11 Si, 2 Ba and the sum $(Ca_{0.5} + Na)$ near to 1 in the unit cell; every K, if present, is probably associated with Ba, as the sum $(Ba + K)$ may surpass 2 only by a small amount.

The synthesis of harmotome was performed for the first time by Barrer and Marshall (1964b) at 250°C from a gel $BaO \cdot Al_2O_3 \cdot 7.5 SiO_2$ in 19 days, hence this phase was alkali-free in contrast with the natural phases. Perrotta (1976) operating in the presence of Na with Na:Ba ratios from 38 to 8 was able to obtain well crystallized harmotomes in three to four days at 95°C; the $SiO_2 : Al_2O_3$ ratio in the batch was 5.25, Ba was added as BaI_2.

It is quite difficult to say who did the first synthesis of phillipsite, because there was some confusion between synthetic phases with the frameworks of phillipsite, merlinoite, gismondine. So phase $K - M$ synthetized by Barrer et al. (1959) and said to be similar to phillipsite, was later shown by Passaglia et al. (1977) to be related to merlinoite, and all phases P, also claimed to be similar to phillipsite, were shown to be related to gismondine (see Chap. 3.1). Batches with a single cation seem to be unsuitable for this synthesis, which on the contrary is easy in the presence of K plus Na and/or Ca. Coombs et al. (1959) reported on the hydrothermal decomposition of a chabazite into phillipsite and wairakite. Starting from an alumino-silicate glass with K, Ca, Na, hydrothermally (~ 250°C) treated with $NaOH + KOH$, Sersale et al. (1965) obtained a phase which is beyond any doubt a phillipsite, as shown by the powder pattern and data; as a matter of fact, all natural phillipsites of hydrothermal origin give a powder pattern with a medium line at 6.4 Å, no matter if with or without Ca (Galli and Loschi Ghittoni 1972); this line distinguishes the zeolite from merlinoite, which has a similar pattern but no such peak; in the powder pattern of Sersale et al. (1965) this peak can be observed, although very weak, with all other reflections of phillipsite. Kühl (1969) used phosphate as a complexing agent for a batch of metasilicate or waterglass plus NaOH and KOH, hence in the absence of Ca; this synthesis was possible at ~ 100°C. The zeolite ZK-19 thus obtained is probably phillipsite, even though the 6.4 Å spacing is absent in the powder data, because of the presence of the 4.13 Å spacing, which is also absent in merlinoite. Pure ZK-19 crystallized from mixtures having silica to alumina molar ratios 4 to 16 and Na:(Na + K) ratios

Table 3.21. Chemical analyses and formulae on the basis of 32 oxygens and lattice constants in Å

	1	2	3	4	5	6	7	8	9	10
SiO_2	41.42	41.53	54.97	52.05	49.30	49.97	50.3	48.10	46.3	46.67
Al_2O_3	23.82	24.33	16.24	18.09	18.97	15.15	17.9	17.12	16.4	16.35
Fe_2O_3	0.07	0.08	0.17	0.08	0.11	3.15	0.2	0.08	0.5	
MgO	0.03	0.05	0.10	0.10	0.04	1.52				
CaO	8.50	2.10	6.41		6.70	0.64	4.4	2.96	0.1	1.78
SrO	0.28	0.03	0.12		0.10	0.10		0.11		
BaO	0.09	0.08	1.24	5.51	4.36	8.75	8.1	12.60	18.8	20.35
Na_2O	0.65	3.16	0.50	7.30	0.31	3.33	0.5	0.22	1.9	
K_2O	6.80	13.00	2.49	16.87	2.50	2.24	2.8	3.41	0.2	0.23
H_2O	17.57	15.06	17.88		17.40	14.88	15.8	15.40	15.8	14.66
Total	99.23	99.42	100.12	100.00	99.79	99.73	100.00	100.00	100.0	100.00
Si	9.51	9.50	11.85	11.38	11.02	11.31	11.28	11.23	11.25	11.29
Al	6.45	6.56	4.13	4.66	5.00	4.04	4.73	4.71	4.69	4.66
Fe	0.01	0.01	0.03	0.03	0.02	0.54	0.07	0.01	0.18	
Mg	0.01	0.02	0.03	0.02	0.01	0.51				
Ca	2.09	0.51	1.48		1.60	0.16	1.06	0.74	0.03	0.46
Sr	0.04		0.01		0.01	0.01		0.02		
Ba	0.01	0.01	0.10		0.38	0.78	0.71	1.15	1.79	1.93
Na	0.29	1.40	0.21	2.34	0.13	1.46	0.22	0.10	0.89	
K	1.99	3.79	0.68	2.04	0.71	0.65	0.80	1.01	0.06	0.07
H_2O	13.54	11.49	12.85	12.31	12.97	11.35	11.82	11.99	12.80	11.83
E%	-1.8	+4.7	-0.1	+4.3	+3.0	-8.9	+0.9	-4.3	-5.1	-3.9

	1	2	3	4	5	6	7	8	9	10
a	9.876	9.938	9.904	9.997	9.914	9.921	9.869	9.887	9.847	9.879
b	14.318	14.321	14.228	14.190	14.247	14.135	14.214	14.169	14.102	14.139
c	8.689	8.754	8.692	8.722	8.706	8.685	8.679	8.695	8.684	8.693
β_m	124°19'	124°35'	124°43'	125°5'	124°39'	124°55'	124°51'	124°38'	124°45'	124°49'
c_0	14.387	14.415	14.289	14.275	14.324	14.244	14.243	14.309	14.270	14.275
β_0	89°31'	90°	90°	90°7'	89°56'	90°5'	90°13'	89°59'	90°14'	90°11'
D	2.20	2.20	2.12			2.21			2.44	2.35

1. Phillipsite, Casal Brunori, Rome, Italy (Galli and Loschi Ghittoni 1972); new lattice constants by the authors.
2. Phillipsite, San Venanzo, Perugia, Italy (Galli and Loschi Ghittoni 1972).
3. Phillipsite, Montegalda, Vicenza, Italy (do); new lattice constants by the authors.
4. Phillipsite, Indian Ocean (no.14 of Sheppard et al. 1970); new lattice constants by the authors.
5. Wellsite, M. Calvarina, Vicenza, Italy (Galli 1972); new lattice constants by the authors.
6. Wellsite, Wikieup, Arizona (Sheppard and Gude 1971).
7. Wellsite, Kurtsy, Crimea (Černý et al. 1977).
8. Wellsite, Selva di Trissino, Vicenza, Italy (Passaglia and Bertoldi 1983).
9. Harmotome, Korsnas, Finland (Černý et al.1977).
10. Harmotome, Andreasberg, Germany (Rinaldi et al. 1974).

Note: sample no.2 was assumed by the original authors to be metrically orthorhombic, although no measurement of β_0 was made by them.

0.3 to 0.85; the phillipsites obtained have a ratio silica to alumina from 4 to 5, equivalent to 67% and 71% of Si in the tetrahedra.

Colella and Aiello (1975) reported on low temperature (~ 80°C) syntheses with several ratios NaOH : (NaOH + KOH) of alkalies reacting with a rhyolitic pumice low in Ca and Mg and high in Na and K: phillipsite (with the 6.4 Å spacing) is obtained when the ratio Na : (Na + K) in the reacting alkali solution is 1, 0.8, 0.6, 0.4, but one has to remember that K is also present as a solution product of the rhyolitic glass.

3.2.4 Optical and Other Physical Properties

Harmotome is optically positive with $2V = 80°$ $\alpha = 1.506$ $\beta = 1.509$ $\gamma = 1.514$; the orientation is $\gamma \rightarrow b$ and $\hat{a\beta} = 23.5°$ in the acute angle; colourless, with white streak, vitreous lustre; Mohs-hardness = 4.5; cleavage {010} perfect, {100} poor; these data for Korsnäs harmotome (Sahama and Lehtinen 1967) are nearly equal to those given by Meier (1939) for a sample from Glen Riddle. For wellsite, full data are available only for the original sample (Pratt and Foote 1897): $2V = +39°$ $\alpha = 1.498$ $\beta = 1.500$ $\gamma = 1.503$ the orientation being $\gamma \rightarrow b$ $\hat{a\beta} = 38°$ in the acute angle; on the whole, despite the lower Ba content, wellsite has the same indices and orientation as harmotome. Phillipsite (Winchell and Winchell 1951) has another orientation $\alpha \rightarrow b$ and $\hat{c\gamma} = 11° - 30°$ in the acute angle, $2V = 60° - 80°$ and even more until negative. Galli and Loschi Ghittoni (1972) give refractive indices for all their analyzed phillipsites, the lower values $\alpha = 1.490$ $\beta = 1.493$ $\gamma = 1.498$ being for Ca-poor alkali-rich phillipsites; still lower mean indices n = 1.480 where measured in high silica high alkali samples from deep-sea cores. Phillipsite is usually milky or yellowish white, transparent with white streak and vitreous lustre; Mohs-hardness = 4 – 4.5, cleavages {010} and {100} perfect.

3.2.5 Thermal and Other Physico-Chemical Properties

Harmotome, with five different water sites in the structure, looses its water in three steps (see DTG of Fig. 3.2 J) at 120°C, 230°C, and 320°C, plus one shoulder at 50°C and its dehydration is complete at 400°C; its structure is destroyed nearly at the same temperature and a new phase crystallizes at 750°C (see DTA of Fig. 3.2 J). The thermal behaviour of wellsite (Fig. 3.2 K) is similar to that of harmotome, with the last water loss shifted to higher temperature (~ 350°C).

Phillipsite, again with five water sites in the structure, looses its water in five steps (see DTG of Fig. 3.2 L) at 70°C, 120°C, 140°C, 180°C, and 320°C, plus two shoulders at 205°C and 280°C; it is destroyed at 400°C (Reeuwijk 1974).

The ion exchange properties of natural phillipsites were first studied by Ames (1964a, 1964b, 1965) and later on by Barrer and Munday (1971) and Shibue (1981). Barrer and Munday give several exchange isotherms (Fig. 3.2 M) for a sedimentary phillipsite from Pine Valley, Nevada, at 25°C. The

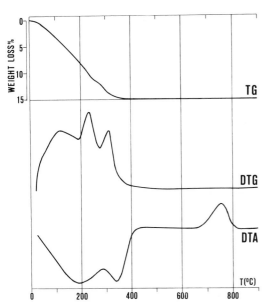

Fig. 3.2 J. Thermal curves of harmotome from Andreasberg, Germany; in air, heating rate 20°C/min (by the authors)

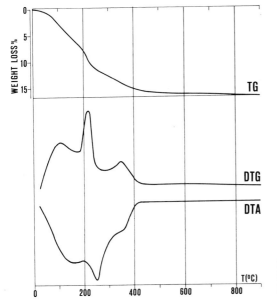

Fig. 3.2 K. Thermal curves of wellsite from Monte Calvarina, Italy; in air, heating rate 20°C/min (by the authors)

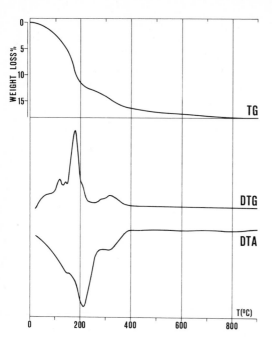

Fig. 3.2 L. Thermal curves of phillipsite from Casal Brunori, Rome, Italy; in air, heating rate 20°C/min (by the authors)

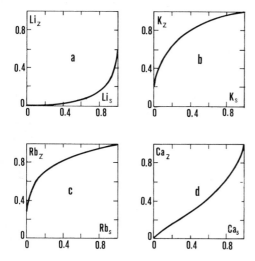

Fig. 3.2 M a – d. Exchange isotherms of phillipsite from Pine Valley, Nevada, at 25°C, total normality 0.1 N (Barrer and Munday 1971): **a** Na – Li; **b** Na – K; **c** Na – Rb; **d** Na – Ca

normal selectivity sequence Cs ~ Rb ~ K > Na > Li and Ba > Ca ~ Na ~ Sr is valid also for phillipsite, but the preference for K over Na is as strong as in chabazite; the exchange capacity was 3.2 meq/g. Shibue (1981) compared the Na – K – Ca exchange properties of a sedimentary phillipsite from Tekopa, California, with those of a hydrothermal phillipsite from Oki Islands, Japan, at 70°C, 35°C, and 5°C, and found that the Si/Al ratio of the sample strongly influences the exchange equilibria. Sherman (1977) emphasized the high selectivity of phillipsite for $(NH_4)^+$. Barrer and Townsend (1976) studied the exchange of the Cu – Zn – Co-ammine ions in the Nevada phillipsite; full exchange is possible in ammoniacal solution.

Maggiore et al. (1981) investigated the activity of a sedimentary phillipsite from the Neapolitan Yellow Tuff, Italy, as a catalyst for the isopropanol dehydration and found values which compare well with those of silica-alumina, gamma-alumina and synthetic 5 A zeolite.

3.2.6 Occurrences and Genesis

Harmotome is known to have both sedimentary and hydrothermal origin; the second one is by far the most common.

Harmotome was found three times in lacustrine formations from the western United States: genesis in a hydrologically closed system is probable for all three.

Charles Milton (written communication in Sheppard and Gude 1971) identified harmotome and barite with X-ray and optical methods in the Green River Formation, more exactly in an oil shale in the Piceance Creek basin of Colorado. A "sodic harmotome", which should more properly be called wellsite, was described in lacustrine Pliocene tuffs, in the Big Sandy Formation, Mohave Co., Arizona, by Sheppard and Gude (1971); it is associated with analcime, chabazite, clinoptilolite, erionite; its chemical analysis is no. 6 in Table 3.21.

Sheppard and Gude (1983) described harmotome from the cement of a basaltic, volcaniclastic sandstone from a lacustrine deposit near Kirkland Junction, Yavapai Co., Arizona; whilst the genesis of a zeolite in such a geological and chemical environment is obviously easy, the problem is to identify the source of barium.

Harmotome has been found several times in deep sea sediments (Kastner and Stonecipher 1978), in areas of the Pacific Ocean with a slow sedimentation rate, commonly associated with phillipsite. The presence of palagonite nuclei in the harmotome crystals gives evidence for a genesis by alteration of palagonite. After Arrhenius and Bonatti (1965), harmotome grows in the early stages of interaction between sea water and volcanic glass; in the later stages, only phillipsite is formed.

An occurrence similar to the sedimentary one is that of the harmotome which is present in the montmorillonite-groundmass of sandstones and porcellanites of Carboniferous red beds of the Ostrava-Karvina basin, Czechoslovakia, but Kralik (1971) stated that it is the product of post-volcanic hydrothermal mineralization, although he attributes a diagenetic origin not only to the montmorillonite but also to analcime, associated with harmotome in the same rock.

Some famous occurrences of harmotome in vugs of massive rocks (with hydrothermal genesis, probably), already listed by Hintze (1897) and Dana (1914), are:

Scotland: in a granite not far from the contact with gneiss from Strontian, Argyllshire, as fine crystals up to 2 cm with barite.

Northern Ireland: in the basalts of Antrim, at Giant's Causeway.

Norway: in the red quartz-syenite from Tonsenas, Oslo.

Germany: as very fine crystals at Andreasberg, Harz, in a hydrothermal metalliferous vein deposit connected with the Brocken granite; in the basalt of Rossberg, Rossdorf, Hessen (see also Deisinger 1973).

Newer descriptions of similar occurrences of harmotome are the following:

Scotland: in veins in a muscovite-biotite-schist and in a pegmatite at Allt Tigh Cumhaig, Inverness-shire, Scotland, as small beautiful crystals on blende, barite or calcite, including minute crystals of pyrite and marcasite (Russell 1946).

Sweden: in a pegmatite at Ultevis as fine crystals intergrown with kaolinite and montmorillonite as alteration products of scapolite (Byström 1956).

Finland: as well developed crystals, up to 1 cm in length, with apophyllite in the lead mine (no geological data available) of Korsnäs (Sahama and Lehtinen 1967).

Germany: in joints of a melilite-nephelinite in Höwenegg, Hegau, with phillipsite and garronite (Walenta 1974).

Czechoslovakia: in a desilicated pegmatite at Hrubsice, Western Moravia, as 12-fold twins (Černý and Povondra 1965a).

Austria: in the basalt of Weitendorf, Styria, crystals up to 2 mm in length, with delessite and calcite (Machatschki 1926).

Italy: in the graphite lenses of a gneiss at Cerisieri, Piemonte, as fairly large crystals with quartz, blende, galena (Pelloux 1931).

Bulgaria: in fissures of latitic lavas North of Iskra, Khaskovo District, with laumontite and heulandite (Kostov 1962).

U.S.A.: in a hyalophane-bearing anorthositic gabbro at Glen Riddle, Pennsylvania, as well developed cruciform twins with hydrargillite (Meier 1939).

Japan: in green tuffs at Udo, Shimane Pref., as typical interpenetration twins, with calcite and laumontite (Kinoshita 1922).

New Zealand: in an andesite dyke, at the contact with limestone, at Rehia and Maungarahu, Tokatoka District, as large crystals with analcime, crystallized at 200°C from a Na + Ba solution; the source of Ba is barite concretions in adjacent limestones (Black 1969).

Some hydrothermal occurrences of wellsite are the following:

In the Buck Creek corundum mine (type locality), Clay Co., North Carolina, as small crystals with chabazite on feldspar or corundum (Pratt and Foote 1897).

- In the oligoclase pegmatite of Vezna, Western Moravia, Czechoslovakia, as twins and fourlings (Černý 1960).
- In a basalt at Monte Calvarina, Verona, Italy, with analcime and chabazite (Galli 1972); in the fissures of the bedded chert associated with ophiolites at Cassagna, Liguria, Italy, with quartz (Lucchetti 1976).
- In the fissures of an igneous rock at Kurtzy, Crimea (Shkabara 1940), as large crystals with phillipsite, analcime, leonhardite, calcite.

Phillipsite is quite a common mineral both in sediments and vugs of massive rocks. Its formation is possible at ambient temperature, because the zeolite is a frequent component of deep sea sediments, and also of the vitric tuffs which were zeolitized in nearly surface conditions. The hydrothermal deposition is possible in a wide temperature range, from ~60°C (hot springs) up to ~200°C.

There is no report of the genesis of phillipsite in weathering surface conditions, viz. in alkaline soils. Hay (1978) considers in the same context the alteration of the tuffaceous and non-tuffaceous sediments of Pleistocene and Holocene age in the vicinity of the Olduvai Gorge, Tanzania (Hay 1963b, 1976), which are very similar in genesis to the zeolites of saline alkaline lakes; in these African sediments, phillipsite is formed together with natrolite, chabazite, analcime; the lower level of the zeolitic alteration represents the water table; the crystallization of zeolites is favoured by the high concentration of sodium carbonate-bicarbonate at the land surface.

There are two main types of sedimentary quasi-surface genesis of phillipsite: in saline lakes with silicic tuffs, and by burial diagenesis of basic tuffs.

Phillipsite with the first genesis is common in the Western United States: an alternation of Pleistocene mudstone with vitreous tuffs ($SiO_2 > 70\%$) is zeolitized mainly to phillipsite, both in monomineralic beds or associated with clinoptilolite, chabazite, erionite, clay minerals, opal, potassium feldspar; the phillipsite is silica-rich calcium-poor, with $Na:K > 1$ and has low refractive indices (~1.475). This typical situation is found for instance at Lake Tekopa, Inyo Co., California, (Sheppard and Gude 1968), in the Barstow Formation, San Bernardino Co., California (Sheppard and Gude 1969a) and in the Big Sandy Formation, Mohave Co., Arizona (Sheppard and Gude 1973). Hay (1964) described several other similar occurrences associated with rhyolitic tuffs: Searles Lake, California; China Lake, California; Teels Marsh, Nevada; and also Lake Magadi, Kenya; saline minerals, which accompany phillipsite in the former occurrence, are lacking in the following: Owens Lake, California; Waucoba Lake beds, California; Olduvai Gorge, Tanganyika.

Sheppard (1971) gives references for another 13 sedimentary lacustrine occurrences in the United States.

The sedimentary phillipsite generated by diagenesis (probably in a hydrologically open system) of basic ($SiO_2 \sim 50\%$) tuffs or ash flow deposits is common in central Italy (Gottardi and Obradovic 1978); well known is the Quaternary "Tufo grigio Campano", an ash flow deposit of potassic alkali-trachytic composition, widely altered to clay minerals (upper level), to phillipsite plus chabazite (medium levels) or to potassium feldspar (low levels); of similar composition and age is the "Tufo giallo napoletano", a tuff also altered to phillipsite plus chabazite; other phillipsite-bearing volcanics are the "tufo lionato" (Rome) and the pyroclastics of Monte Vulture, Potenza. Passaglia (pers. comm.) analysed these phillipsites and found a cation content similar to that of normal hydrothermal phillipsites, and a Si/Al ratio just a little higher than the normal values; their refractive indices are also high (~ 1.495): so on the whole, these sedimentary phillipsites are chemically quite different from those from lacustrine deposits.

The basaltic tuffs of the Honolulu Group on Oahu, Hawaii, were first palagonitized, and later the palagonitized tuffs were widely transformed into zeolites, the most abundant being a phillipsite, whose chemical composition is not known, although it is probably near to that of common hydrothermal phillipsite, because of its high (~ 1.500) refractive index, and hence similar to sedimentary phillipsites of central Italy and different from those of lacustrine deposits (Hay and Iijima 1968a, b). Please note that the term "palagonite" is used here in a wide sense, for alteration of sideromelane (basic glass), and not strictly for volcanic glass hydrated by reaction with sea water. Iijima and Harada (1969) used the word palagonite in the same sense when describing the transformation of glassy basaltic tuffs into phillipsite, gismondine, chabazite, gonnardite, natrolite, analcime and faujasite at Oahu, Hawaii.

The presence of phillipsite in deep-sea sediments, after the first find by Murray and Renard (1891) has been the object of so many publications that only a brief account of them can be given here; there are two modes of occurrence: in true sediments and in manganese nodules.

Review articles on the first are due to Bonatti (1965), Venkatarathuam and Biscaye (1973), Stonecipher (1976), Bernat and Church (1978), Honnorez (1978), Kastner and Stonecipher (1978).

The complex distribution of phillipsite and clinoptilolite in deep-sea sediments may be summarized as follows:

Phillipsite more frequent	Clinoptilolite more frequent
1. in clayey sediments	in calcareous and other sediments
2. where accumulation rate is less than $10\,m/10^6$ years	where accumulation rate is more than $10\,m/10^6$ years
3. in Pacific Ocean	both in Atlantic and Pacific Oceans
4. between latitudes 50°N and 50°S	at all latitudes

| 5. in Upper Miocene to recent sediments | in Lower Miocene to Cretaceous sediments |
| 6. in sediments containing basic volcanic glasses | in sediments containing rhyolitic glasses or from mafic glasses in the presence of excess silica |

After Petzing and Chester (1979), there is only one factor responsible for all differences, namely the last mentioned basicity or acidity of the glass shards the zeolites crystallized from. It must be emphasized that most zeolites from deep sea sediments grew at the expense of glass fragments spread in the oceans; ocean ridge basalts are not an important source of volcanic material for zeolite formation, which happened at the expense of glass shards spread over large areas in two ways: first, silicic explosive activity disperses into the oceans, via the atmosphere, huge amounts of acid glassy particles, and second, rapid chilling in sea water of hot basic lava from submarine central volcanoes (not related to ridges) causes granulation or pulverization of the hot lava itself, and turbidity currents disperse these basic glassy fragments; a basic-alkalic explosive activity does certainly also exist, but its contribution is minor when compared with the previous two mentioned sources. Zeolites reported as vein or cavity fillings have a (submarine) hydrothermal genesis and obviously are not considered here. The arguments of Petzing and Chester (1979) are:

1. Clay-carbonate.

Carbonate deposition is possible only above the carbonate compensation depth which means only at moderate depth, hence near to islands or continents where the volcanism is essentially acid and favours the formation of clinoptilolite; by contrast, deep sea clays are deposited at great depth in regions remote from continents where a silica undersaturated alkaline volcanic activity is possible and favours the subsequent formation of phillipsite.

2. High-low accumulation rate.

Near the continents or to islands, the rate of deposition is high and for the reasons given under 1., clinoptilolite formation is favoured.

3. Pacific-Atlantic.

As ocean ridge basalts cannot be considered here, there is no basic volcanism in the Atlantic, capable of spreading basic glass shards in the ocean; on the contrary acid volcanism exists at the ocean margins and its debris is also brought to the ocean by rivers; in the Pacific Ocean, both types of volcanism are present, as are phillipsite and clinoptilolite.

4. Latitude higher or lower than 50°.

No explanation is given for this division.

5. Recent-Cretaceous.

Phillipsite prevails over clinoptilolite in recent sediments because basic glasses break down to phillipsite faster; the presence of basic glasses in ancient sediments was minor for causes related to the geological history of the earth; acid glasses need a longer geological time to be altered to clinoptilolite.

On the whole, Petzing and Chester's arguments are fascinating because of their simplicity, but some other phenomena must also be active in the formation of the two zeolites; for instance it is well known that phillipsite never occurs in sediments older than Jurassic. It is difficult to explain this fact without invoking the transformation of phillipsite into clinoptilolite or some other mineral during geological times. The formation of clinoptilolite from basic glasses in the presence of high silica activity, due to other sedimentary components, is admitted by Petzing and Chester too. The crystallization of phillipsite in a solution where calcium carbonate is precipitating, could also be impossible for some other chemical reason.

Kastner and Stonecipher (1978) gave a detailed, prudent and realistic picture of the deep-sea occurrences of phillipsite and clinoptilolite; this paper also contains an appendix with the list of deep-sea sedimentary facies in which clinoptilolite and/or phillipsite occur. Apart from the general distribution rules already mentioned, they emphasize the dissolution of phillipsite at some depth (in some cases only 4 m) under the sediment surface, as indicated by etched crystal faces; this fact may be related to the contrasting increase of the clinoptilolite content with depth. Phillipsite is often associated, apart from the basaltic glass shards, with smectite and Fe- and Mn-oxides and/or -hydroxides. The growth of phillipsite from basaltic glass probably always takes place after the hydration or alteration of the glass to palagonite; anyway the relation of phillipsite crystallization to the presence of a basaltic glass is more definite than that of clinoptilolite to the presence of a silicic glass. Kastner and Stonecipher (1978) consider also the thermodynamic stability of the two zeolites, as some authors (e.g. Hay 1966) suggested that they are unstable under surface conditions; this view is supported at least for phillipsite by its disappearance in favour of clinoptilolite with depth and age; but thermodynamical calculations by Glaccum and Boström (1976) suggested a small but definite stability field for phillipsite under the conditions of the deep-sea floor.

Honnorez (1978) described in detail the phenomenon of palagonitization of the basaltic glass, which preceeds the crystallization of phillipsite, on the basis of his research on the samples from Palagonia, Sicily; he distinguishes three stages (initial, mature, final) of palagonitization, divided by the two following major reactions:

fresh glass + sea water → "palagonite" + intergranular authigenic minerals

"palagonite" + sea water → intragranular authigenic minerals .

Palagonite is an amorphous phase originated by hydration and oxidation of the fresh glass, with Fe and Ti passively accumulated in the residual glass (= palagonite) and Mg, Ca, K, Si, Al, Na (in decreasing order) leached from the glass to form authigenic minerals (zeolites, smectites) as intergranular cement. The final stage of palagonitization is reached when the palagonite has been replaced by intragranular authigenic minerals (zeolite, smectite, Fe – Mn-oxide aggregates).

The presence of phillipsite in deep-sea manganese nodules is well known; Bonatti and Nayudu (1965) gave a full report on this kind of occurrence, a more recent article with a literature review by Burns and Burns (1978). It is now generally accepted that these nodules are generated in the vicinity of submarine basic volcanoes; evidence hereof is given by the presence of palagonitic glass in the nodules, hence this glass is the more obvious precursor of the phillipsite. The complex mineralogical and chemical composition of the nodules is still a puzzle; the presence of phillipsite is less problematic, as the growth of this zeolite from basaltic glass via palagonite is a generally accepted process. Yoshikawa et al. (1982) analyzed the phillipsite from the Mn-nodules of the South Pacific Ocean, and found a composition intermediate between those of normal deep sea sediments and those from saline lake deposits.

Phillipsite is obviously rare as a constituent of marine sediments now on land (Iijima 1978), because it is less and less frequent as older rocks are considered. Sameshima (1978) reported only one occurrence of phillipsite, at Parnell Grit, in his study of the zeolites of the Waitemata Group, New Zealand, an Early Miocene marine sedimentary sequence including volcaniclastic deposits. Sersale et al. (1963) were the first to report on a phillipsite-rich rock layer at Garbagna, Alessandria, Italy, which was later recognized by Gianello and Gottardi (1969) as a marine cineritic deposit (tuffite) at the boundary between Oligocene and Miocene; the tuffite layer can be recognized all over the Northern Apennines, but it is strongly zeolitized in one place only.

Phillipsite is quite common as a (probably hydrothermal) deposit in vugs and geodes of basic volcanics like basalts, but not in other kinds of rocks; in particular phillipsite does not occur in vugs of acid rocks; neither rhyolite nor granite nor gneiss, which however, are common host rocks for other zeolites. On the whole, hydrothermal genesis repeats the preference of phillipsite for basic rocks, already seen for deep-sea phillipsites.

In view of the common presence of the mineral in many basic lavas, only some classic occurrences described in Hintze (1897) and Dana (1914) will be repeated here:

Iceland: at Dyrefjord and many other localities (see also Betz 1981).
Northern Ireland: in the basalts of Antrim, in the Giant's Causeway tholeiites, with gmelinite at Island Magee, and in many other localities (see also Walker 1960a).
Germany: with chabazite in the basalt from Annerod, Giessen, and from Nidda, both Hessen; in the limburgite from Sasbach, Kaiserstuhl, Baden

(see also Lorent 1933; Fricke 1971; Rinaldi et al. 1975a) with faujasite and offretite.
Czechoslovakia: in the basalt of Zálezly (= Salesl), Bohemia.
Italy: in the leucitites from Capo di Bove (type locality of gismondine) and from many other localities near Rome (Osa, Acquacetosa, Casal Brunori, Civitella, Vallerano; see also Galli and Loschi Ghittoni 1972) with chabazite; all these samples have a remarkable high Al-content; in the basalts from Aci Castello and Isola dei Ciclopi, Sicily (see also Di Franco 1933).

More recently described occurrences are:

France: in several basaltic rocks of the Central Massif (Pongiluppi 1976; Pillard et al. 1980).
Germany: in the basalt near Rossdorf, Hessen, with natrolite, scolecite, harmotome (Deisinger 1973); in the melilite-nephelinite at Höwenegg, Hegau, with harmotome (Walenta 1974).
Hungary: in the basalts at Mt. Halyagos and at Savarly, Balaton Lake area, as large crystals with apophyllite, thaumasite, heulandite (Mauritz 1931); in the cavities of the basalts of Halap and Gulacs, Lake Balaton area, with stilbite, natrolite, gismondine, thaumasite in the first locality, and with stilbite, natrolite, mesolite, scolecite, chabazite in the second (Mauritz 1938; Koch 1978).
Italy: in a differentiated venanzite at San Venanzo, Perugia, as an Al-rich variety, with willhendersonite (Morbidelli 1964); in the metagabbro of the "Gruppo di Voltri", Liguria, with thomsonite, chabazite, gismondine, saponite (Cortesogno et al. 1975).
U.S.A.: in veins and cavities of a nepheline-melilite basalt at Moilili quarry, Honolulu, Hawaii, with chabazite, hydronepheline, allophane (Dunham 1933).
Japan: in the amygdales of an altered basalt at Mazé, Niigata Pref., with heulandite and natrolite (Harada et al. 1967).
Taiwan: in the amygdales of the Penghu basalts, with thomsonite, scolecite, chabazite (Lin 1979).
Australia: in the Tertiary basalt of Tasmania, with apophyllite, chabazite, gonnardite, natrolite, stilbite, tacharanite and tobermorite (Sutherland 1965); in vesicles of a Quaternary alkali-basalt at Clunes, Malmsbury, Tylden and Trentham, Victoria, with chabazite (Hollis 1979).

3.2.7 Uses and Applications

The interesting property of phillipsite as an ammonia adsorber (Sherman 1977) suggests an application for the removal of ammonia from wastewater; from this point of view phillipsite is superior to clinoptilolite and inferior only to merlinoite or Linde W, but the first is not available in nature as large masses, and the synthetic Linde W is certainly more expensive than a sedimentary phillipsite. Subsequently Sherman and Ross (1982) patented a process of this type based on natural and synthetic phillipsites.

A similar application, ammonia removal during coal-gasification, was proposed by Hayhurst (1978) who demonstrated that phillipsite was better than mordenite, chabazite, clinoptilolite, or ferrierrite for this particular task.

3.3 Merlinoite

$(K, Na)_5(Ba, Ca)_2(Al_9Si_{23}O_{64}) \cdot 24 H_2O$
Immm
$a = 14.12$
$b = 14.23$
$c = 9.95 (\overset{\circ}{A})$

3.3.1 History and Nomenclature

Passaglia et al. (1977) described a new zeolitic mineral from the nepheline melilitite from Cupaello, Rieti, Italy, and named it in honour of Stefano Merlino, professor of crystallography at the University of Pisa.

3.3.2 Crystallography

The structure of merlinoite was resolved and refined by Galli et al. (1979) (see Chap. 0.2.3). There is no (Si, Al)-order in the framework, which is topologically tetragonal I4/mmm. The framework contains double octagonal rings, which deviate from regularity, the dimensions in the two main directions being 7 and 8 Å, although in the unit cell a and b are nearly equal. The deviation from the higher symmetry is due to the different arrangement of two (K, Ba) sites, which would be equivalent in the tetragonal space group. There are 8 framework oxygens around each (K, Ba) site, but they are a little too far from the cation site on the average: so four are pushed away and four are pulled nearer in the first site, and two are pushed away and six are pulled nearer in the second site (see also Gottardi 1979). There are also three other cation sites filled with (Ca, K, Na), in the centre of the largest cage and on the side of the same cage, in a niche of the "double crankshaft". The water molecules are mainly in the same large cage too.

The morphology is known only in a rough way: there is a pseudo-tetragonal prism with a pseudo-tetragonal bipyramid of the same order: their indices could be {100} {010} {011} {101} or {110} {111}.

Fig. 3.3 A. Merlinoite from Rieti, Italy (SEM Philips, × 900)

Fig. 3.3 B. Merlinoite from Rieti, Italy (SEM Philips, × 2000)

3.3.3 Chemistry and Synthesis

The two available chemical analyses are listed in Table 3.3I: both are rather near to the average formula of the title page, although a high balance error corresponds to the second analysis, which hence is unreliable.

The synthesis of merlinoite-like phases was perfomed long before the mineral was recognized as a new species. Barrer and Baynham (1956) reported on the synthesis of phase $K-M$, with a powder pattern similar to merlinoite; it was obtained hydrothermally from $K_2O \cdot Al_2O_3 \cdot nSiO_2$ with $1 \leqslant n \leqslant 8$ from 85 °C to 450 °C, but the best yield with $n = 3$ and $T = 250$ °C. Barrer et al. (1959) reported on the synthesis and properties of the "orthorhombic form of species P", also with a powder pattern similar to merlinoite; it grew hydrothermally from gels $Na_2O \cdot Al_2O_3 \cdot 5SiO_2$ with 200% alkali excess at 250 °C. Milton (1961) patented zeolite Linde W, which later on was shown by Sherman (1977) to be the same zeolite as $K-M$.

Belitskiy and Pavlyuchenko (1967) synthesized a barium silicate $Ba_8(Al_8Si_{24}O_{64})Cl_8 \cdot 12H_2O$, which was shown by Soloveva et al. (1972) to have a framework with double octagonal rings of tetrahedra, hence the same framework which later on was attributed to merlinoite. This silicate is tetragonal proper and contains also Cl as extraframework anion.

Therefore, on the whole, phases with the merlinoite framework were obtained with either K or Ba, both cations being present in merlinoite. The mineral was synthesized also using natural products as starting materials: so $K-M$ was obtained by Colella and Aiello (1971) from rhyolitic pumice, and by Barrer and Mainwaring (1972) from kaolinite plus amorphous silica. Colella et al. (1977) give as best conditions for merlinoite synthesis: rhyolitic pumice (1 g) + NaOH + KOH with Na : (Na + K) = 0.12, at 140 °C, total alkali concentration 1 N, water (10 g).

3.3.4 Optical and Other Physical Properties

Passaglia et al. (1977) give only an average refractive index, n = 1.494 due to the minute size of their crystals. Khomyakov et al. (1981) give full optical data: $\alpha = 1.499$ $\beta = 1.500$ $\gamma = 1.501$ $2V = 56°$, $\alpha \rightarrow b$ $\beta \rightarrow c$ $\gamma \rightarrow a$. An IR spectrum is also presented.

3.3.5 Thermal and Other Physico-Chemical Properties

The thermal curves are given in Fig. 3.3C. There are four and perhaps even five peaks in the DTG between 25 °C and 450 °C; please note that there are eight water sites in the structure with the following occupancies in decreasing order: 100, 100, 60, 44, 21, 20, 20, 20 (all %). A small peak at 650 °C cannot be interpreted with certainty. The lattice of the synthetic analogue is destroyed at 800 °C – 850 °C (Colella et al. 1977).

Barrer and Baynham (1956) found an easy Na- and Ca-exchange of phase $K-M$. Barrer and Mainwaring (1972) obtained the Na-, Li-, Ba- and Ca-

Table 3.3I. Chemical analyses and formulae on the basis of 64 oxygens and lattice constants in Å

	1	2
SiO_2	51.82	44.21
Al_2O_3	18.04	23.66
Fe_2O_3	0.72	
CaO	3.18	
SrO		1.13
BaO	2.49	9.57
Na_2O	0.65	1.29
K_2O	7.53	7.11
H_2O	15.57	13.00
Total	100.00	100.00
Si	22.68	20.15
Al	9.31	12.71
Fe	0.24	
Ca	1.49	
Sr		0.30
Ba	0.43	1.71
Na	0.55	1.14
K	4.21	4.15
H_2O	22.74	19.76
E%	+11.0	+36.6
a	14.116	14.05
b	14.229	14.10
c	9.946	9.99
D	2.14	2.27

1. Cupaello, Rieti, Italy (Passaglia et al. 1977).
2. Khibira Massif, Kola Peninsula, U.S.S.R. (Khomyakov et al. 1981).

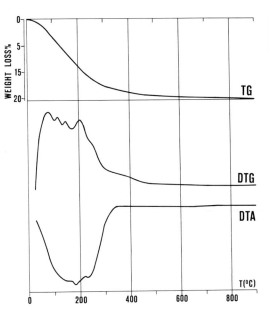

Fig. 3.3C. Thermal curves of merlinoite from Rieti, Italy; in air, heating rate 20°C/min (by the authors)

forms from K – M. Sherman (1978a) gives the following selectivity series for Linde W: $Mg < Ca \simeq Na \ll NH_4 \lesssim K$. He also emphasized that Linde W gives by far the best results in NH_4 capacity in a mixed cation exchange test. On this basis, Sherman (1978b) proposed IONSIV W-85 (a similar phase) as ion exchanger for removal of urea nitrogen from blood serum.

Barrer and Baynham (1956) found that K – M and its Na- and Ca-forms did not show any appreciable sorptive power toward oxygen or argon at −183 °C when dehydrated at 350°C; however, at 20°C, all the dehydrated forms sorbed ammonia quite rapidly. K – M has been shown to sorb considerable amounts of Ar and Kr, at high temperature and pressure (Barrer and Vaughan 1971); the limited sorption of oxygen at −195 °C is explained as due to activated diffusion too slow to measure at this temperature (Barrer and Mainwaring 1972).

3.3.6 Occurrences and Genesis

The type locality is Cupaello, Rieti, Italy, where the mineral occurs in vugs of a nepheline melilitite. It was found for the second time in the same kind of rock at Höwenegg, Hegau, Germany, with amicite (Alberti et al. 1979). Khomyakov et al. (1981) found merlinoite in an altered, cataclastized pegmatitic rock from the Khibira Massif, Kola Peninsula, U.S.S.R., Parodi (in Stoppani and Curti 1982, p. 195) found fine crystals, up to 4 mm in size, in pyroxene-rich ejecta from the Sabatino volcano complex, occurring at Valle Biachella, Sacrofano, Rome, Italy.

3.4 Mazzite

$K_3Ca_{1.5}Mg_2(Al_{10}Si_{26}O_{72}) \cdot 28\,H_2O$
$P6_3/mmc$
$a = 18.39$
$c = 7.65\,(\text{Å})$

3.4.1 History and Nomenclature

Mazzite is the first zeolite to be found as a natural mineral, when a synthetic counterpart with the same framework (zeolite Omega and ZSM-4) was already known. Galli et al. (1974) described the mineral from Mont Semiol, the type locality of offretite, known in the past as Mont Semiouse, near Montbrison, Loire, France, and named it for prof. Fiorenzo Mazzi, a distinguished mineralogist at Pavia University.

3.4.2 Crystallography

The framework of mazzite (Chap. 0.2.3) was determined by Galli (1974), who later (1975b) published a refinement down to 5%: it may be described as columns of gmelinite cages (Chap. 0.3) assembled with glide planes between them. There are two kinds of tetrahedral sites in the unit cell, the first T(1) on the 6-rings which are the common element between two successive gmelinite cages of the same column, the second T(2) which builds up the sides of the cage. Galli (1975b) found average distances $T(1)-O = 1.653\,\text{Å}$ and $T(2)-O = 1.639\,\text{Å}$, the difference being only 0.014 Å. Alberti and Vezzalini (1981b) gave some evidence, based on crystal energy calculations, showing that this small difference is significant and that more Al is in T(1) (33%) than in T(2) (24%). Mg is at the centre of the gmelinite cage, completely surrounded by water and with no contact to framework oxygens (type IV of Chap. 0.4). K is in a site between the gmelinite cages, and coordinates six framework oxygens and only two water molecules, which fill only 50% of the sites. Ca is in the centre of the large channels and coordinates only the water molecules which coat the wall of these channels. Rinaldi et al. (1975b) studied mazzite dehydrated at 600°C: at this temperature, it still retains some water ($9\,H_2O$) and Mg jumps from the centre of the gmelinite cage to the centre of the 6-ring at the top or the bottom of the cage, where Mg coordinates three oxygens plus some of the residual water which lies on the same inversion hexad. After Rinaldi et al. (1975b), K and Ca interchange their sites with dehydration, and this would be the second example, after offretite, of an "internal ion exchange" following Mortier et al. (1976a). Alberti and Vezzalini (1981b) were not able to support this "internal ion exchange" with crystal energy calculations, but they showed that Mg and H_2O must be alternated along the inversion hexad, as is predictable.

Fig. 3.4A. Mazzite from Mt. Semiol, France (×20)

Fig. 3.4B. Mazzite from Mt. Semiol, France (SEM Philips, ×120)

Fig. 3.4C. Mazzite from Mt. Semiol, France (SEM Philips, ×120)

Fig. 3.4D. Mazzite from Mt. Semiol, France (SEM Philips, ×950)

The morphology of mazzite is given by long and thin hexagonal prisms (order unknown), terminated with the basal pinacoid; the crystals are needle-like.

3.4.3 Chemistry and Synthesis

The analysis of the only known sample is as follows: SiO_2 58.10, Al_2O_3 18.14, MgO 2.92, CaO 2.75, Na_2O 0.03, K_2O 3.27, H_2O 18.42, total 103.63 (all %, the sum is well over 100, because of the water loss in the vacuum chamber of the electron microprobe). The calculated formula is

$$Na_{0.03}K_{1.91}Mg_{1.99}Ca_{1.35}(Al_{9.77}Si_{26.53}O_{72}) \cdot 28.06\,H_2O$$

balance error $= 13.58\%$ $a = 18.392$ $c = 7.646\,(Å)$ $D_{calc} = 2.11$.

A phase named zeolite Omega with the same framework may be synthesized in the presence of TMA ($=$ tetramethylammonium) (Flanigen and Kellberg 1967, 1968, 1970; Aiello and Barrer 1970). The reaction mixtures $t(TMA_2O) \cdot s(Na_2O) \cdot Al_2O_3 \cdot nSiO_2 \cdot pH_2O$ with $t \approx 1.6$, $s \approx 6.4$ and $8 \leqslant n \leqslant 20$, $120 \leqslant p \leqslant 320$, must be treated at 80°C for 7 days hydrothermally. The Si : (Si + Al) is slightly higher (77% − 80%) than the natural one (74%). Also the phase named ZSM-4 (Ciric 1966, 1968) has the same framework and may be synthesized only in the presence of TMA.

3.4.4 Optical and Other Physical Properties

Mazzite is uniaxial negative with $\omega = 1.506$ $\varepsilon = 1.499$. Transparent, colourless, streak white. Mohs-hardness probably 4.

3.4.5 Thermal and Other Physical Properties

Thermal curves are not known for mazzite. Rinaldi et al. (1975b) showed that, after dehydration in vacuum at 600°C, mazzite still retains 9 of the original $28\,H_2O$. Zeolite Omega (Aiello and Barrer 1970) first loses the water around 200°C, then the organic component burns at 600°C; there is no X-ray powder pattern over ca. 700°C. After burning the organic component in air, the adsorption properties of zeolite Omega are quite interesting, in accordance with the presence of channels surrounded by 12-tetrahedra rings: at 25°C and P = 700 torr, 81 mg of n-butane or 29 mg of isobutane per 1 g of zeolite are adsorbed (Breck 1974).

3.4.6 Occurrences and Genesis

Mazzite has been found so far only at Mont Semiol, Montbrison, Loire, France, where it occurs as hydrothermal deposition in vugs of an olivine basalt with offretite, chabazite, phillipsite (Galli et al. 1974; see also Betz 1982).

3.5 Paulingite

$Na_{12}K_{68}Ca_{41}(Al_{162}Si_{500}O_{1344}) \cdot 705 H_2O$
Im3m
$a = 35.09 (\text{Å})$

3.5.1 History and Nomenclature

Kamb and Oke (1960) described the new mineral from Rock Island Dam, Washington, and named it in honour of the Nobel laureate Linus Carl Pauling.

3.5.2 Crystallography

Gordon et al. (1966) determined the crystal structure (see Chap. 0.2.3) which is characterized by an extraordinarily complex framework with double 8-rings, gmelinite cages and truncated cubo-octahedra or alpha cages; as the refinement could be carried only to R = 14%, no (Si, Al) order was detected. One cation is in the gmelinite cage and is completely surrounded by water molecules. Two other cations are near the lobes of the double 24-ring, and coordinate also framework oxygens.

The crystals are clear rhombic dodecahedra.

3.5.3 Chemistry and Synthesis

All the reliable data on the chemical composition are due to Tschernich and Wise (1982) and are listed in Table 3.5 I. There is some variability between the four analyses, with a Si: (Si + Al) ratio ranging from 74% to 77%, the number of Ca per cell from 30 to 60, the number of K per cell from 36 to 74; there are ca 10 Na per cell, and also some Ba. So, like many phillipsites, paulingite is a (K, Ca)-zeolite, but with more silica. The measured density is not available.

We are not aware of any report on paulingite synthesis.

3.5.4 Optical and Other Physical Properties

Isotropic with $n \simeq 1.48$ (see Table 3.5 II), clear colourless to yellowish to bright orange, bright vitreous lustre. No cleavage, Mohs-hardness 5.

3.5.5 Thermal and Other Physico-Chemical Properties

The thermal curves (see Fig. 3.5 C) show a three-step water loss, at 100°C, 200°C, and 380°C; this can hardly be related with the 14 water sites of Gordon et al. (1966).

Fig. 3.5 A. Paulingite from Rock Island Dam, Washington (SEM Philips, ×60)

Fig. 3.5 B. Paulingite from Rock Island Dam, Washington (SEM Philips, ×120)

Table 3.5 I. Chemical analyses and formulae on the basis of 1344 oxygens, and lattice constants in Å

	1	2	3	4
SiO_2	63.53	61.99	61.13	55.11
Al_2O_3	16.43	18.04	17.95	16.16
Fe_2O_3	0.16	0.18	0.04	0.07
MgO	0.02	0.14	0.05	
CaO	3.46	4.30	4.41	6.10
SrO		0.32	0.07	
BaO	0.90	4.11	1.58	0.44
Na_2O	0.97	0.54	0.43	0.78
K_2O	6.87	5.53	7.19	3.14
H_2O				18.20
Total	92.34	95.15	92.85	100.00
Si	515.4	499.7	499.5	499.4
Al	157.1	171.4	172.8	172.6
Fe	1.0	1.1	0.3	0.5
Mg		1.8	0.6	
Ca	30.1	37.1	38.6	59.2
Sr		1.4	0.3	
Ba	2.9	13.0	5.1	1.6
Na	15.2	8.5	6.7	13.7
K	71.0	56.8	74.9	36.3
H_2O				550.0
E%	+3.9	+0.3	+1.5	+0.9
a	35.093	35.049	35.059	35.088
D				2.09

1. Rock Island Dam, Washington (Tschernich and Wise, 1982).
2. Chase Creek, Canada (do).
3. Riggins, Idaho (do).
4. Ritter, Oregon (do; H_2O from the TG of fig. 3.5 C ; D is calculated).

Table 3.5 II. Refractive indexes

	1	2	3	4
n	1.473	1.484	1.480	1.484

Numbers refer to the same samples as in Table 3.5 I.

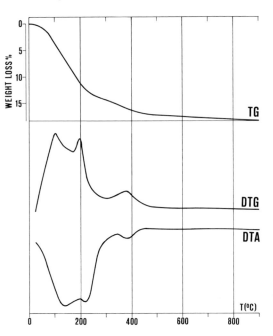

Fig. 3.5C. Thermal curves of paulingite from Ritter, Oregon; in air, heating rate 20°C/min (by the authors)

3.5.6 Occurrences and Genesis

The zeolite is known only from vugs and cavities of basalts.

Canada: in rounded vesicular basalt fragments from Chase Creek, north of Falkland, British Columbia, as fine, coloured crystals with erionite, harmotome, heulandite (Tschernich and Wise 1982).

U.S.A.: in a vesicular basalt occurring as boulders at the Rock Island Dam (the type locality) on the Columbia River, Washington, as fine crystals light yellow to colourless with erionite, offretite, phillipsite, clinoptilolite (Kamb and Oke 1960; Tschernich and Wise 1982); in a vesicular basalt occurring near the junction of Papoose Creek and Squaw Creek, west of Riggins, Idaho, as transparent crystals with phillipsite (Tschernich and Wise 1982); in the olivine basalt from Three Mile Creek, Grant Co., Oregon, as transparent crystals with heulandite, phillipsite, chabazite (Tschernich and Wise 1982).

4 Zeolites with 6-Rings

4.1 Gmelinite

$Na_8(Al_8Si_{16}O_{48}) \cdot 22\,H_2O$

$P6_3/mmc$

$a = 13.75$

$c = 10.06\,(\text{Å})$

4.1.1 History and Nomenclature

Brewster (1825a) proposed the name for the minerals occurring at Montecchio Maggiore, Vicenza, Italy, and at Glenarm, Ireland, in honour of the chemist G. C. Gmelin. Samples from the first locality had been incompletely described earlier under other names now obsolete. No other synonyms have any importance other than historical.

4.1.2 Crystallography

The structure was determined by Fischer (1966) and refined by Galli et al. (1982). There is no Si – Al order in the framework. Out of the two cation sites, the first is on a 6_3, which goes through the center of the double 6-ring: this cation is outside this cage, but in contact on one side with three framework oxygens of the cage. The cation coordinates three water molecules on the other side, hence the coordination is of type III of Chap. 0.4. The second cation is in the main channel, which is parallel to c and is surrounded by a 12-tetrahedra ring; these cations are attached to the wall of the channel, again with coordination of type III: three framework oxygens on one side, two water molecules on the other. These cation sites are the same whether the prevailing cation is sodium, as is most common, or calcium; the occupancy factors are obviously different in the two cases.

The morphology (Fig. 4.1 B) is quite simple, with hexagonal dipyramid $\{10\bar{1}1\}$ and prism $\{10\bar{1}0\}$, the pinacoid $\{0001\}$ being less common; other forms are rare. The twinning with $\{10\bar{1}1\}$ twin- and composition-plane is rare.

Intergrowths gmelinite-chabazite are common and not easy to recognize at a glance. If a powder or a Gandolfi pattern proves the presense of both species

Fig. 4.1A. Gmelinite from Flinders, Australia (×5)

Fig. 4.1B. Gmelinite from Flinders, Australia, with analcime (SEM Philips, ×15)

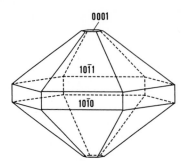

0001
10Ī1
10Ī0

Fig. 4.1C. Common morphology of gmelinite

in the same specimen, a clue to distinguish the domains of the two kinds may be found in the colour, more pink or reddish in gmelinite, white or even yellowish in chabazite (but chabazite may also have red shades).

The presence of stacking-faults on the {0001} planes is not only probable in view of the structure type, but also supported by some chemical properties (see Chap. 4.1.5).

4.1.3 Chemistry and Synthesis

A review of the chemistry of gmelinites was given by Passaglia et al. (1978a), who listed chemical formula and unit cell data for 21 specimens; sample no. 21 has a formula very near to the stoichiometric one given in the title page. The $Si:(Si+Al)$ ratio is nearly constant in all 21 samples, $68\% \pm 2\%$ to be compared with 67% of the stoichiometric formula. The ratio $(Na+K):(Mg+Ca+Sr)$ nearly coincides with the $Na:Ca$ ratio, because K, Mg, and Sr are normally very low and Ba is absent; this ratio ranges from 259 (in sample no. 21) to nearly 1; in only one sample (no. 1) is this ratio decidedly smaller than unity. The number of water molecules oscillates around 22. Some selected analyses and formulae from this source are given in Table 4.1 I along with the lattice constants, which anyway do not show a large variation; there is a positive correlation of the Sr-content with a, and of $(Na+K):(Mg+Ca+Sr)$ with c.

The synthesis of gmelinite is possible mainly in a sodium bearing hydrothermal environment, but has been performed also in the absence of sodium.

Barrer et al. (1959) reported on the first synthesis of crystals 1 to 12 µm in diameter, from 60°C to 100°C, with $SiO_2:Al_2O_3 = 9$ and $Na_2O:Al_2O_3 = 4$ in the batch. The product of the synthesis was probably a phase with chabazite and gmelinite domains; its chemical formula was $Na_7(Al_7Si_{17}O_{48}) \cdot 18\,H_2O$, not far from the stoichiometric one in the title page. Similar results were obtained by Pereyron et al. (1971) and by Barrer and Mainwaring (1972). Gmelinite-like phases were synthetized also in presence of some K, or Li (Barrer and Mainwaring 1972; Pereyron et al. 1971), or Ba (Barrer et al. 1974).

Table 4.1I. Chemical analyses and formulae on the basis of 48 oxygens and lattice constants in Å

	1	2	3	4
SiO_2	46.89	50.00	49.91	49.52
Al_2O_3	19.19	19.17	19.08	18.87
CaO	5.55	2.95	0.68	0.08
SrO	6.72	0.58		
Na_2O	1.17	7.20	10.12	11.79
K_2O	0.24	0.10	0.49	0.37
H_2O	20.16	20.00	19.70	19.37
Total	100.00	100.00	100.00	100.00
Si	16.21	16.62	16.59	16.49
Al	7.82	7.51	7.48	7.41
Ca	2.06	1.05	0.24	0.03
Sr	1.35	0.11		
Na	0.78	4.64	6.52	7.61
K	0.11	0.04	0.21	0.16
H_2O	23.23	22.17	21.85	21.51
E%	+1.01	+7.19	+3.34	-5.36
a	13.805	13.750	13.760	13.756
c	9.974	10.055	10.065	10.064
D	2.11	2.04	2.01	2.01

1. Montecchio Maggiore, Vicenza, Italy (no.1 of Passaglia et al. 1978a) with 0.08% FeO.
2. Montecchhio Maggiore, Vicenza, Italy (no.2 of Passaglia et al. 1978a).
3. Victoria, Australia (no.20 of Passaglia et al. 1978a) with 0.02% MgO.
4. Queensland, Australia (no.21 of Passaglia et al. 1978a).

The synthesis of gmelinite in the absence of Na was performed with $Ca(OH)_2 + N(CH_3)_4OH$ (Barrer and Denny 1961), but with a low yield. Barrer and Marshall (1964a) obtained a good yield of hexagonal plates from a batch with a gel $SrO \cdot Al_2O_3 \cdot 4.4SiO_2$ at 205 °C after 6 days; the product was certainly a gmelinite-like phase, but the procedure was not reproducible.

4.1.4 Optical and Other Physical Properties

The optical properties of gmelinites are not well known, but they are similar to those of chabazite. In our opinion, they may be summarized as follows: average index ranging from 1.475 to 1.485, nearly always uniaxial positive (but sometimes negative), birefringence 0.002 – 0.005; biaxial crystals both positive and negative, with 2 V small, are not rare. The colour may be white, transparent or opaque, but usually crystals are pink to salmon red, rarely dark red; in thin section always transparent colourless. This tendency of the colour towards red shades may be an easy distinguishing character from chabazite, which also sometimes features reddish colour, but normally is white or yellowish. In chabazite-gmelinite intergrowths, the colour shade may indicate which parts pertain to the former and which to the latter.

Gmelinite is piezoelectric (Ventriglia 1953) in spite of the centric structure.

4.1.5 Thermal and Other Physico-Chemical Properties

Figure 4.1 D shows the thermal curves for a typical sodian gmelinite from Victoria, Australia (no. 3 in Table 4.1 I). Water is lost in three steps: after a shoulder at 100 °C, there are two peaks at 175 °C and 300 °C in the DTG; this behaviour is in accordance with the presence of three water sites in the structure: the first loss at 100 °C of 8 or $9H_2O$ probably corresponds to W(3), which is less tightly bound being in the large channels; it is not possible to deduce which one of W(1) and W(2) with $6H_2O$ each, is emptied first at the two higher temperatures.

Passaglia et al. (1978a) give the DTG for a Ca-rich gmelinite, which is not essentially different from the curve of the normal sodian gmelinite here described.

Ion exchange is certainly quite easy in gmelinite, as several authors describe ion exchanged phases, but no equilibrium diagrams on gmelinite have been published.

Kühl and Miale (1978) present data on the gas absorption of cation exchanged samples, and in particular of the ammonium form from which the hydrogen form was obtained by calcination. The gas absorption was found to be very low in comparison with the value which should be expected in view of the presence of large channels built by 12-tetrahedra rings: this result confirms previous data by Barrer (1944) who first concluded that stacking faults of the chabazite type block the channels, thus reducing gas absorption to the values proper for chabazite. Kühl and Miale (1978) found also a surprising result

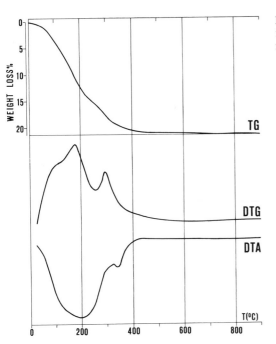

Fig. 4.1 D. Thermal curves of gmelinite from Flinders, Australia; in air, heating rate 20°C/min (by the authors)

with the K-form, which is the most temperature stable, as is common for many other zeolites; this form, after calcination at 550°C, rehydrates immediately to the original water content, but water continues to be sorbed at a slow rate, up to 24 g H_2O/100 g zeolite; with an uncalcined K-gmelinite, a hydration up to 38 g H_2O/100 g zeolite was shown to be possible in a suitable atmosphere!

4.1.6 Occurrences and Genesis

All known occurrences are probably of hydrothermal origin, namely all gmelinites occur as crystals in geodes or fissures, and there is no evidence of reaction or equilibrium with the wall rock.

Well known occurrences, also described in Hintze's (1897) and Dana's (1914) handbooks, are in the following localities:

Northern Ireland: in the basalts from Co. Antrim, notably in Glenarm, Portrush, Island Magee.
Germany: in the metalliferous vein deposit at Andreasberg, Harz.
Italy: in the basalt at Montecchio Maggiore, Vicenza, with analcime, natrolite, chabazite, phillipsite, stilbite, heulandite; also in nearby localities.
Cyprus: in the basalts at Pyrgo, Mansoura and Malounda (see also Passaglia et al., 1978).

U.S.A.: at Bergen Hill, New Jersey.

There are only a few more recent descriptions of gmelinite occurrences:

Northern Ireland: Walker (1959) noticed that in the Antrim basalt, the presence of gmelinite is restricted to a narrow zone along the eastern seaboard (less than 1% of the area of the basalts), although zeolites are very common in the whole trap. The gmelinite is associated with chabazite, analcime, natrolite, thomsonite and phillipsite, and frequently forms overgrowths on chabazite. The narrowness of the gmelinite zone was considered as evidence for the chemistry of the hydrothermal fluids being of a very particular composition (mixing of seawater with meteoric water?).

Germany: in the basalt of Zeilberg, Unterfranken, Jakob (1982) found gmelinite with chabazite, analcime, laumontite.

Hungary: the basalts North of Lake Balaton are long known for occurrences of gmelinite and many other zeolites: Koch (1978) gave a review on this topic.

Italy: Passaglia (1966) found gmelinite in the basalts of Monte Baldo, Trentino, with natrolite, analcime, chabazite, phillipsite and other silicates.

U.S.S.R.: Shkabara (1940) described gmelinite with analcime and wellsite and other silicates in the fissures of an igneous rock near Kurtzy, Crimea; a similar nearby occurrence was described by Kopteva (1955).

Greenland: Karup-Møller (1976) described gmelinite associated with herschelite (sodium rich chabazite) as hypogene alteration of ussingite, a hydrothermal silicate found in the Ilimaussaq intrusion.

Canada: in the traps of Nova Scotia, northeast of a line drawn from Cape d'Or to Cape Split, Walker and Parsons (1922) found frequent gmelinite with abundant chabazite and analcime, and frequent natrolite; Aumento (1964) gave more data on these occurrences.

Australia: Birch (1976) observed the association gmelinite, chabazite, natrolite in the basalts of Flinders; the intergrowth gmelinite-chabazite is common in these rocks.

Bass et al. (1973) found gmelinite in a DSDP-core, as a hydrothermal replacement of plagioclase of a basalt; this is certainly not a sedimentary gmelinite.

4.2 Chabazite, Willhendersonite

Chabazite	Willhendersonite
$Ca_2(Al_4Si_8O_{24}) \cdot 12 H_2O$	$K_2Ca_2(Al_6Si_6O_{24}) \cdot 10 H_2O$
$P\bar{1}$	$P\bar{1}$
$a = 9.41$	$a = 9.20$
$b = 9.42$	$b = 9.18$
$c = 9.42 (Å)$	$c = 9.49 (Å)$
$\alpha = 94°11'$	$\alpha = 92°36'$
$\beta = 94°16'$	$\beta = 92°26'$
$\gamma = 94°21'$	$\gamma = 90°3'$
or at the most	
$R\bar{3}m$	
$a_{rh} = 9.41 (Å)$	
$\alpha_{rh} = 94°19'$	
or	
$a_{hex} = 13.80$	
$c_{hex} = 15.02 (Å)$	

4.2.1 History and Nomenclature

The first certain description of chabazite was given by von Born (1772), who named it "zeolithus crystallisatus cubicus Islandiae"; Romé de l'Isle (1783) wrote of a "zéolite en cube". The modern name was, more or less, proposed by Bosc d'Antic (1792) as "chabasie", derived from the Greek χαβάσιος, "un mot grec, qui désignait un certain espêce de pierre". Breithaupt (1818) changed it to Chabasit, and in this form is still used in Germany; the Italians write cabasite. The English-speaking mineralogists use chabazite, and we were not able to discover who first substituted the z for the s. The French now use chabazite, but wrote chabasite in the 19th century.

Two more names are still used in connection with this mineral species. Lévy (1825) introduced the name Herschelite, in honour of J.F.W. Herschel, for a chabazite from Aci Castello, Sicily, Italy, with a special pseudo-hexagonal habit; Mason (1962) redefined herschelite as a zeolite with the chabazite framework and Na > Ca, because this chemical characteristic is typical for the samples from Sicily. Breithaupt (written communication in Tamnau, 1836) introduced the name phakolite (from the Greek φακός = lentil) for a chabazite from Leipa, Bohemia, with a particular pseudo-hexagonal habit, similar to that of herschelite; the word is still used as a varietal name to denote those chabazites with "phakolitic (or phacolitic) habit".

Other synonyms, now fully obsolete, are acadialite, seebachite, haydenite.

Peacor et al. (1984) proposed the name willhendersonite for a zeolite with the same framework occurring at San Venanzo, Terni, Italy, in honour of

William A. Henderson, Jr., of Stamford, Connecticut, who first noticed the uniqueness of this species.

4.2.2 Crystallography

The crystal structure was first determined in space group $R\bar{3}m$ by Dent and Smith (1958) and by Nowacki et al. (1958), as an AABBCCAA... stacking of 6-rings (see Chap. 0.4) forming large cages (see Fig. 0.3 A h).

The maximum symmetry compatible with this framework is $R\bar{3}m$; this is also the highest symmetry which has been attributed to chabazite, although many mineralogists long ago claimed that the real symmetry was lower, mainly on the basis of optical observations (see e.g. Winchell and Winchell 1951; Majer 1953; Akizuki 1980, 1981 b). Two of the early refinements (Smith 1962; Smith et al. 1963) were performed in $R\bar{3}m$, but the third one (Smith et al. 1964) was done in $P\bar{1}$, the deviation from trigonal symmetry being ascribed to the position of the Ca-atom.

Gottardi (1978) emphasized that the twinning observed with the polarizing microscope was not compatible with symmetry $\bar{3}m$ and that a structure refinement performed on an untwinned fragment of a chabazite with well-developed twins might be successful in explaining both the symmetry lowering and the Si/Al order. This kind of research was done by Mazzi and Galli (1983) who investigated four chabazites and found small but significative deviations from trigonal symmetry, leading to space group $P\bar{1}$. These deviations are due to asymmetric Al-enrichment in the tetrahedra and to the cation sites whose occupancy is correlated to the Al-fraction in nearby tetrahedra. The Al-fraction oscillates from 0.20 to 0.38 in an irregular way, which is interpreted by Mazzi and Galli as due to the presence of domains with ordered Si/Al distributions.

Calligaris et al. (1982, 1983) refined a natural crystal and also K- and Ag-exchanged forms; Calligaris and Nardin (1982) the Ba- and Cd-forms; Alberti et al. (1982b) refined another natural crystal and its Na-, K-, Ca-, Sr-modifications; from all the data, the cation sites may be listed in the following schematic way (first code after Alberti et al. 1982b, second code by Mortier 1982):

Site C1 – A: on the triad, in the centre of the double 6-ring; sometimes empty, as in Na- and K-forms, but partially occupied in natural, Ca- and Sr-forms.

Site C2 – C: on the triad, outside the double 6-ring, in contact with three oxygens of the ring; nearly always with high occupancy.

Site C3 – K: on the triad, nearly at the centre of the large cage; with low occupancy.

Site C4 – G: nearly at the centre of the 8-tetrahedra ring which is the port between the large cages; with low occupancy.

Water molecules have one site with high occupancy in all forms, near the cation site C2 – C, the remainder being spread over many sites with medium to low occupancy.

Fig. 4.2 A. Chabazite from Heinekeim, Germany (×10)

Fig. 4.2 B. Chabazite from Nidda, Hessen, Germany (SEM Philips, ×120)

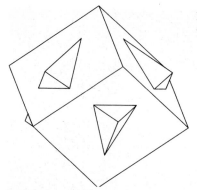

Fig. 4.2C. Common morphology of [0001]-twinned chabazite

Fig. 4.2D. Chabazite, variety "herschelite", from Aci Castello, Sicily, Italy (SEM Philips, ×60)

Cation sites do not change appreciably in dehydrated chabazites, except for site C3 – K which is obviously empty because not in contact with framework oxygens; in the Ca-form only sites C1 – A and C2 – C are occupied, but in the Na-form also site C4 – G is partially occupied and the framework undergoes some distortion assuming symmetry C2/m (Mortier et al. 1977).

Willhendersonite (Tillmanns and Fischer 1981; Peacor et al. 1984) has the same framework as chabazite, but an ordered (Si, Al)-distribution with an alternation of the two atoms in the tetrahedra, as Si : Al = 1 in this mineral.

The most frequent morphology of chabazite is quite simple, as pseudocubic rhombohedra corresponding to the shape of the unit cell are usually

Fig. 4.2 E. Chabazite with "phacolitic" habit, from Richmond, Australia (SEM Philips, ×8)

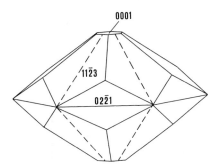

Fig. 4.2 F. Morphology of the "phacolitic" habit

found. The following forms with hexagonal indices are known in more complex combinations: $\{10\bar{1}1\}$, $\{01\bar{1}2\}$, $\{02\bar{2}1\}$, the first being always dominant. Twinning in chabazite is complex; pseudomerohedral twins of chabazite $P\bar{1}$ with the pseudo-mirror plane as twin-plane, not observable with the normal goniometric method, can be detected by reflection-type interference contrast microscopy (Akizuki 1980, 1981 b). The triad (or pseudo-triad) is a frequent twin-axis, with both $\{10\bar{1}0\}$ and $\{0001\}$ as contact planes (Fig. 4.2 C). The result may be a complex twin with lens shape ("phacolitic habit", see Fig. 4.2 F) or rounded shape. The lens shape is due to the presence of the rhombohedra $\{02\bar{2}1\}$ and the hexagonal bipyramid $\{11\bar{2}3\}$ plus the pinacoid $\{0001\}$ from absent to small, or also dominant as in the original herschelite. The rounded

Fig. 4.2G. Chabazite with another "phacolitic" habit from Trial Harbour, Tasmania (SEM Philips, ×8)

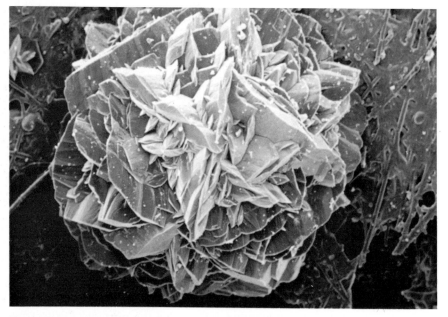

Fig. 4.2H. Rosette aggregates of Sr- and Al-rich chabazite from Osa, Rome, Italy (SEM Philips, ×60)

Fig. 4.2I. Sedimentary chabazite, with phillipsite, in the "tufo grigio Campano" from San Mango, Avellino, Italy (SEM Philips, ×900)

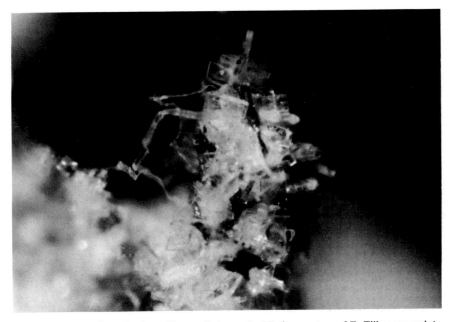

Fig. 4.2J. Willhendersonite from Mayen, Germany (×50) (by courtesy of E. Tillmanns and A. Queisser)

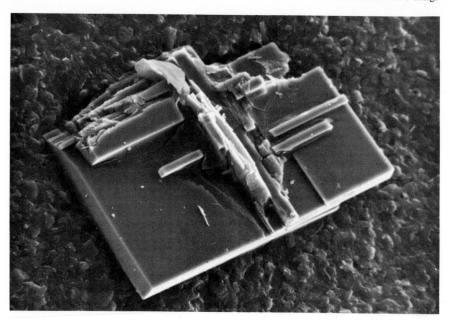

Fig. 4.2 K. Willhendersonite from Mayen, Germany (SEM, ×500) (by courtesy of E. Tillmanns and A. Queisser)

shape is due to the presence of more complex forms, the {0001} being absent; re-entrant angles are possible. Both types of twin are always correlated to complex, mostly alkali-rich, chemical compositions (Passaglia 1970).

Chabazite-gmelinite intergrowths are frequent (see Chap. 4.1.2).

Crystals of willhendersonite from Terni form trellis-like intergrowths, twinned nearly at 90° (twin axis is probably [111], the pseudo-triad); crystals from Terni and Mayen are mostly lamellar {001}, the other faces being {100} and {010}.

4.2.3 Chemistry and Synthesis

The chemistry of chabazite shows large variations from the schematic formula of the title page, both in Si:(Si + Al) ratio and in cation content (see Table 4.2 I). The percentage of the tetrahedra occupied by silicon oscillates from 59% to 80%, but the most frequent value is just 66%. The most Al-rich samples are high in Sr (by chance?); the most Si-rich have a sedimentary origin, the highest Si-percentage in tetrahedra for a hydrothermal chabazite being 73%. The most frequent extraframework cation is just Ca, up to nearly two atoms per rhombohedral unit cell, but Na can be as high as 3.1 (analysis no. 2), K as high as 2.1 (analysis no. 3) and usually not less than 0.4 − 0.3, Sr up to 0.5 (no. 4), while Ba is always very low.

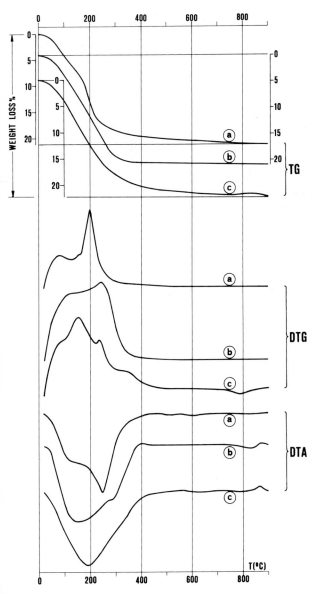

Fig. 4.2 L. Thermal curves of (a) Sr- and Al-rich chabazite from Vallerano, Rome, Italy, (b) "herschelite" (Na-rich) from Poggio Rufo, Palagonia, Sicily, Italy, and (c) typical Ca-rich chabazite from Keller, Nidda, Germany; in air, heating rate 20°C/min (by the authors)

Table 4.21. Chemical composition and formulae on the basis of 24 oxygens and lattice constants in Å

	1	2	3	4	5	6	7	8	9
SiO_2	46.40	42.45	43.44	38.21	59.68	60.86	52.94	49.82	34.▪
Al_2O_3	19.30	22.06	21.18	22.66	13.11	12.37	16.47	17.91	28.▪
Fe_2O_3	0.06	0.11	0.60	0.10	0.13	0.39	1.28	0.08	
MgO	0.06	0.24	0.39	0.04	0.79	0.31	0.54	0.36	
CaO	10.04	1.01	2.46	6.57	1.13	5.29	4.14	7.07	10.▪
SrO	0.32	0.47	0.14	5.27			0.05	0.28	
BaO	0.02		0.08	0.90				0.06	
Na_2O	0.10	9.20	2.90	0.48	5.30	0.55	0.78	0.24	
K_2O	0.92	4.70	9.21	3.89	0.62	1.09	5.80	2.68	8.▪
H_2O	22.80	19.73	19.58	21.44	19.01	17.27	18.00	21.50	18.▪
Total	100.02	99.97	100.09	99.56	99.86	98.32	100.00	100.00	100.▪
Si	8.03	7.40	7.60	7.05	9.51	9.65	8.69	8.44	6.▪
Al	3.94	4.53	4.37	4.93	2.46	2.31	3.19	3.58	5.▪
Fe	0.01	0.01	0.08	0.01	0.02	0.05	0.05	0.01	
Mg	0.02	0.06	0.10	0.01	0.19	0.07	0.13	0.09	
Ca	1.86	0.19	0.46	1.30	0.19	0.90	0.73	1.28	2.▪
Sr	0.03	0.05	0.01	0.56			0.01	0.03	
Ba				0.07					
Na	0.03	3.11	0.98	0.17	1.64	0.17	0.25	0.08	
K	0.20	1.05	2.06	0.92	0.13	0.22	1.21	0.58	1.▪
H_2O	13.16	11.47	11.43	13.20	10.10	9.13	9.86	12.15	10.▪
E%	-2.78	-4.30	-6.46	-0.45	-2.05	-1.29	1.3	3.7	-0.▪
a_{hex}	13.790	13.863	13.849	13.773	13.705	13.721	13.789	13.807	
c_{hex}	15.040	15.165	15.165	15.389	14.870	14.795	14.994	15.009	
D	2.008	2.09	2.10	2.20					2.▪

1. Col del Lares, Val di Fassa, Italy (Passaglia, 1970).
2. "Herschelite", Acitrezza, Sicily, Italy (Passaglia, 1970).
3. Vesuvio, Italy (De Gennaro and Franco, 1976) with 0.10% Rb_2O.
4. Casal Brunori, Roma, Italy, (Passaglia, 1970); new lattice constants by authors.
5. Fossil Canyon, California, with FeO 0.02%, TiO_2 0.04%, P_2O_5 0.02%, CO_2 ▪ (Gude and Sheppard, 1966).
6. Grant Co., Oregon (Sheppard and Gude, 1970) with FeO 0.09%, TiO_2 0.05%, 0.01% each P_2O_5, MnO, CO_2, Cl, F.
7. San Mango, Avellino, Italy (Passaglia and Vezzalini, 1985).
8. Stigliano, Viterbo, Italy (Passaglia and Vezzalini, 1985).
NOTE: chabazites 1 to 4 have a hydrothermal genesis, 5 to 8 a sedimentary ▪
9. Willhendersonite, San Venanzo, Terni, Italy (Peacor et al., 1984); latt constants a=b=9.16 c=9.45(A) $\alpha = \beta$ =92°30' γ =90°.

The hydrothermal chabazites show the widest variation in chemistry; the sedimentary ones often are "acid", i.e. silica-rich, and sodic, but the silica percentage can also be not so high (70% of the tetrahedra with Si in no. 8), and Ca can be the dominant cation (analyses nos. 6 and 8).

Passaglia (1978a) studied the lattice constant variations in cation exchanged chabazites.

Willhendersonite is quite near to the schematic formula of the title page. The lattice constants of the sample from Terni are metrically monoclinic (see Table 4.2I); the sample from Mayen gives $a = 9.23$, $b = 9.21$, $c = 9.52 (\text{Å})$, $\alpha = 92°42'$, $\beta = 92°24'$, $\gamma = 90°6'$; the lattice constants in the title page are average values of the two.

Barrer and Baynham (1956) were the first to document a synthesis from gels of composition $K_2O \cdot Al_2O_3 \cdot nSiO_2$ with $1 \leqslant n \leqslant 8$ and $60 < T < 100°C$, in the presence of excess KOH so as to have pH ~ 10.5; bayerite co-crystallized. The product gave the powder pattern of chabazite; ion-exchanged forms of Li, Na, NH_4, Rb, Cs, were also obtained.

Chabazite-gmelinites were obtained also from $Na_2O \cdot Al_2O_3 \cdot nSiO_2$ gels with $7 \leqslant n \leqslant 9$, 200% NaOH excess at ca. 80°C (Barrer et al. 1959) and from $CaO \cdot Al_2O_3 \cdot 2SiO_2$, $3N(CH_3)_4OH$ gels at 200°C (Barrer and Denny 1961).

Aiello and Franco (1968) were the first to use natural materials, halloysite and montmorillonite, for a synthesis in the presence of KOH at 80°C. Franco and Aiello (1969) synthesized chabazite in similar conditions but starting from a volcanic glass. Tomita et al. (1969) were able to crystallize chabazite by boiling a powdered volcanic glass (low in Mg, medium in Ca, Na, K) with 5% NaOH or 5% KOH. With sodium hydroxide, chabazite appeared after 7 h, but was later gradually converted to faujasite and then to Na – P (see Chap. 3.1.3). With potassium hydroxide, the crystallization could be detected after 13 h, and proceeded to completeness with persistence up to 200 h. Similar results were obtained on a pumice low in Ca and high in alkali by Höller et al. (1974), who did their test on wider pH and T ranges; they were able to synthesize chabazite also at very low pH (7 to 8), using only H_2O as reacting fluid. Colella and Aiello (1975) used a pumice of similar composition and searched for the influence of a changing Na/K ratio; they obtained good yields of true chabazite only if the ratio $K : (K + Na)$ was 0.80 or higher; if this ratio was in the range $0.8 - 0.5$, a chabazite-gmelinite intergrowth crystallized; if the range was lower than 0.5, no chabazite appeared. Strangely enough, although the commonest hydrothermal chabazite has Ca as dominant cation, no chabazite has been synthesized so far from Ca-rich batches, except in the presence of $N(CH_3)_4OH$, which is not a probable constituent of the natural fluids from which zeolites grow.

Information on the stability of chabazite is scanty. Aoki (1978) studied the hydrothermal persistence of chabazite in the presence of water only, or with silica gel (1/3 or 2/3 of the chabazite weight), the water pressure being up to 1 Kbar. Chabazite begins to alter when heated for several days (1 to 14) at nearly 250°C in all cases. In the presence of water only, wairakite was the main transformation product, gonnardite appearing in a small intermediate temperature interval. If silica gel was also added, heulandite was the main transformation product, with wairakite and mordenite, the latter in the runs with more silica.

4.2.4 Optical and Other Physical Properties

The optical properties of chabazite are not well known; as a matter of fact, a small change in one index may change the whole scheme of the optical properties because of the low birefringence, which may be negative or positive. Table 4.2 II gives the (mostly uniaxial) optical data for some samples whose analyses are given in Table 4.2 I. As said before, crystals may be uniaxial, but very often are biaxial; full optical data of this kind are rare. Majer (1953) gave: $\alpha = 1.4848$, $\beta = 1.4852$, $\gamma = 1.4858$, $\gamma \rightarrow c$, $2V =$ from $+67$ to $+75°$ for a chabazite from Bor, Serbia, with formula $Ca_{1.78}Na_{0.32}K_{0.13}(Al_{3.61}Si_{8.29}O_{24})$ $\cdot 12.49 H_2O$. He measured similar $2V$ angles for other chabazites. As is easily predictable, lower indices correspond to sodian silicic crystals, medium indices to calcian stoichiometric crystals, high indices to strontium-bearing silica-poor crystals. From colourless transparent, to white, but also pinkish and rarely brick-red; streak white, vitreous lustre; Mohs-hardness $= 4 - 5$. Cleavage parallel to the (pseudo)rhombohedron.

Willhendersonite from Terni has $\alpha = 1.507$, β not known, $\gamma = 1.517$, optical sign $+$, $2V$ large; willhendersonite from Mayen has $\alpha = 1.505$, $\beta = 1.511$, $\gamma = 1.517$, $2V = +87°$; for the orientation, see Peacor et al. (1984); colourless with vitreous lustre; streak white. The cleavages parallel to (100), (010) and (001) are perfect.

Rabinowitsch and Wood (1933b) measured the electrical resistance which is lower for zeolites than for other common ionic crystals, ranging from 10^7 to 10^{10} ohm per cm^3 for chabazites of different occurrences, but being a definite quantity for crystals of a given origin. The resistance decreases with rise of temperature under conditions which exclude loss of water. Wetzlar (1938) found a dielectric constant $\varepsilon = 10$ (~ 3000 Hz) for chabazite, which is twice as much as for gypsum.

4.2.5 Thermal and Other Physico-Chemical Properties

The thermal curves (Fig. 4.2 J) are influenced by the dominant cation present in the chabazite. "Herschelite" (Na-rich chabazite) shows a very broad area in the DTG, with two peaks at 170°C and 250°C; a typical Ca-rich chabazite gives as many as 4 peaks in the DTG, at 100°C, 150°C, 220°C and 280°C, in accordance with the presence of many possible water sites in the structure; an anomalous Sr – Al-rich chabazite loses water in two main steps at 80°C and 200°C, plus perhaps a minor one at 160°C; the dehydration is nearly complete at 400°C for all samples, and continues at a very low rate up to 800°C, rehydration being possible under this limit.

Table 4.2 II. Optical data

	1	2	3	4	5	6
ω	1.4994	1.4773	1.474	1.5166	1.462	1.461
ε	1.4962	1.4750		1.5142	1.460	1.465
Sign	−	−		−	−	+

NOTE: samples bear the same numbers as in Table 4.2 I.

The literature on ion exchange and gas absorption on chabazite is quite large and abundant, so that only a short summary is given here; important literature sources may be found in the references of the quoted papers; the pioneering work of Nacken and Wolff (1921) and Rabinowitsch and Wood (1933a) is worth mentioning.

A full set of ion exchange diagrams on a natural chabazite from Nova Scotia is given by Barrer et al. (1969), who also measured the heat of exchange. The ion pairs, which exhibit almost complete exchange, are here arranged in the affinity series:

- monovalent ions $Tl > K > Ag > Rb > NH_4 > Pb > Na$
- divalent ions $Ba > Sr > Ca$.

Similar results were obtained on a synthetic chabazite AW-500 by Ames (1964a, b). The rate of Na/K exchange was studied by Duffy and Rees (1974, 1975).

A summary of the adsorption properties is given by Breck (1974); calcian chabazite from Nova Scotia readily adsorbs small molecules (numbers give mg per 1 g of dehydrated zeolite, at P = 700 torr) CO_2 240 at 298 K, O_2 200 at 90 K, N_2 200 at 77 K; a somewhat larger molecule like propylene C_3H_6 is adsorbed in 95 mg at 298 K. Ion exchange strongly influences the adsorption properties: nitrogen adsorption is smaller in the sodium form and absent in the potassium form; the adsorption of hydrocarbons is similarly affected.

4.2.6 Occurrences and Genesis

Chabazite is a very common zeolite; the genesis may be sedimentary, but no crystallization in weathering or surface conditions is known, and its presence in deep sea material is associated with basalts, hence this origin is certainly hydrothermal (Bass et al. 1973).

Sedimentary chabazite generated by alteration of glass-bearing tuffs is quite common; two environments for zeolitization are well known: continental zones with normal water drainage (hydrologic open systems) and saline lakes (hydrologic closed systems).

Large tuffaceous masses, mainly Quaternary, occurring in central Italy, are widely transformed to chabazite plus phillipsite as open systems (Gottardi and Obradovic 1978): the first evidence was given by Norin (1955) and by Sersale (1958) for the "tufo giallo napoletano", by Sersale (1959a) for the "tufo lionato" in the region of Rome, by Colella et al. (1973) for the "tufo grigio campano", an ash-flow tuff in the region of Naples. Also the "tufo chiaro", a reworked ash-flow tuff of the Monte Vulture, Lucania, Italy, was found to contain remarkable concentrations of chabazite by Sersale (1960b). Sersale (1960a) and Lenzi and Passaglia (1974) found a high content of chabazite in the "tufo rosso a pomici nere", an ash-flow deposit in the region of the Bracciano Lake and Orvieto.

Passaglia and Vezzalini (1984) performed chemical analyses of many of these sedimentary chabazites carefully separated from the matrix and found that all had a Si/Al ratio equal to or slightly higher than that of typical hydrothermal chabazites, the cationic content being either calcian only or mixed calcian potassian. The genesis of these chabazites is believed to be due to the slow percolation of water through the still loose volcanic materials, hence in an open system; this hypothesis is certainly true for the "tufo grigio Campano", which often features an upper zone with glass and clay minerals (smectites), a median zone with chabazite and/or phillipsite, and a lower zone with potassium feldspar; the "tufo lionato" has an irregular zeolite dispersion in the rock, the crystals being a little larger (30 μm instead of 10 μm of the before mentioned tuffs) so that in this last case the genesis is probably connected to a post-volcanic low hydrothermalism.

Lenzi and Passaglia (1974) proposed for the "tufo rosso a pomici nere" a particular kind of genesis (called "geoautoclave" by Aleksiev and Djourova 1975): when the surface water entered the pore space of the ash-flow deposit, the temperature was high enough to form a hydrothermal environment which homogeneously zeolitized the rock. Quite similar to this Italian occurrence is that of Laacher See, Eifel, Germany (Gottardi and Obradovic 1978), where the "Trass", a zeolitized ash-flow deposit, is built up homogeneously by chabazite plus phillipsite (Sersale 1959b); the geoautoclave genesis may probably be accepted here too.

The main product of palagonitization is phillipsite (see Chap. 3.2.6): chabazite is sometimes a by-product, as in Oahu, Hawaii (Hay and Iijima 1968a, b; Iijima and Harada 1969). These authors interpreted the formation of phillipsite and chabazite, plus analcime and montmorillonite as due to the slow percolation of ground water through the alkali basalt glass (sideromelane) of the tuffaceous deposits of Koko Craters, Hawaii; the process is called palagonitization; this sodian chabazite has a ratio Si/Al slightly lower than 2. Sodium-rich chabazite (herschelite) was found in Sicily by Capola (1948) and Sturiale (1963) in the palagonitic tuffs (hyaloclastites) of Palagonia, and by Irrera (1949) in a palagonitic basalt at Ficarazzi.

Genesis in hydrologically closed systems is different, as are the chabazites thus formed. "Closed system" is more or less a synonym for a salt lake basin with no effluent streams, in this context; most occurrences of this type are in the western United States. Gude and Sheppard (1966) and Sheppard and Gude (1969a) described a high silica, alkali chabazite of the Barstow Formation, San Bernardino Co., California, where monomineralic beds of chabazite are rare, the zeolite being usually associated with clinoptilolite, erionite, mordenite; its genesis is related to the alteration of a vitric glass high in silica and alkali.

Regis and Sand (1967) found a similar chabazite in bedded Tertiary lake deposits at Bowie, Arizona; it was formed by alteration of bedded pyroclastics, with analcime, erionite, clinoptilolite.

Sheppard and Gude (1970) described another high silica but calcic chabazite from the John Day Formation, Grant Co., Oregon, as an alteration product of a silicic tuff. Sheppard and Gude (1973) found chabazite sometimes as a monomineralic bed, more frequently associated with erionite, clinoptilolite and clay minerals in the Big Sandy Formation, Mohave Co., Arizona; the parent material is the glass, rhyolitic or dacitic, of a tuff deposit.

A relatively pure chabazite deposit of several million tons was found by Gude and Sheppard (1978) in the siliceous tuff of a Pliocene lacustrine deposit near Durkee, Baker Co., Oregon, not far from the erionite type locality; the sodium-rich and silica-rich chabazite is an early product of reaction between glass (rhyolitic to dacitic) with alkaline (pH > 9) saline lake waters.

Lower concentrations of sedimentary chabazite have been found in the following saline lake basins:

- Olduvai Gorge, Tanzania (Hay 1964) with erionite and phillipsite; Lake Tekopa, Inyo Co., California (Sheppard and Gude 1968) with phillipsite, clinoptilolite, erionite;
- Nevada Test Site (Hoover 1968) with clay minerals;
- Lake Magadi Region, Kenya (Surdam and Eugster 1976) with erionite.

Some similar occurrences of chabazitized rhyolitic to dacitic tuffs in the U.S.S.R. have been described by Mikhailov et al. (1975).

Chabazite is never present in the zeolite facies of zones of low burial metamorphism, although some series of chabazite-bearing sediments have been described where the zeolite type changes with depth: in all these cases, chabazite occurs always at the top of the sedimentary sequence. So Iijima and Utada (1966) studied alteration of Cenozoic and Mesozoic pyroclastic rocks of Japan, finding chabazite with analcime, natrolite and phillipsite at the highest level, where syngenesis or diagenesis is more probable. Similar conclusions were reached by Shimazu and Kawakami (1967) for the basalt and tuff Yahiko formation (Middle Miocene), Niigata Pref., Japan, and by Kristmannsdóttir and Tómasson (1978) when studying the basalt and hyaloclastite sequence in Icelandic geothermal areas: chabazite grew at the expenses of volcanic glass when T < 70°C, the maximum T in the field being 190°C. Sameshima (1978), too, considered low hydrothermal alteration as the genesis for the chabazite in the Miocene marine volcano-sedimentary sequence of the Waitemata Group, Auckland, New Zealand; this chabazite has a chemistry with high alkali and Si/Al = 2, hence different from the same mineral from lacustrine deposits, and similar to those of central Italy and of Sicily; this occurrence in New

Zealand is the only known sedimentary occurrence in a submarine tuff (tuffite).

A zoning in hydrothermal deposition of zeolites, including chabazite, was also found by Kostov (1970) in Bulgaria, where the distribution of zeolites correlates with the sites of sulphide mineralization and sites of granodioritic to monzonitic intrusions.

The large majority of fine chabazite crystals, which embellish the showcases of museums and collections, were found in vugs of massive rocks, and hence probably had a hydrothermal genesis. This kind of origin is so common, that only a small survey of these occurrences can be given here; many others can be found in Hintze's (1897) and Dana's (1914) handbooks.

The occurrences in basic lavas are reviewed first.

Iceland: there are a hundred or so occurrences in the basic lavas of this island; Walker (1960b) gave a good report for the occurrence in the Berufjördur; Betz (1981) lists 17 zeolite locations, mostly with chabazite, and quotes an article in Icelandic by Jakobsson (1977) where more information can be found.

Greenland: Bøggild (1953) lists ca. 30 occurrences; a recent detailed study was dedicated to the alteration of ussingite to gmelinite + chabazite at Ilimaussaq (Karup-Møller 1976).

Faröes: very common in all basalts (Currie 1907); Betz (1981) reviews all localities with zeolites: the best crystals, up to 4 cm, have been collected in Vidoy and Sandoy, with heulandite, stilbite, apophyllite.

Scotland: well known are the crystals in the basalt of Storr, Skye; associated zeolites are levyne and stilbite.

Northern Ireland: Walker (1951, also 1959) states that chabazite is ubiquitous in the Tertiary basalt of the Garron Plateau area; he gives detailed data on the distribution of the different crystal habits.

France: in several localities of the Central Massif (Pongiluppi 1976).

Germany: famous occurrences of the mineral are in the basalts of Annerod and Nidda, Hessen, and also Vogelsberg, Hessen (Mörtel 1971; Hentschel and Vollrath 1977); in the tephrite at Eichert, Sasbach, Kaiserstuhl, Fricke (1971) found crystals with phillipsite.

Czechoslovakia: a strontian (3.01% SrO) chabazite was found by Černý and Povondra (1965b) in an andesite of Komňa, Moravia, and another one (1.64% SrO) at Řepčice, Bohemia (Ulrych and Rychly 1981).

Hungary (a general review is in Koch 1978): in the basalts near the Balaton Lake, with phillipsite (Mauritz 1931, 1938, 1955); in the basalts of Dunabogdány with analcime and stilbite (Reichert and Erdélyi 1935); in an amphibole-andesite at Sátoros, with heulandite, stilbite, epistilbite, laumontite (Erdélyi 1942); in the altered andesites of Matra Mts. with laumontite, heulandite, stilbite (Mezösi 1965).

Italy: in the basalts of Acitrezza and Aci Castello, Sicily (Mason 1962; Passaglia 1970); this is the type locality for the Na-rich "herschelite"

(Lévy 1825); in the leucitites of Casal Brunori, Vallerano, Acquacetosa, Osa, all in the vicinity of Rome (Passaglia 1970); all these are Al- and Sr-rich (up to 5.34% SrO) and with phacolitic habit; in the porphyrites of Alpe di Siusi, Bolzano, with fibrous zeolites, heulandite and clinoptilolite, mordenite, dachiardite and laumontite (Vezzalini and Alberti 1975); in the metagabbros of the "Gruppo di Voltri" with fibrous zeolites, gismondine, phillipsite (Cortesogno et al. 1975).

Yugoslavia: in the andesite of Bor, eastern Serbia, with stilbite (Majer 1953).

Bulgaria: in several basic rocks (basalts but also tuffs) Kostov (1962) found chabazite, often brick red in colour.

U.S.S.R.: Fersman (1922) reviewed 200 occurrences with 22 zeolitic species, including chabazite.

Kerguelen Islands: in the decomposed basalts with heulandite and fibrous zeolites (Lacroix 1915).

Canada: in the Triassic basalts of Nova Scotia with many other zeolites (Aumento 1964).

U.S.A.: in the basalts of Table Mt., Colorado, with fibrous zeolites, levyne, stilbite, laumontite (Johnson and Waldschmidt 1925); in the Tertiary basalt of Ritter Hot Springs, Grant Co., Colorado, with mesolite, analcime, stilbite (Hewett et al. 1928); in the Hawaian basalts (Dunham 1933) with phillipsite; in the basalts near the Skookumchuck River, Washington, fine crystals with mesolite, heulandite, mordenite (Tschernich 1972).

Brazil: in the basalts of southern Brazil (Franco 1952; Mason and Greenberg 1954).

Australia: in the basalt of Richmond and Collingwood, Victoria, and at Kyogle, New South Wales (Hodge-Smith 1929); all these are Na-rich; in the Tertiary basalts of Tasmania, with gonnardite, natrolite, stilbite, phillipsite (Sutherland 1965).

Japan: in the amygdales of the altered basalt at Chojabaru, Nagasaki Pref., with levyne, erionite and stilbite (Shimazu and Mizota 1972); in the amygdales of a Pliocene trachybasalt at Dōzen, Oki Island, with levyne, thomsonite, cowlesite (Tiba and Matsubara 1977).

India: very famous occurrences of zeolites, including chabazite, are in the Deccan Traps of Western India (Sukheswala et al. 1974).

Hydrothermal chabazite occurs also as late hydrothermal deposits in granites, gneisses and other feldspar bearing rocks.

In the vugs of the granitoid rocks of the Aar and St. Gotthard Massifs, Parker (1922) found chabazite associated with adularia, apophyllite and calcic zeolites, like stilbite, heulandite, scolecite. Weibel (1963) described fine crystals on a heulandite layer coating fissures of a syenite near Tavetsch, Switzerland. In the fissures of the granite of Baveno, Lago Maggiore, Italy, Pagliani (1948) found the zeolite with heulandite, stilbite, laumontite.

In the volcanoclastic leucitic rocks (welded tuffs and lahars) at Ercolano, Naples, Italy, De Gennaro and Franco (1976) found a chabazite with potassium prevailing over all other cations.

In an amphibolite at Červena Hora, Moravia, Novotná (1926) found the mineral with several other zeolites.

In a gneiss from Mt. Medvezhyi, Siberia, Shkabara (1941) described chabazite with heulandite and stellerite.

In the gneiss of Jones Falls, Maryland, Shannon (1925) found chabazite with heulandite and stilbite. Brownell (1938) found natrolite and chabazite in the cavities of the gneiss or between gneiss and a vein of massive sulphide in the Sherritt Gordon mine, Manitoba, Canada.

Rao and Cunha e Silva (1963) found chabazite with stilbite and heulandite in the druses at the contact of tactites with mica schists in the scheelite mines of northeastern Brazil.

Willhendersonite was found in the S. Venanzo quarry, Terni, Italy, with thomsonite, phillipsite, in a basic decomposed rock rich in olivine (probably the melilite-leucitite well known in this locality), and also in a limestone inclusion in a basalt at Ettringer Bellerberg, Eifel, Germany, with gismondine, chabazite, phillipsite, thomsonite, ettringite, thaumasite (Peacor et al. 1984).

4.2.7 Uses and Applications

The sedimentary chabazite from Bowie, Arizona, is used not only for the removal of H_2O, CO_2, H_2S in industrial gases, but also for the separation CO_2/CH_4 from gas produced by decaying garbage in landfills: a plant of this type with chabazite beds is active in Los Angeles (Mumpton 1975).

Maggiore et al. (1981) compared the catalytic activity of two sedimentary chabazites from central Italy with the activity of silica-alumina, gamma-alumina and synthetic 5 A zeolite in isopropanol dehydration: the measured activities are more interesting if referred to units of active surface rather than units of weight.

4.3 Levyne

$NaCa_{2.5}(Al_6Si_{12}O_{36}) \cdot 18 H_2O$
$R\bar{3}m$
$a_{rh} = 10.84\,(\text{Å})$
$\alpha_{rh} = 75°57'$
or
$a_{hex} = 13.35$
$c_{hex} = 22.90\,(\text{Å})$

4.3.1 History and Nomenclature

Brewster (1825b) proposed the name levyne for a zeolite found at Dalsnipa, Sandoy, Faröe Islands, in honour of A. Lévy of Paris University. The variant "levynite" has also been used. "Mesoline" is an obsolete synonym.

4.3.2 Crystallography

A model for the framework of levyne proposed by Barrer and Kerr (1959) was found to be correct by Merlino et al. (1975), who refined the structure finding all cation and water sites. The framework can be described as an AABCCABBCAA... sequence of 6-rings (see Chap. 0.2.4) which build up a typical cage (see Fig. 0.3 A j). There is no Al – Si ordering in the tetrahedra. All cations are aligned along the triad at the centre of the cage, and distributed in five sites (first code after Merlino et al. 1975, second after Mortier 1982):

C1 – B: outside the double 6-ring, but in contact with three oxygens of the same, and coordinating also three water molecules; fully occupied by Ca.
C2 – C: nearly in the centre of the cage, without contact with the framework oxygens, with a low mixed occupancy.
C5 – E: in the centre of the single ring, sparsely occupied by all cations.
(C3 + C4) – D: two nearby sites with low occupancy, which cannot be occupied simultaneously, nor if C5 is occupied; they are set between C2 – C and C5 – E.

The morphology is characterized by {0001} lamellae with a hexagonal outline given by the rhombohedra {$10\bar{1}1$} {$01\bar{1}2$} {$30\bar{3}2$}: please note that Hintze (1897) and Dana (1914) give a value of c which is halved with respect to that given in the title page, hence their indices must be modified so as to double l; moreover their h and k have been interchanged, to have as most frequent form the rhombohedron corresponding to the unit cell in obverse setting, which is very probable, even if not certain. Penetration twins (Fig. 4.3 B) with the triad as twin axis and (0001) as contact plane, are frequent, the forms being {0001} and {$10\bar{1}1$}. Sheaf-like aggregates are frequent as associations of two interpenetrating crystals forming an angle of nearly 30° with their (0001) planes, so as to form a capital X (an undescribed twin?).

Intergrowths of levyne with erionite and/or offretite are rather common and were discovered by Shimazu and Mizota (1972) who described a levyne-erionite intergrowth from Iki Island, Japan, and several years later also by Passaglia et al. (1974), who described the same intergrowth from Sardinia, Italy, and by Sheppard et al. (1974) who described an intergrowth with offretite from Beech Creeck, Oregon (but later on Bennett and Grose (1978) showed that it was in fact erionite). Wise and Tschernich (1976b) described in detail intergrowths of levyne with erionite and with offretite from several localities of the United States. Levyne-offretite intergrowths from Australia were described by England and Ostwald (1979) and by Birch (1980). As a matter of fact, both erionite and offretite are probably present in all these intergrowths, because of the similarity of their structures and of their powder patterns, so that the mineral given by the different authors as grown on levyne probably is the most abundant one but not the only one. The fibrous offretite/erionites build up a coating on the {0001} faces of the levyne lamellae; the fibres are normal to the plane of the lamellae; (0001) layers inside the levyne

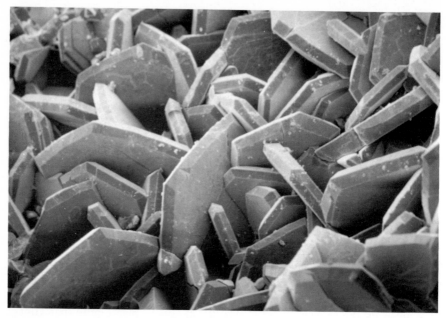

Fig. 4.3A. Levyne from Nurri, Sardinia, Italy (SEM Philips, ×60)

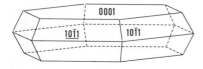

Fig. 4.3B. Common morphology of levyne twins; faces of different individuals are distinguished by the presence or absence of underlining of their indices

are also possible. The same intergrowth had been observed by Walker (1951), who was however not able to reveal the nature of the overgrown fibrous material.

4.3.3 Chemistry and Synthesis

The chemical composition of levynes (see Table 4.3I) is usually quite near to the schematic formula in the title page, the sodium content being sometimes much higher; the Si/Al ratio does not show large variations. A larger set of analyses may be found in Galli et al. (1981), who also detected a correlation between (Na + K) and the lattice constant a_{hex}.

The synthesis of levyne by St. Claire Deville in 1862 (quoted by Breck 1974) is often considered the first synthesis of a zeolite; as a matter of fact, he enclosed a mixture of potassium silicate and sodium aluminate in a glass tube and, after hydrothermal treatment, obtained hexagonal lamellae, which were analysed and found to have a chemical composition similar to levyne (but probably it was not).

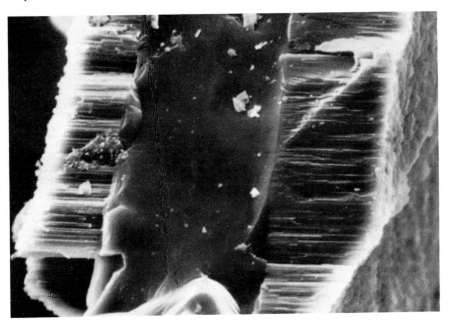

Fig. 4.3 C. Levyne-erionite over- and inter-growth from Montresta, Sardinia, Italy (SEM, ×2500)

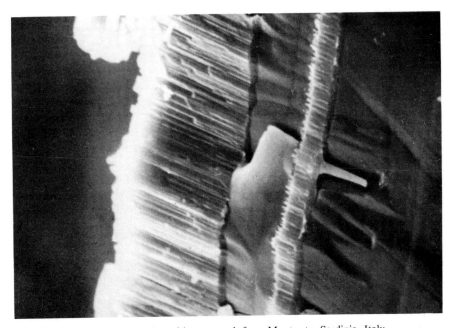

Fig. 4.3 D. Levyne-erionite over- and inter-growth from Montresta, Sardinia, Italy (SEM, ×4500)

Table 4.3I. Chemical analyses and formulae on the basis of 36 oxygens and lattice constants in Å

	1	2	3	4	5
SiO_2	43.88	47.33	47.80	49.76	47.45
Al_2O_3	22.73	21.66	20.75	19.09	21.26
CaO	10.57	10.30	9.33	8.88	3.90
Na_2O	2.01	1.36	1.74	1.71	7.26
K_2O	0.80	0.65	0.19	1.05	1.79
H_2O	20.54	20.22	20.37	19.51	18.15
Total	100.53	101.52	100.18	100.00	100.00
Si	11.15	11.69	11.94	12.35	11.78
Al	6.81	6.31	6.11	5.58	6.22
Ca	2.88	2.73	2.50	2.36	1.04
Na	0.99	0.65	0.84	0.82	3.50
K	0.26	0.20	0.06	0.33	0.57
H_2O	17.40	16.66	16.98	16.14	15.03
E%	-2.81	-0.02	3.60	-4.99	0.35
a_{rh}	10.867	10.870	10.818	10.855	10.829
α_{rh}	75°43'	75°42'	76°22'	75°48'	76°37'
a_{hex}	13.338	13.338	13.376	13.337	13.426
c_{hex}	23.004	23.014	22.728	22.953	22.686
D	2.04				2.11

1. Dōzen, Oki Island, Shimane Pref., Japan (Tiba and Matsubara 1977).
2. Nurri, Sardinia (Passaglia et al. 1974).
3. Padria, Sardinia (do).
4. Montresta, Sardinia (do).
5. Island Magee, Co. Antrim, Northern Ireland (Galli et al. 1981), with 0.01% Fe_2O_3, 0.03% MgO, 0,15% SrO.

Kerr (1969) reported on the first synthesis proven by X-ray data. He started from mixtures of $(0.1$ to $0.2) R_2O \cdot (0.8$ to $0.9) Na_2O \cdot Al_2O_3 \cdot (4$ to $5) SiO_2 \cdot (20$ to $50) H_2O$ where R = 1-methyl-1-azonia-4-azabicyclo-[2.2.2]-octane and heated at 100°C from nearly 2 to 25 h; the crystalline product, called ZK-20, gives a powder pattern similar to levyne; this phase may be changed to the hydrogen form by heating at 600°C.

Wirsching (1981) obtained levyne in a simulated "open system", by hydrothermal reaction of a basaltic glass with a frequently renewed 0.1 to 0.01 N $CaCl_2$ solution, the temperature being 100°C, the time needed 200 days.

Table 4.3 II. Refractive indices

	1	2	3	4
ω	1.497	1.510	1.498	1.489
ε	1.494	1.502	1.498	1.487

Source of data as in Table 4.3 I.

4.3.4 Optical and Other Physical Properties

Levyne is nearly always uniaxial negative, but may be biaxial negative or nearly isotropic. There is some correlation between calcium content and the refractive indices (see Table 4.3 II). Crystals are usually transparent colourless, sometime reddish or yellowish. Lustre vitreous, streak white. Mohshardness 4 to 4.5. Cleavage $\{10\bar{1}1\}$.

4.3.5 Thermal and Other Physico-Chemical Properties

The thermal curves (Fig. 4.3 E) show two main water losses at 70°C and 300°C, and perhaps a minor one at 180°C; there are many more water sites in the structure.

No other physico-chemical data are known for natural levyne, but Kerr (1969) described several properties of his synthetic levyne ZK-20, such as ion-exchange with every kind of cation and catalytic activity (for instance in hydrocarbon cracking).

4.3.6 Occurrences and Genesis

Levyne always has a hydrothermal genesis.

Faröes: the type locality, Dalsnipa, Sandoy, is not the only occurrence in the islands, as the mineral has been found in many other localities: Midvagur in the island Vagar (Hintze 1897). Torvadalsa, Bordoy and Nolsoy (Galli et al. 1981), Nesvik, Streymoy, and Nordepil, Bordoy (Betz 1981; see also Currie 1907, and Görgey 1910).

Greenland: Bøggild (1953) lists the following occurrences in the Upernavik district, at Ingnerit Fjord; in Umanak district, at Aumarutigssat; in Ritenbeck district, at Ujaragssuit and at Igdlukunguaq; in Godhavn district, at Puilasoq, at Lyngmarken, at Akiarut, at Kavnerssuaq, at Unartorssuaq, at Bluafjeld, at Qarusuit, at Iginiarfik, at Ivisarqut; in East Greenland, at Iluileq, at Henry Gletscher, at Gaasefjord, at Kap Bror Ruys; Hintze (1897) mentions also Disko, Godhavn district.

Iceland: Hintze (1897) lists the following localities: Onundar Fjordur, Dyra Fjordur, Mossfell on the bank of Leiruvogsa, Reynivellir, Thyrill, Hruni on the bank of the river Laxa, at the entrance of the Rodefjord. Galli et al. (1981) list four other localities: Onondafjordur; Hvalfjordur, 100 m

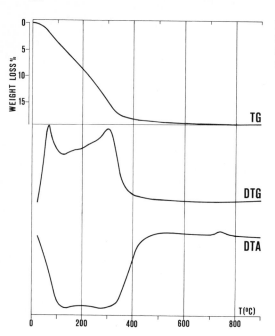

Fig. 4.3E. Thermal curves of levyne from Pargate quarry, Northern Ireland; in air, heating rate 20°C/min (by the authors)

above the coastal range near "Fossa" waterfall; Fagridalur Valley, Mt. Saudahlidarfiall; Reydarfjordur, North of Budareyri, Mt. Teigargerdistindur. Betz (1981) adds some more occurrences: Hvalfjordur (on the southern shore, near Hvaleyri); Mjöadalsá Canyon near Sanddalstunga, about 6 km North of Hvammur; Skagafell; Skessa and Eyrarfjall Mountains on the Southwest head of Reydarfjordur; Walker (1960b) found many zeolites, including levyne, in the famous Breiddalur-Berufjordur area.

Northern Ireland: Walker (1951) describes the following occurrences in Tertiary basalts: Garron Plateau and Island Magee in Co. Antrim; Cluntygeeragh, Dungiven and Cabragh, Termoneeny and Benyevenagh in Co. Londonderry. Galli et al. (1981) list some other localities: Little Deer Park, Pargate Quarry and Dunseverich, all Co. Antrim.

Scotland: Hintze (1897) lists Hartfield Moss, Glasgow, and the famous Storr in the Island of Skye.

Germany: Hentschel and Vollrath (1977) described a levyne-offretite intergrowth (see Chap. 4.3.2) in a Tertiary basalt of Ober-Widdersheim, Vogelsberg; Hentschel (1980b) found the mineral in the Ca-rich ejecta of the Bellerberg volcano, Eifel.

Bohemia: Hintze (1897) reports on the very old finds in Horni Kamenice (= Ober-Kamnitz) and also in Triebsch (= ?).

Hungary: Erdélyi (1940) found levyne with epistilbite, heulandite, chabazite, stilbite, laumontite, scolecite and mesolite at the andesite-granite contact in the quarry of Nadap, Comitat Fejer.

Italy: besides the old known occurrence of Montecchio Maggiore, Vicenza (Galli et al. 1981), three newly described occurrences are in Sardinia at Padria, Montresta and Nurri (Passaglia et al. 1974); two of these are levyne-erionite intergrowths.

U.S.A. and Canada: the occurrence at Table Mt., Golden, Colorado, already mentioned by Hintze (1897), was described in detail by Johnson and Wald-schmidt (1925), who found also analcime, apophyllite, chabazite, thomsonite, fibrous zeolites, stilbite and laumontite in the cavities of the Tertiary basalt. Hewett et al. (1928) were not certain of the identification of levyne in the Tertiary basalt of Ritter Hot Spring, Oregon. Sheppard et al. (1974) found a levyne-erionite (not offretite, as later on shown by Bennett and Grose 1978) intergrowth in vesicles of the Picture Gorge basalt, near Beech Creek, Grant Co., Oregon; associated zeolites are offretite, chabazite, analcime, thomsonite (White 1975). Wise and Tschernich (1976b) found a complex intergrowth of levyne-offretite-erionite in some vesicles of a basalt in a small quarry at the corner of Lava and River Roads, Milwaukie, Oregon, and a levyne-offretite intergrowth southwest of Westwold, along the road to Douglas Lake, Southern British Columbia, and along Queen Creek 6 miles West of Superior, Pinal Co., Arizona. The chemical analyses show that each of the intergrown minerals has a different Si/Al-range.

Argentina: Cortelezzi (1973) found levyne with wairakite and epistilbite in the basalt of Cerro China Muerta.

Japan: Shimazu and Mizota (1972) found a levyne-erionite intergrowth with chabazite and stilbite in the amygdales of an altered basalt at Chojabaru, Iki Island, Nagasaki Pref.. Tiba and Matsubara (1977) found it in the amygdales of a Pliocene trachybasalt lava near Kuniga, Nishi-no-Shima, Dozen, with chabazite, thomsonite and cowlesite.

India: Chatterjee (1971) found it with heulandite in the Deccan Traps near Bophal.

Australia: Thin hexagonal plates of levyne with chabazite were described in a basalt from Clunes, Victoria, by Birch (1979). A levyne-offretite intergrowth was found in the Tertiary basalt of the Merriwa district, Hunter valley, New South Wales, by England and Ostwald (1979) and two others in the cavities of basalts from Flinders, Victoria, by Birch (1980) and by Coulsell (1980).

4.4 Erionite

$NaK_2MgCa_{1.5}(Al_8Si_{28}O_{72}) \cdot 28 H_2O$
$P6_3/mmc$
$a = 13.15$
$c = 15.05 (Å)$

4.4.1 History and Nomenclature

Eakle (1898, 1899) proposed the name for a fibrous mineral occurring in seams in a rhyolitic tuff at Durkee, Oregon, from the Greek $\varepsilon\rho\iota o\nu$ = wool, because of its woolly aspect. Staples (1957) and Staples and Gard (1959) identified as type locality the old Durkee opal mine, Swayze Creek, Baker Co., Oregon. Deffeyes (1959) improved the crystallographic data. Hey and Fejer (1962) cast some doubts on erionite being an independent species from offretite, but Bennett and Gard (1967) and Harada et al. (1967) re-stated the rank of species for erionite.

4.4.2 Crystallography

A model for the erionite framework (see Chap. 0.2.4) as a stacking AABAAC... of 6-rings was first proposed by Staples and Gard (1959), and later on confirmed and refined by a photographic method by Kawahara and Curien (1969), but only to R = 17% and later on by Gard and Tait (1973) to R = 15%. The framework is characterized by the presence of three types of cavities, double 6-rings, cancrinite and erionite cages (see Chap. 0.3). No (Si, Al)-order was detected. Potassium is at the centre of the cancrinite cage, the double 6-rings are empty and Ca, Na and Mg are dispersed in several sites of the erionite cage. A crystal from the same occurrence (Mazé, Japan) was refined in the dehydrated form down to R = 6% by Schlenker et al. (1977a), who found Ca in the cancrinite cage, Mg (plus some Ca and/or Na) at the centre of the single 6-rings, and K at the center of the boat shaped 8-rings which form the walls of the erionite cage. So it is probable, also on the basis of ion exchange data, that in both offretite and erionite, K is in the cancrinite cage if the crystal is hydrated, but it is substituted by Ca in the dehydrated form. This phenomenon is termed "internal ion exchange" by Mortier et al. (1976a).

　　The morphology is quite simple, with rare hexagonal prisms (order unknown) terminated with the basal pinacoid, but usually erionite occurs as thin fibres, often forming a compact felt, but sometimes tufts with delicate woolly aspect, as in the holotype.

　　The occurrence of intergrowth with offretite is quite common in view of the similarity of the two structures; also the single crystals really contain some

Fig. 4.4 A. Fibrous erionite from Cape Lookout, Oregon (SEM Philips, ×30)

Fig. 4.4 B. Fibrous erionite from Cape Lookout, Oregon (SEM Philips, ×3900)

Fig. 4.4C. Sedimentary erionite from Durkee, Oregon (SEM Philips, ×500)

Fig. 4.4D. Sedimentary erionite from Durkee, Oregon (SEM Philips, ×950)

Fig. 4.4E. Sedimentary erionite with chabazite from Bowie, Arizona (SEM Philips, ×3900)

stacking faults of the offretite type as shown by TEM technique by Kokotailo et al. (1972); 6% of the single 6-rings are stacked in the offretite way in the Mazé specimen according to Schlenker et al. (1977a). Kerr et al. (1970) investigated the erionite-offretite intergrowths in synthetic samples. Macro-intergrowths have been described by Rinaldi (1976). Fibrous erionite-offretite intergrowths often built up a coating over levyne lamellae (see Chap. 4.5.3 and 4.5.6).

4.4.3 Chemistry and Synthesis

The formula in the title page gives only an idea of the composition, but real erionites show large variations, although not so large as a survey of the literature data may lead one to suppose, because of the unreliability of many analyses of sedimentary erionites, which were unsufficiently purified from the rock matrix. Some data are listed in Table 4.4I, more analyses may be found in Sheppard and Gude (1969b) and in Wise and Tschernich (1976b). The ratio $Si:(Si+Al)$ is a little higher (78%) in sedimentary samples, than in the hydrothermal ones (75%). The cationic content varies considerably with Mg usually not far from unity, Ca near to 1.5 in hydrothermal erionites, but much lower in some sedimentary ones, in which alkalies may be quite high.

The synthesis of an erionite-like phase was described by Aiello and Barrer (1970): gels of composition $mM_2O \cdot Al_2O_3 \cdot 15\,SiO_2$, where M is TMA (= tetra-

Table 4.4I. Chemical analyses and formulae on the basis of 72 oxygens and lattice constants in Å

	1	2	3	4	5	6
SiO_2	57.40	58.02	54.72	62.52	58.89	60.67
Al_2O_3	15.60	14.47	15.24	16.48	14.23	12.90
Fe_2O_3		0.22	1.04	0.01	0.40	1.44
MgO	1.11	0.42	1.17	1.12	1.16	1.09
CaO	2.92	3.29	4.32	2.49	2.67	0.65
Na_2O	1.45	1.08	1.00	1.72	0.64	4.39
K_2O	3.40	4.64	2.46	4.84	4.85	4.09
H_2O	17.58	17.82	19.12		16.59	14.63
Total	99.46	100.00	99.07	89.18	99.58	100.00
Si	27.42	27.84	26.90	27.56	27.96	28.19
Al	8.79	8.19	8.83	8.56	7.96	7.07
Fe		0.08	0.39		0.14	0.50
Mg	0.79	0.30	0.86	0.74	0.82	0.75
Ca	1.50	1.69	2.28	1.18	1.36	0.32
Na	1.34	1.01	0.95	1.48	0.59	3.96
K	2.07	2.84	1.54	2.72	2.94	2.42
H_2O	28.01	28.53	31.35		26.27	22.68
E%	+10.0	+5.6	+5.2	+6.9	+2.8	-11.3
a	13.254	13.186	13.24	13.245	13.218	
c	15.100	15.055	15.12	15.13	15.060	
D	1.90	2.11	2.08		2.07	

1. Durkee, Oregon (no.2 of Staples and Gard 1959); lattice constants of Sheppard and Gude (1969b).
2. Reese River, Nevada (Gude and Sheppard 1981); all Fe as Fe^{3+}, plus 0.04% TiO_2; D is calculated.
3. Mazé, Japan (no.4 of Harada et al. 1967).
4. Yaquina Head, Oregon (no.16 of Wise and Tschernich 1976b).
5. Jersey Valley, Nevada (no.9 of Sheppard and Gude 1969b); plus minor P_2O_5, Cl, F; D is calculated.
6. Lake Tekopa, California (Sheppard and Gude 1968) plus minor TiO_2, P_2O_5, MnO.

methylammonium) or Na and m = 8 or 10, with TMA : Na = 4 or 1. The
formula of the synthetic phase is $TMA_3Na_6Al_9Si_{27}O_{72} \cdot 17H_2O$. Erionite was
synthetized also in the absence of TMA, but using natural crystals as seeds, by
Nikolina et al. (1981) starting from sodium aluminosilicate and an aluminium
salt; better results were obtained if some KOH or K_2CO_3 was added.

4.4.4 Optical and Other Physical Properties

Erionite is uniaxial positive (in contrast to offretite which is negative), some
refractive indices being listed in Table 4.4II. Transparent colourless, streak
white. Mohs-hardness 4; a cleavage parallel to c is probable.

Sediments rich in erionite may be recognized by a simple field test
(Sheppard, oral communication): every splinter of such a rock, if hit with the
pointed side of a geologist's hammer, sticks to the hammer.

4.4.5 Thermal and Other Physico-Chemical Properties

Figure 4.4F presents the thermal curves of an erionite of hydrothermal
genesis, hence a sample without contaminants. Three peaks (100°C, 140°C,
170°C) and a shoulder (250°C) may be seen in the DTG, in partial agreement
with the structure refinements of Kawahara and Curien (1969) and by Gard
and Tait (1973), which both give the presence of three water molecules; the
water loss is nearly complete at 300°C, but continues at a low rate up to
700°C. Figure 4.4G gives the thermal curves of a sedimentary zeolite rich in
erionite: the behaviour has the same trend, but only one broad water loss
between 25°C and 350°C may be seen.

The erionite-like phase synthetized by Aiello and Barrer (1970) first loses
its water, then its TMA is decomposed by oxidation and eventually at 360°C it
inverts to sodalite, namely the stacking sequence becomes ABC...: this
framework change at such a low temperature is remarkable; the sodalite
lattice is destroyed at 700°C.

Ames (1961) found that the normal replacement series

$$Cs > K > Na > Li \quad \text{and} \quad Ba > Sr > Ca > Mg$$

is valid for erionite too. Ames (1964b) gives exchange isotherms at 25°C, total
normality 1N, for Na ↔ Ca and Na ↔ Sr, both not far from straight lines, and
Ca ↔ Sr, a typical isotherm with selectivity reversal; Sr is favoured in the

Table 4.4II. Refractive indices

	1	2	3	5	6
ω	1.468	1.468	1.477	1.467	1.458
ε	1.473	1.472	1.480	1.471	1.462

Numbers refer to the same samples of Table 4.4 I.

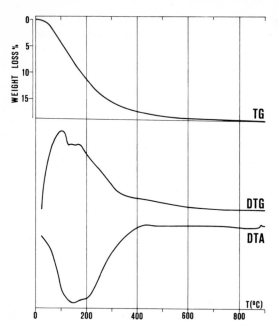

Fig. 4.4F. Thermal curves of erionite from Agate Beach, Oregon; in air, heating rate 20°C/min (by the authors)

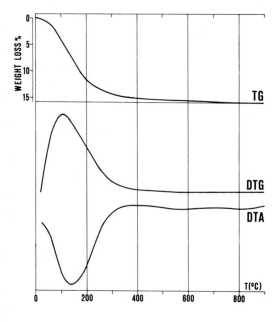

Fig. 4.4G. Thermal curves of sedimentary erionite from Eastgate, Churchill Co., Nevada; in air, heating rate 20°C/min (by the authors)

zeolite at low concentration in the solution, not at high concentration. Ames (1965) gives self-diffusion coefficients (cm^2/sec) for several cations at 50°C and 70°C. Sherry (1979) determined the exchange isotherms at 5 and 25°C, total normality 0.1 N, for the couples (Li, Na), (K, Na), (Rb, Na), (Ca, Na), (Cs, Na), (Sr, Na), (Ba, Na) and proposed a selectivity series: Rb > Cs ⩾ K > Ba > Sr > Ca > Na > Li valid at this lower concentration.

Eberly (1964) gives adsorption data at 98.4°C and P = 300 mm (Hg?) in mmoles of sorbed compound per 1 g of natural erionite: n-pentane 0.26, n-hexane 0.29, n-heptane 0.15; the values for 2-methyl-pentane, 2,2-dimethyl-butane and benzene are very low or zero; this kind of adsorption corresponds to an effective pore diameter of 4.5–5.5 Å in agreement with the diameter calculated from the crystal structure. The adsorptions are not substantially varied by ion exchange, and are about one fourth of the corresponding values in the synthetic zeolite 5 A or CaA.

4.4.6 Occurrences and Genesis

Erionite is more common in sedimentary rocks than as hydrothermal deposition, although its frequency as a surface mineral was recognized only 50 years after its discovery.

It is not known as a weathering mineral, or from soils.

Deffeyes (1959) described the following sedimentary occurrences: in a 4 m thick tuffaceous rhyolitic bed, zeolitized to erionite in its upper part, at the Miocene-Pliocene boundary, in northern Jersey Valley, Sonoma Range Quadrangle, Nevada; in early Pliocene sediments, sometimes with clinoptilolite, from Shoshone Range and from the valley of Reese River, Nevada; in the lacustrine beds of Blancan (Plio-Pleistocene) from Pine Valley, Nevada; with clinoptilolite, in thin bedded altered tuff (upper and middle Eocene) along Beaver Divide, east of Sand Draw, Wyoming; also with clinoptilolite in the White River (Oligocene) formation, South Dakota. Although the available data are insufficient to define the genesis, by analogy with many other sedimentary zeolites in the western United States, it is probable that at least some of these zeolitized tuffaceous beds underwent a crystallization in a saline lacustrine environment (hydrologically closed system). Genesis in closed systems was attributed with certainty to many occurrences in the western states: with chabazite and clinoptilolite in rhyolitic tuff of a saline lacustrine formation at Durkee, Baker Co., Oregon (Sheppard et al. 1983) and at Rome, Oregon; with clinoptilolite, in the altered tuffs of the Barstow Formation, San Bernardino Co., California (Sheppard and Gude 1969a); in nearly mono-mineralic beds or in association with clinoptilolite and many other zeolites in the tuffs of the Big Sandy formation, Mohave Co., Arizona (Sheppard and Gude 1973); as monomineralic beds or in association with other zeolites in rhyolitic tuffs of Lake Tekopa, Inyo Co., California (Sheppard and Gude 1968); other occurrences of this kind can be found in Sheppard (1971). Similar is the genesis by alteration from trachytic glass in the Lake Magadi Region, Kenya (Surdam and Eugster 1976).

Genesis in a hydrologically open system, namely by slow percolation of meteoric water through the porous tuffaceous sediments, was invoked by Barrows (1980) for explaining the presence of erionite with other zeolites in the thick Miocene volcaniclastic sequence of Southern Desatoya Mts., Nevada.

Erionite is never present in deep sea sediments; its rare presence in deep sea samples, is always due to hydrothermal deposition in basalts (Kastner and Stonecipher 1978).

The "geoautoclave" genesis (Aleksiev and Djourova 1975) has been attributed to the erionite occurring in the acid welded tuff from Ashio, Tochigi Pref., Japan, by Matsubara et al. (1978a), who consider the mineral to be a product of glass transformation under the high vapour pressure conditions which prevailed after the accumulation and welding of hot ash.

There is no definite evidence for the genesis of erionite by burial diagenesis, although it cannot be excluded in some cases, when the zeolite from tuffaceous sediments could have crystallized immediately after sedimentation or after burial at some depth, or later on when the rock was again nearly at surface. The erionite occurring in a 90 cm thick tuff bed of the Timber Bay marine formation, Kaipara, New Zealand (Sameshima 1978) probably was formed by burial diagenesis.

Hydrothermal genesis is obviously possible for erionite, as for all zeolites. Diffuse hydrothermalism may cause the formation of microscopic crystals dispersed in a rock, like typical sedimentary zeolites, as in the Yellowstone National Park, where erionite occurs with a complete collection of zeolites (Bargar et al. 1981).

The holotype, occurring with opal as woolly filling of seams in a rhyolitic tuff at Durkee, Oregon (Eakle 1898, 1899; Staples and Gard 1959) could be a re-deposition from low hydrothermal solutions which dissolved the locally abundant sedimentary erionite. The only other woolly erionite, which has been found near Reese River, Nevada, probably has the same origin (Gude and Sheppard 1981). Erionite is associated with chalcedony, opal and melanophlogite in the amygdales in Columbia River lavas, near Freedom, Idaho (Reed 1937): the paragenesis is similar to the two previously described.

The paulingite occurring in a basalt by Rock Island Dam, Washington (Kamb and Oke 1960) is associated with a zonal growth with erionite and offretite (Wise and Tschernich 1976b); these authors described also the following other examples of deposition in vugs of lavas in the United States: an intricate intergrowth of erionite, offretite and levyne in a basalt near Milwaukie, Oregon; erionite-offretite intergrowths from Winkleman, Pinal Co., Arizona, and from Horseshoe Dam, Maricopa Co., Arizona; pure erionite with clinoptilolite, mordenite, phillipsite in olivine basalts from Clifton, Greenlee Co., and Thumb Butte, Graham Co., both Arizona; with mordenite, clinoptilolite, phillipsite, dachiardite in the tholeiitic basalt from Cape

Lookout, Tillamook Co., Oregon; with clinoptilolite at Yaquina Head, Lincoln Co., Oregon.

Other occurrences outside the U.S.A. are:

Faröes: as compact fibrous material with chabazite (Hey 1959).

Germany: as zonal growths of erionite and offretite, with faujasite, phillipsite in a limburgite at Sasbach, Kaiserstuhl (Rinaldi 1976); as a fibrous intergrowth with offretite in the basalt of Gedern, Vogelsberg (Betz and Hentschel 1978).

Czechoslovakia: as intergrowth with offretite in a basalt from Prackovice nad Labem (Rychly et al. 1982).

Italy: as fibrous coating on levyne in an andesite from Sardinia (Passaglia et al. 1974).

U.S.S.R.: with heulandite and laumontite in lower Triassic lavas from Nidym River, Siberia (Belitskii and Bukin 1968); with chabazite, mordenite, stilbite and opal in a volcanic rock from Shurdo, Georgia (Batiashvili and Gvakhariya 1968).

Japan: with heulandite and chabazite in a basalt from Mazé, Niigata Pref. (Harada et al. 1967; Shimazu and Kawakami 1967); as fibrous coating on levyne, with chabazite and stilbite in an altered basalt from Chojabaru, Iki Island (Shimazu and Mizota 1972).

4.5 Offretite

$KCaMg(Al_5Si_{13}O_{36}) \cdot 15 H_2O$
$P\bar{6}m2$
$a = 13.29$
$c = 7.58\,(\text{Å})$

4.5.1 History and Nomenclature

Gonnard (1890) proposed the new name offretite for a hexagonal mineral from Mont Semiouse (called Mont Semiol in all modern maps), near Montbrison, Loire, France, in honour of prof. Offret of Lyon, France.

4.5.2 Crystallography

A model for the framework of offretite was proposed by Bennett and Gard (1967) (see Chap. 0.2.4) and later on confirmed and refined by Gard and Tait (1972). The framework is characterized by a sequence AAB... which gives rise to three types of cavities: cancrinite cages, gmelinite cages (see Chap. 0.3) and large channels surrounded by 12 tetrahedra, parallel to c. All Al is present in the double rings AA, whereas the single rings B are Al-free. K is at the centre of cancrinite cages (site B of Mortier 1982) with 100% occupancy, and coordinates framework oxygens only (type I of Chap. 0.4); Ca(2) alternately occupies one of two equivalent sites inside the double 6-ring (6% occupancy in each of the two) and coordinates framework oxygens only (type I of Chap. 0.4); Mg is at the centre of the gmelinite cage (86% occupancy; site F of Mortier 1982) with five water molecules interposed between the cation and the cage (type IV of Chap. 0.4); Ca(1) is on the axis, a triad, of the large channels (39% occupancy; site G of Mortier 1982) and coordinates water molecules only (type IV of Chap. 0.4). After dehydration (Mortier et al. 1976a), Mg becomes 3-coordinated, jumping from the centre of the gmelinite cage to the centre of the distorted single 6-ring, which is its port to the upper or lower cage of the same kind; Ca is in the cancrinite cage and K is shifted nearly to the centre of one of the 8-rings of the gmelinite cage; the large channel is empty. After adsorption of CO in the dehydrated form (Mortier et al. 1976b), K and Ca do not change their position, but Mg is now out of the plane of the single 6-ring and is four coordinated, the fourth ligand being the oxygen of the CO molecule.

The morphology is characterized by the presence of hexagonal prisms (order unknown) with the basal pinacoid and also by barrel-shaped crystals whose forms are not known.

Offretite is always intergrown with erionite: in some cases with macro-domains as observed by Rinaldi et al. (1975a) and by Rinaldi (1976), or as

Fig. 4.5A. Offretite from Mt. Sémiol, France (SEM Philips, ×60)

cryptodomains detected via single crystal study (Mortier et al. 1976a; Pongiluppi 1976). The fact can be easily explained because of the similarity in the 6-ring stackings: AAB... in offretite, AABAAC... in erionite.

Offretite-erionite intergrowths occur frequently also as overgrowth on both sites of levyne lamellae (see Chap. 4.3), thus forming typical sandwiches; sometimes offretite-erionite layers occur also inside the levyne crystals.

4.5.3 Chemistry and Synthesis

The chemical composition of offretite is usually quite near to the stoichiometric formula of the title page, with $(Ca + Mg) \sim 2$ but variable Ca/Mg ratio (see Table 4.5 I and also Wise and Tschernich 1976 b).

Rubin (1969) obtained an offretite-like phase (probably with some intergrown erionite) from mixtures

$$(1.1 + 0.4) M_2O \cdot Al_2O_3 \cdot (5 - 10) SiO_2 \cdot (0 - 8) H_2O$$

where M is TMA ($= $ tetramethylammonium) plus Na and/or K.

Aiello and Barrer (1970) synthetized an offretite-like phase "O", without any intergrowth with other phases from gels

$$mM_2O \cdot Al_2O_3 \cdot nSiO_2$$

where M is TMA or Na or K with $4 \leqslant m \leqslant 10$ and $10 \leqslant n \leqslant 20$; the synthesis was not possible in the absence of either TMA or K, the hydrothermal treatment lasting one week at 80°C; this phase is much more silica-rich than the natural ones. Kerr et al. (1970) showed that synthetic offretite often occurs as intergrowths not only with erionite, but also with zeolite L, a synthetic phase without any natural counterpart. Barrer and Sieber (1977) obtained offretite-like phases also from gels containing Li and Cs, as well as TMA, but the product was always associated with other zeolites, mainly ZK-5.

4.5.4 Optical and Other Physical Properties

Optical data are scanty; the sample of Mont Semiol is uniaxial negative with $\omega = 1.489$, $\varepsilon = 1.486$ (Sheppard and Gude 1969 b), transparent colourless, streak white. Mohs-hardness probably 4. There is probably some cleavage parallel to c.

4.5.5 Thermal and Other Physico-Chemical Properties

The thermal curves show (see Fig. 4.5 B) two main water losses at 125 °C and 195 °C, with a nearly complete dehydration at 400°C; a correspondence with the presence of seven different water sites, with occupancies ranging from 90% to 14%, is quite hard to find. The temperature of lattice breakdown is

Table 4.5I. Chemical analyses and formulae on the basis of 36 oxygens and lattice constants in Å

	1	2	3
SiO_2	53.0	50.64	56.44
Al_2O_3	18.1	18.64	20.04
Fe_2O_3		0.11	0.10
MgO	2.0	0.99	0.26
CaO	4.1	6.29	8.04
SrO		0.26	
Na_2O		0.18	0.73
K_2O	3.6	3.68	3.22
H_2O	18.8	19.19	
Total	99.6	99.98	88.83
Si	12.92	12.55	12.83
Al	5.20	5.45	5.31
Fe		0.02	0.02
Mg	0.73	0.37	0.09
Ca	1.07	1.67	1.94
Sr		0.04	
Na		0.09	0.32
K	1.12	1.16	0.92
H_2O	15.29	15.86	
$E\%$	+10.3	+1.1	+0.6
a	13.291	13.314	13.315
c	7.582	7.567	7.556
D	2.09	2.10	

1. Mont Semiol, Loire, France (Sheppard and Gude, 1969b); D is calculated.
2. Araules, Haute-Loire, France (Pongiluppi, 1976).
3. Owihee Dam, Oregon (no.10 of Wise and Tschernich, 1976b).

not known for a natural sample, but the synthetic offretite-like phase, which is more silica-rich, gives its powder pattern up to 820°C (Aiello et al. 1970).

Aiello et al. (1970) tested synthetic offretite for the sorption of n-hexane and c-hexane, which is very low in the unmodified synthetic sample, whose large channels are blocked by TMA, but after burning at 650°C in air both molecules are readily imbibed. Even better results are obtained when the synthetic offretite is exchanged with ammonium nitrate, so as to remove nearly

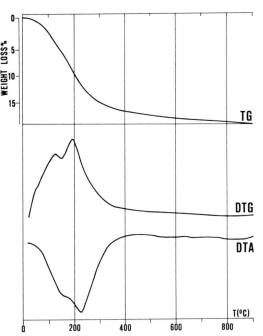

Fig. 4.5B. Thermal curves of offretite from Mt. Sémiol, France; in air, heating rate 20°C/min (by the authors)

half the potassium, and then heated in air to 550°C: at this point the cationic content per unit cell is KH_3, and the phase adsorbs 120 mg of c-hexane per 1 g of sorbent at P = 60 mm Hg and T = 22 °C, or also 120 mg of m-xylene per 1 g of sorbent at P = 8 mm Hg and T = 40 °C; the larger molecules of tri-methyl-benzene are not adsorbed.

Mirodatos et al. (1975) (see also the references quoted by them) established a catalytic activity of the synthetic offretite with K_3H as cations per unit cell: iso-octane is cracked mainly to iso-butane, because the small gmelinite cages are blocked and the large channels only are active. If the unit cell cation content is KH_3, the small gmelinite cages are also active and the cracking gives a higher quantity of smaller molecules, like propane.

4.5.6 Occurrences and Genesis

Offretite is known to occur only in vugs of massive rocks, mostly basic volcanics, in a restricted number of localities; the hydrothermal genesis is probable.

France: the type locality is Mont Semiol, Montbrison, Loire, where it occurs with phillipsite, chabazite and mazzite in the vugs of an olivine basalt (Gonnard 1890; Galli et al. 1974); Pongiluppi (1976) found it at Araules, Haute-Loire, with chabazite and phillipsite in a basalt.

Germany: as intergrowth with erionite in the amygdales of the limburgite from Sasbach, Kaiserstuhl, with faujasite and phillipsite (Rinaldi et al.

1975a; Rinaldi 1976); in the basalts of Grossen Buseck, Annerod, Herbstein, all in Vogelsberg, Hessen, with chabazite, phillipsite, faujasite (Hentschel and Vollrath 1977; Hentschel 1980a); as an intergrowth with erionite in the basalt of Gedern, Vogelsberg, Hessen (Betz and Hentschel 1978); in the basalt of Arensberg, Eifel (Hentschel 1982).

Czechoslovakia: as an intergrowth with erionite, associated with wellsite at Prackovice nad Labem (probably in a basalt) (Rychly et al. 1982).

Canada: as fibrous overgrowths on levyne lamellae (in a basalt?) southwest of Westwold, British Columbia (Wise and Tschernich 1976b).

U.S.A.: as a fibrous intergrowth with erionite, coating levyne lamellae in the vesicles of the Picture Gorge basalt, Beech Creek, Grant Co., Oregon (Sheppard et al. 1974; Bennet and Grose 1978); as intricate intergrowths with erionite and levyne, in vesicles of a basalt at Milwaukie, Oregon, with stilbite and chabazite (Wise and Tschernich 1976b); as fibrous overgrowth on levyne lamellae (in a basalt?) along Queen Creek, Pinal Co., Arizona (Wise and Tschernich 1976b); as tiny hexagonal prisms (0.3 × 1.0 mm) of pure offretite at Owyhee Dam, Oregon (Wise and Tschernich 1976b); as a complex levyne-erionite-offretite intergrowth in the vesicles of an olivine basalt at Malpais Hill, Pinal Co., Arizona, with heulandite and phillipsite and at Horseshoe Dam, Maricopa Co., Arizona (Wise and Tschernich 1976b).

Australia: as fibrous aggregates (probably an intergrowth with erionite) coating levyne lamellae in the Tertiary basalt in the Merriwa District, New South Wales (England and Ostwald 1979); as fibrous aggregates (probably an intergrowth) coating levyne lamellae in a vesicular basalt at Flinders, Victoria (Birch 1980); in the same area Coulsell (1980) found offretite with chabazite, gmelinite, analcime, natrolite, thomsonite, phillipsite, levyne.

4.6 Faujasite

$Na_{20}Ca_{12}Mg_8(Al_{60}Si_{132}O_{384}) \cdot 235 H_2O$
Fd3m
$a = 24.60\,(\text{Å})$

4.6.1 History and Nomenclature

Damour (1842) described the new mineral from Sasbach, Kaiserstuhl, Germany, previously considered as apophyllite, and named it in honour of Faujas de Saint-Fond.

Faujasite is exceptional amongst natural zeolites, because its synthetic counterparts have found important applications as sorbents and catalysts, so that the literature available on these synthetic phases is so wide that the reference list thereof would be longer than the already long reference list of this book. Therefore, only a very short review on these synthetic phases will be given here. More data on the faujasite-like phases may be found in Breck (1974) and in Barrer (1978, 1982).

4.6.2 Crystallography

The structure, resolved by Bergerhoff et al. (1956, 1958), corresponds to the most open framework of all natural zeolites, with 51% of the cell space void in the dehydrated modification; there are nearly 80 $\mathring{A}^3$ per tetrahedron, to be compared with 38 $\mathring{A}^3$ per tetrahedron in quartz. The structure was refined down to R = 13% by Baur (1964) (see Chap. 0.2.4) and to an R = 8% by Rinaldi (oral communication). Baur (1964) located 40% of the cations inside the sodalite cage (or β-cage, or cuboctahedron) in a site called I' by Smith (1976), outside the double 6-ring, but in contact with three oxygens of this double ring, a position similar to the main cation site in chabazite and other zeolites of this group. Baur (1964) also found the position of 16% of the water molecules; the remaining majority probably have a disordered distribution in the large cavities, also called supercages. Rinaldi (oral communication) found a larger fraction (50%) of the cations in the same sites, and nearly the same fraction of the water molecules.

Mortier (1982) lists 67 refinements of synthetic phases with the same framework and different Si:(Si+Al) ratios, variously ion exchanged and modified. These results cannot be considered here; a large review may be found in Smith (1971, 1976).

The only common form of the crystals is the octahedron {111}. Twinning on {111} is possible, also as penetration twins.

4.6.3 Chemistry and Synthesis

The chemistry of faujasite (see Table 4.6I) is rather variable, and sometimes quite far from the schematic formula of the title page. Si occupies normally $70 \pm 2\%$ of the tetrahedra, but in one case this value is 64%. The most abundant cation may be Na (up to 35 per cell) or Ca (up to 20 per cell); Mg may be absent, but normally is present in relevant amount: so faujasite is one of the few Mg-rich zeolites, with offretite, mazzite and ferrierite.

The first synthesis of a zeolite with a faujasite framework was announced by Breck et al. (1956), but full details were published only by Milton (1959), who coded it (Linde) X; in the same year, Barrer et al. (1959) also described a synthetic phase R, which was said to be related to "Linde Molecular Sieve X", hence they recognized the priority of Milton (1959). Breck (1962, 1964) synthesized zeolite Y, which was considered as a distinct species at that time, because of its particular physico-chemical properties: for instance Ca-exchanged Y adsorbs triethylammine, Ca-exchanged X does not. A full report of the synthesis of these two phases was given by Breck and Flanigen (1968);

Fig. 4.6 A. Faujasite from Sasbach, Germany (SEM Philips, ×30)

in short, starting from gels $5\,Na_2O \cdot Al_2O_3 \cdot 4\,SiO_2 \cdot 320\,H_2O$, there is a visible crystallization at $100\,°C$ with separation of zeolite X near to

$$Na_{86}(Al_{86}Si_{106}O_{384}) \cdot 264\,H_2O$$

lattice constant $a = 24.93$ Å; starting from gels $10\,Na_2O \cdot Al_2O_3 \cdot 20\,SiO_2$, which are best aged for some time at room temperature, and then heated at $100\,°C$, one obtains the crystallization of zeolite Y with formula $Na_{56}(Al_{56}Si_{136}O_{384}) \cdot 250\,H_2O$, hence very near to natural faujasite in composition; these gel compositions are appropriate if silica is added as sodium silicate; the use of silica gel requires higher SiO_2 concentrations. Stamires (1973) showed that zeolites X, Y and faujasite are terms of a continuous series with $Si:(Si+Al)$ ranging from 52% to 73%, and lattice constants from 24.98 to 24.62 Å; all physico-chemical properties vary with continuity as a function of the $Si:(Si+Al)$ ratio: as do triethylammine adsorption in Na-forms, and catalytic activity for cumene cracking.

4.6.4 Optical and Other Physical Properties

Faujasite is isotropic with n $\simeq 1.47$ (see Table 4.6 II), transparent colourless or white, streak white, lustre vitreous, Mohs-hardness 5; cleavage {111} perfect.

Table 4.6I. Chemical analyses and formulae on the basis of 384 oxygens and lattice constants in Å

	1	2	3	4	5	6	7
SiO_2	49.54	42.13	50.8	49.75	46.80	46.82	45.72
Al_2O_3	15.93	19.73	18.2	16.93	17.42	17.42	18.12
Fe_2O_3			0.48	0.04	0.03	0.02	
MgO	2.10	0.84		0.64	1.29	0.76	0.96
CaO	3.57	3.75	4.1	4.59	6.00	6.55	6.85
SrO		0.06			0.13	0.15	
BaO		0.14		0.05			
Na_2O	2.35	6.19	4.0	3.16	1.66	1.46	1.61
K_2O	0.51	0.92	1.7	1.43	0.17	0.30	0.16
H_2O	26.00	26.24					
Total	100.00	100.00	79.48	76.59	73.50	73.48	73.42
Si	139.08	123.43	135.73	137.20	133.60	133.92	131.04
Al	52.72	68.14	57.23	54.92	58.56	58.72	61.28
Fe			0.96	0.08			
Mg	8.79	3.67		2.72	5.44	3.20	4.16
Ca	10.74	11.77	12.37	13.68	18.40	20.16	21.12
Sr		0.10			0.32	0.32	
Ba		0.16		0.08			
Na	12.79	35.16	20.69	16.88	9.28	8.16	8.96
K	1.83	3.44	5.76	4.64	0.64	1.12	0.64
H_2O	243.44	256.39					
E%	-1.8	-2.7	+13.7	+1.0	+0.5	+3.7	+1.9
a	24.60	24.783	24.69	24.638			
D	1.93			1.93			

1. Sasbach, Germany (average of 8 analyses by Rinaldi et al. 1975a; H_2O, lattice constant and density by Fischer and O'Daniel 1956).
2. Aci Castello, Italy (Rinaldi, written communication).
3. Oahu, Hawaii (Iijima and Harada 1969).
4. San Bernardino, California (average of 4 analyses, Wise 1982).
5. Annerod, Germany (do).
6. Grossen Buseck, Germany (do).
7. Hasselborn, Germany (do).

Table 4.6 II. Refractive indices

	1	3	4
n	1.480	1.471	1.466

Numbers refer to the same samples as in Table 4.6 I

4.6.5 Thermal and Other Physico-Chemical Properties

The thermal curves (see Fig. 4.6 B) reveal a major water loss at 175 °C and a minor one at 300 °C: the first one probably corresponds to the (disordered) water molecules in the supercages and the second one to the more tightly bound molecules in the sodalite cages. The water loss continues at a very low rate up to 800 °C.

Ion-exchange is certainly very easy in natural faujasite, although no data for natural samples are known to us, because there is a lot of experimental evidence on the exchange capacities of the synthetic zeolites X and Y (see for instance Sherry 1966, 1967, 1968 a, b; Olson and Sherry 1968; Breck 1974); in zeolite X, the behaviour varies with the degree of exchange: up to nearly 50% exchange, which broadly corresponds to the cations in the supercage, the selectivity series is

$$Ag \gg Tl > Cs > Rb > K > Na > Li .$$

Over 50%, hence if also the cations in sodalite cage are exchanged, the selectivity series is

$$Ag \gg Tl > Na > K > Rb > Cs > Li$$

so Cs and Rb and solvatated Li enter with difficulty into the small cage through 6-rings. Some typical exchange isotherms are shown in Fig. 4.6 C; the Ca $-$ Na isotherm is bimodal, hence there are two different ways of exchange depending on the site, inside or outside the sodalite cage. The Sr $-$ Na isotherm is anomalous, and this fact is probably due to the formation of two phases.

Zeolite Y, which is chemically more similar to faujasite than zeolite X, has a lower exchange capacity because of its lower Al-content. Some exchange isotherms are shown in Fig. 4.6 D. The selectivity series for monovalent ions, up to 68% exchange, is

$$Tl > Ag > Cs > Rb > NH_4 > K > Na > Li .$$

Na is not exchanged over 68% at room temperature neither by monovalent ions, like Cs, NH_4, Tl, nor by any divalent or trivalent ions.

The hydrogen forms have been obtained by NH_4 exchange followed by a suitable thermal treatment (Breck 1974): the Na-free form of zeolite X cannot be obtained in this way, at least at room temperature, and a partial hydrogen form is the result of the ignition; if the ion exchange is forced to completion

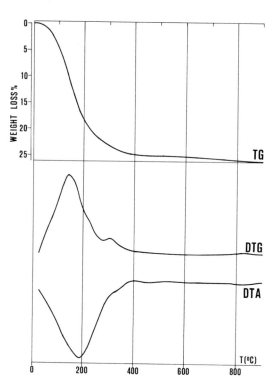

Fig. 4.6B. Thermal curves of faujasite from Sasbach, Germany; in air, heating rate 20°C/min (by the authors)

at higher temperature, the subsequent ignition destroys the lattice; the complete Na – NH$_4$ exchange is possible in zeolite Y, whose pure hydrogen form may be obtained easily and is thermally quite stable.

A summary of the adsorption properties of the synthetic zeolites with the faujasite framework is given in Table 4.6III. The adsorption properties are the highest among natural zeolites; higher values are obtained for polar molecules like H$_2$O or CO$_2$, whereas non-polar molecules like O$_2$ and N$_2$ give lower values; the large pore structure allows the adsorption of large molecules such as benzene and tri-propyl-ammine, up to values over 20% of the weight of the activated zeolite; molecules with a kinetic diameter over 8.5 Å are not sorbed.

The literature on the catalytic properties of zeolites X and Y is so wide, and the number of possible reactions so high, that we can only advise those interested in this matter to read the review articles by Venuto (1971), Rabo and Poutsma (1971), Poutsma (1976), Naccache and Ben Taarit (1980), Imelik et al. (1980) and Gallezot (1984).

4.6.6 Occurrences and Genesis

A sedimentary genesis has been invoked for faujasite from only two occurrences: the palagonite tuffs from Oahu, Hawaii (Iijima and Harada 1969),

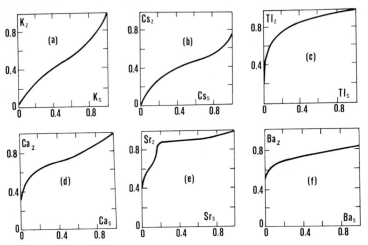

Fig. 4.6 C a – f. Ion-exchange isotherms for zeolite X, at 25 °C and 0.1 N total normality. **a** Na – K (Sherry 1966). **b** Na – Cs (Sherry 1966). **c** Na – Tl (Sherry 1966). **d** Na – Ca (Sherry 1968a). **e** Na – Sr (Sherry 1968a). **f** Na – Ba (Sherry 1968a)

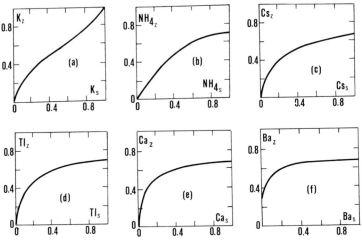

Fig. 4.6 D a – f. Ion-exchange isotherms for zeolite Y, at 25 °C and 0.1 N total normality. **a** Na – K (Sherry 1966). **b** Na – NH$_4$ (Sherry 1966). **c** Na – Cs (Sherry 1966). **d** Na – Tl (Sherry 1966). **e** Na – Ca (Sherry 1968a). **f** Na – Ba (Sherry 1968a)

Table 4.6III. Adsorption (g of sorbed fluid per g of activated zeolite) in zeolites NaX and NaY (Breck 1974)

Compound	NH_3	H_2O	CO_2	O_2	N_2	Kr	Xe	benzene	iso-octane	$(C_4H_9)_3N$	$(C_4F_9)_3N$
Kinetic diameter (Å)	2.6	2.7	3.3	3.5	3.6	3.6	4.0	5.9	6.0(?)	6.1	10.2
Pressure (torr)	700	20	700	700	700	700	700	10	40	1	25
T(°C)	25	25	25	-78	-78	-78	25	25	25	25	25
Zeolite NaX	0.19	0.35	0.26	0.054	0.12	0.41	0.25	0.26	0.21	0.227	0.031
Pressure (torr)	700	20	700	700	700			70	40	1	25
T(°C)	25	25	25	-78	-78			25	25	25	25
Zeolite NaY	0.18	0.35	0.23	0.033	0.072			0.26	0.20	0.228	0.018

where faujasite is present in the cement with phillipsite, gismondine, chabazite, gonnardite, natrolite, analcime; sedimentary origin by slow percolation of meteoric water in these tuffs was attributed to these zeolites by Hay and Iijima (1968a and b), who did not, however, find faujasite. Similar is the origin of the zeolites (water clear mm-sized octahedra of faujasite and phillipsite) derived from the palagonitized basaltic glass of a cinder cone at Halloran Spring Quadrangle, San Bernardino Co., California (Wise 1982).

A hydrothermal origin is probable for all other occurrences in vugs of massive rocks:

Germany: in the vugs of the limburgite from Sasbach, Kaiserstuhl, Baden (the type locality), as fine crystals up to 2 mm with phillipsite and offretite (Hintze 1897; Rinaldi et al. 1975a); in basalts from Annerod, Giessen, Hessen, with chabazite, offretite, phillipsite (Hintze 1897; Hentschel 1980a); in a basalt north of Grossen Buseck, Vogelsberg, Hessen, as crystals up to 4 mm with chabazite, phillipsite and offretite (Hintze 1897; Hentschel 1980a); in an altered basalt from Stempel, Marburg, Hessen (Hintze 1897).

Switzerland: in the granitic rocks of the Aar Massif (Parker 1922).

Italy: in the basalts of Aci Castello, Sicily (Rinaldi, oral communication).

Canada: as milky white crystals up to 2 mm with quartz and fluorite in the Daisy mica mine, Ottawa Co., Quebec (Hoffmann 1901).

5 Zeolites of the Mordenite Group

5.1 Mordenite

$Na_3KCa_2(Al_8Si_{40}O_{96}) \cdot 28 H_2O$
Cmcm
$a = 18.11$
$b = 20.46$
$c = 7.52 (\mathring{A})$

5.1.1 History and Nomenclature

The name was introduced by How (1864) for a mineral found at Morden, King's Co., Nova Scotia. Cross and Eakins (1886) named a supposedly new mineral from Green and Table Mts., Jefferson Co., Colorado, ptilolite from $\pi\tau\iota\lambda o\nu$ = dawn, because of its plumose aspect. Later Davis (1958) showed that ptilolite was identical to mordenite; other obsolete synonyms are flokite, arduinite and pseudonatrolite.

5.1.2 Crystallography

The structure of mordenite (see Chap. 0.5) was resolved by Meier (1961) and refined by Graemlich (1971). Subsequently a series of ion-exchanged modifications of a natural mordenite were refined, mostly in the dehydrated form: dehydrated Ca, dehydrated H, rehydrated Ca, dehydrated and rehydrated K, by Mortier et al. (1975, 1976, 1978); dehydrated Rb, Ba, Cs, Na, H (via acid exchange), Ca + H, by Schlenker et al. (1978, 1979).

As far as the (Si, Al)-distribution is concerned, there is a small but definite Al-enrichment in the 4-rings which connect the hexagonal sheets; as a matter of fact, the tetrahedra of these 4-rings are just a little larger than the tetrahedra of the sheets, but this small difference is systematically present in all mentioned refinements. The cations are present mainly in three sites, coded A, D, E by Mortier (1982) and I, IV, VI by Mortier et al. (1975, 1976, 1978) and by Schlenker et al. (1978, 1979). To specify these sites, let us remember that there are two sets of channels parallel to c in this structure, a larger and a smaller one; let us now consider one of the above mentioned 4-rings and its

Fig. 5.1 A. Mordenite from Valle dei Zuccanti, Vicenza, Italy (×4)

Fig. 5.1 B. Mordenite from Stevenson, Washington (SEM Philips, ×240)

Fig. 5.1 C. Mordenite from Stevenson, Washington (SEM Philips, ×2000)

Fig. 5.1 D. Sedimentary mordenite from Ponza, Italy (SEM Philips, ×2000)

equivalent by a *c*-translation; the two 4-rings and the two hexagonal sheets connected by them define a cage which, in the *b*-direction, has two exits through two 8-rings; site A is near the centre of the 8-ring which connects the cage with the smaller channel, site D is near to the centre of the 8-ring which connects the cage with the larger channel, site E is in the larger channel, far from the 4-rings. These are the sites preferred by most cations; only Rb and Ba were found in sites normally occupied by water.

The morphology is always characterized by needles, fibres with *c*-elongation; radial aggregates, from loosely bound to compact, are common.

5.1.3 Chemistry and Synthesis

The chemistry of mordenites was reviewed by Passaglia (1975). On the whole, there are no large variations in comparison with the stoichiometric formula given in the title page; the ratio Si: (Si + Al) ranges from 80% to 85% only, the Ca-atoms range from $1.6-2.5$, Na $2.0-5.0$, K $0.1-0.8$ per unit cell. Table 5.1 I contains the chemical analyses, formulae and lattice constants of 5 mordenites. The sedimentary mordenites have the Si-content at the highest level. As a matter of fact, sedimentary mordenites are difficult to purify for analysis but their lattice constants are just a little smaller than the most silica-rich hydrothermal mordenites: in view of the negative correlation of *b* with Si:(Si + Al) found by Passaglia (1975), this fact excludes a strong difference in chemistry.

The first synthesis of mordenite is due to Barrer (1948), who treated hydrothermally at $\sim 290°C$ a gel of composition $Na_2O \cdot Al_2O_3 \cdot 10SiO_2$ (or nearly so) for 2 days. High temperature and long duration of crystallization are therefore characteristic for this synthesis of mordenite, which is the equilibrium phase under the synthesis conditions. Ames and Sand (1958) were able to synthesize a mordenite with a gel $CaO \cdot Al_2O_3 \cdot 10SiO_2$, but the temperature needed was a little higher, $\sim 360°C$. Experimental conditions for the synthesis were given also by Senderov (1963). All these were found to be small-port (SP) mordenites, namely to adsorb molecules smaller than expected on the basis of the crystal structure. Sand (1968) gave a full report on how to drive the process towards the formation of SP- or LP (= large port)-mordenites; in short, if the gel-batch was on the join $(1/3\,Na_2O \cdot SiO_2)$ − $(Na_2O \cdot Al_2O_3 \cdot 9-10SiO_2)$, the product is a LP-mordenite, otherwise it is a SP-mordenite; temperature may be as low as 65°C, but a quicker crystallization occurs at 270°C. The distinction was based on benzene adsorption, which is 7% in weight of dried zeolite for the LP-phase, 0.7% for the SP-phase. A hydrochloric acid treatment allows conversion of the SP-phase into an LP one. This fact supports the idea that in SP-mordenites the channels are blocked not by stacking faults, but by extraframework aluminium atoms, which may be removed by acid treatment, but Raatz et al. (1984) found good evidence supporting the hypothesis that stacking faults disrupt the continuity of the channels in SP-mordenites.

Domine and Quobex (1968) gave data on the influence of chemical parameters and starting materials on the kinetics of mordenite formation. Whitter-

Table 5.1I. Chemical analyses and formulae on the basis of 96 oxygens and lattice constants in Å

	1	2	3	4	5
SiO_2	64.01	69.91	64.65	67.10	69.86
Al_2O_3	12.38	10.70	13.10	11.66	10.23
Fe_2O_3	1.17	0.37	0.06	0.13	0.13
MgO	0.79	0.12	0.02	0.02	0.04
CaO	3.68	2.50	2.36	3.12	2.95
SrO	0.10		0.05	0.05	0.05
BaO	0.04		0.03	0.11	0.06
Na_2O	1.71	3.25	4.37	3.08	2.75
K_2O	1.05	0.40	0.83	0.14	0.23
H_2O	15.00	12.96	14.59	14.33	13.30
Total	100.32	100.21	100.06	99.74	99.60
Si	38.63	40.30	38.82	39.90	40.89
Al	8.81	7.60	9.27	8.17	7.06
Fe	0.53	0.17	0.03	0.06	0.06
Mg	0.71	0.11	0.02	0.03	0.03
Ca	2.38	1.61	1.52	1.99	1.85
Sr	0.03		0.02	0.02	0.02
Ba	0.01		0.01	0.03	0.01
Na	2.00	3.80	5.09	3.55	3.12
K	0.81	0.31	0.64	0.11	0.17
H_2O	30.19	26.03	29.22	28.42	25.96
E%	-1.42	+2.91	+5.14	+6.15	-0.12
a	18.145	18.089	18.115	18.125	18.100
b	20.496	20.451	20.505	20.467	20.446
c	7.520	7.512	7.523	7.519	7.511

1. Monte Civillina, Recoaro, Italy (Passaglia, 1975) with 0.36% FeO.
2. Teigarhorn, Iceland (do).
3. Morden, King's Co., Nova Scotia, Canada (do).
4. Coyote, Garfield Co., Utah (do).
5. 20 miles East of Durangabad, Maharashtra State, India (do).

Table 5.1II. Refractive indices

	2	4
α	1.471	1.473
β	1.474	1.475
γ	1.476	1.478

2. Teigarhorn, Iceland
(Bauer and Hřichová
1962).
4. Coyote, Garfield
Co., Utah (Schaller
1932a).

more (1972) succeeded in synthesizing silica-rich mordenite, with up to 91%
of the tetrahedra occupied by Si, to be compared with the normal maximum
value of 85%. In oxide ratios, normal mordenites are nearly $Na_2O \cdot Al_2O_3$
$\cdot 10SiO_2$, whereas acid mordenites were synthesized up to $Na_2O \cdot Al_2O_3$
$\cdot 19.5SiO_2$. Wirsching (1975) showed that the crystallization of mordenite
from natural rhyolitic glasses is possible, both in open and closed systems.
Ueda et al. (1980) studied the crystallization of mordenite from clear aqueous
solution obtained by dissolving an alumo-silica-gel in water, both in seeded
and unseeded media, thus contributing to the knowledge of the nucleation of
mordenite crystals, which takes from 2 to 4 days to start. Nakajima (1973)
synthesized mordenites in the whole range $Na_2(Al_2Si_{10}O_{24}) - Ca(Al_2Si_{10}O_{24})$.

5.1.4 Optical and Other Physical Properties

The optical properties of mordenites have been determined only by a few
authors, because crystals develop only as needles which are not easy to handle.
Orientation is $\alpha \rightarrow c$, $\beta \rightarrow a$, $\gamma \rightarrow b$, with $\alpha = 1.471$, $\beta = 1.474$, $\gamma = 1.476$ for a
sample from Teigarhorn, Iceland (Bauer and Hřichová 1962), and $\alpha = 1.473$,
$\beta = 1.475$, $\gamma = 1.478$ for a sample from Coyote, Garfield Co., Utah, U.S.A.
(Schaller 1932a), with 2V large and optical sign positive or negative.
Elongation is always negative, since the fibre axis is c.

5.1.5 Thermal and Other Physico-Chemical Properties

The thermal behaviour of natural mordenite is quite simple (see Fig. 5.1E; see
also Reeuwijk 1974): there are two peaks in the DTG, at 60°C and 160°C; the
total water loss of 15% is nearly complete at 300°C. After heating at 800°C,
rehydration is almost complete.

The easy ion-exchange of mordenite with all alkalies and alkali-earths was
demonstrated by Barrer (1948). Ames (1961) showed that the replacement
series $Cs > K > Na > Li$ and $Ba > Sr > Ca > Mg$ are valid for mordenite, as

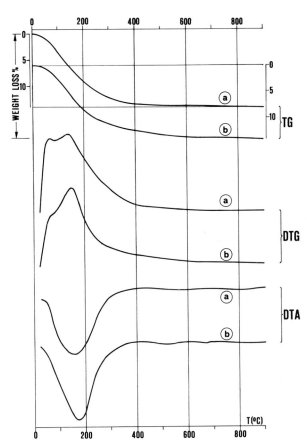

Fig. 5.1E. Thermal curves of mordenite from **a** Morden, Canada, and **b** Poona, India; in air, heating rate 20°C/min (by the authors)

for other zeolites as well. Monovalent ions are in general preferred to divalent ones. A similar study was conducted by Wolf et al. (1980) on mordenite which was dealuminated by HCl. Dealumination decreases the selectivity for K, and increases that for alkaline earth and the natural preference of mordenite for monovalent ions may be reversed in favour of the divalent ones.

Barrer and Townsend (1976) studied the exchange of cuprammine, zinc-ammine and cobaltammine in ammoniacal solution (pH > 10) in mordenite; exchange is easy and selectivity is higher than for the corresponding aquo-ions; each Cu^{2+} or Zn^{2+} has three NH_3 ligands, each Co^{3+} six NH_3 ligands inside the mordenite crystal.

Hydrogen- and hydronium-mordenites are important modifications. Studies in this field are due to Shikunov et al. (1971), Barrer and Klinowski (1975), Kühl (1977). The hydrogen form may be obtained by ammonium exchange followed by calcination at a temperature high enough to expel ammonia but not too high to cause the loss of OH-groups as water: a temperature of ~500°C is proper. This hydrogen form hydrates to the hydronium form under room conditions. The hydronium form can be obtained in a simpler way also by ion-exchange with a 0.1 N HCl solution, but the forms obtained in the two different ways are not equivalent; both contain some aluminium which has been displaced from the tetrahedra of the framework into the channels, and hence is no longer active in the exchange process. H^+ can be neutralized with NaOH and this fact allows the determination of the ratio H^+/Al, which is variable between 50 and 80%, the higher values being given by the form obtained via NH_4-substitution.

5.1.6 Occurrences and Genesis

The genesis of mordenite may be sedimentary or hydrothermal. All crystallization which fills veins, fissures or amygdales of igneous rocks is considered hydrothermal, whereas the massive transformation of the glassy components of volcanic rocks is considered sedimentary; a sedimentary genesis in the absence of volcanic glass has also been identified. Intermediate situations between these two extremes may be recognized in the diffuse presence of mordenite in volcaniclastic rocks which have been subjected to low grade hydrothermal action, such as in a geothermal area.

The mordenite associated with magadiite and trona in the Liwa salt deposit, on the north shore of Lake Chad, certainly crystallized at very low temperature in the absence of any volcanic glass: it is considered a neoformation deriving from pedogenetic reaction between clay, diatomite substance and a concentrated sodium carbonate milieu (Maglione and Tardy 1971). A similar genesis may be attributed to the minor amounts of mordenite, with major analcime, in the Pleistocene to Holocene sediments in the Magadi basin, Eastern Rift Valley, Kenya (Surdam and Eugster 1976).

The first discovery of a sedimentary mordenite derived from volcanic glass is due to Harris and Brindley (1954), who showed that the glass of a pitchstone from Tormore, Arran, Scotland, was altered to mordenite, whose chemical composition was not much different from that of the unaltered glass.

Sheppard and Gude (1964) described Tertiary tuffaceous rocks of the Mojave desert, California, partly or wholly altered to zeolites: mordenite is a minor constituent of some beds, clinoptilolite, analcime, erionite and phillipsite being frequent. A similar zeolitized tuff, rich in clinoptilolite with some mordenite was found at Itaya, Yamagata Pref., Japan by Minato and Takano (1964). In the Green River Formation, Wyoming, Surdam and Parker (1972) and Goodwin (1973) reported the presence of mordenite with clinoptilolite, analcime and potassium feldspar, which all crystallized by reaction between the glass of the tuffs and the water of Lake Gosiute (from fresh to hypersaline). Large deposits of Miocene tuffs, altered up to 70% to zeolites (analcime, mordenite, clinoptilolite), were studied by Suprychev (1976) in the Solt trough, Ukraine, U.S.S.R. Sheppard (1971) lists another ten occurrences in the western United States of tuffs altered to mordenite, mostly in saline lakes.

Mordenite has never been found in deep sea sediments.

Other tuffaceous rocks were altered to mordenite by burial diagenesis, so that different zones were developed. A mordenite zone, overlying a heulandite zone and a laumontite one (the deepest) was found in a tuffaceous sequence of Neogene age in East Tanzawa Mts., central Japan by Yoshitani (1965). In the same mountains, similar rocks of the same age were studied by Seki et al. (1972) in the Oyama-Isehara District; they found thin beds of fine grained pumice tuff interlayered in a more abundant tuff-breccia, which was poorer in glassy materials before the diagenesis; host rock and thin beds share the same chemical composition, but the pumice tuff is altered to mordenite, the tuff-breccia to laumontite; these authors conclude that the higher content of glass and lower permeability of pumices favoured the mordenite crystallization, whereas the higher permeability and the presence of calcic pyroxenes and plagioclases favoured the crystallization of laumontite.

Utada (1970), in his review on zeolitized volcaniclastic rocks of Neogene age in Japan, recognized the presence of five zones with diagenetic minerals: starting from the top, after a fresh glass zone I, there is a zone II, characterized by the presence of clinoptilolite and mordenite, this last mineral being particularly abundant at Minase and Ten'ei, Tohoku District, and at Maji San'in District. The following three zones are characterized by analcime-heulandite, laumontite, albite.

Mordenite and clinoptilolite are the most abundant minerals in the Neogene volcaniclastics of Middle and East Slovakia (Šamajová 1979).

Kostov et al. (1967) and Kostov (1970) recognize a zoning in the hydrothermal deposition of zeolites in volcanic rocks and tuffs of Upper Cretaceous ages in the Srednogorie, Bulgaria; a quartz-mordenite zone occurs locally in the upper part of the formation. Seki et al. (1969a) described a mordenite zone in upper part of the Katayama geothermal area, Onikobe, northeast Japan; the host rock may be an andesitic lava or tuff, or a dacitic tuff; the present temperature of fumarolic gases ranges from 65 °C to 100 °C. Seki (1970) found an extensive transformation into mordenite of a large volcaniclastic andesitic-dacitic formation corresponding with the Atosanupuri active geothermal area, Hokkaido, Japan; also here the temperature never rises over 100 °C, even as deep as 1 km. Honda and Muffler (1970) investigated the cores from a drill-hole in an active hot-spring system down to 65 m, where the temperature was 171 °C. The original sediments (rhyolite detritus) were transformed into mordenite, clinoptilolite and silica minerals, these last being the only phases in the more intensely altered cores.

Occurrences in vugs of massive rocks with probable hydrothermal origin are rather common:

Northern Ireland: As aggregates of small fibres in the basalts of Antrim, (Walker 1960a).
Czechoslovakia: as acicular and fine fibrous crystals in the cracks of a pyroxene andesite in the central Vihorlat Mts. (Šamajová 1977).

Italy: as radially fibrous reddish masses, with quartz in a porphyrite in the Valle dei Zuccanti, Vicenza: this is the locality of "arduinite"(Davis 1958; Passaglia 1966); as radial aggregates associated with dachiardite and clinoptilolite at Alpe di Siusi, Bolzano (Vezzalini and Alberti 1975); as fine acicular crystals ("pseudonatrolite") with dachiardite and other zeolites, as late hydrothermal deposition in the pegmatitic dykes on the island of Elba (Gottardi 1960).

U.S.S.R.: as radially fibrous masses with quartz in a basalt at Mydzk, Volhynia (Thugutt 1933; Shashkina 1958); this mordenite has a remarkable high K content, 5.32% K_2O; as acicular crystals in chalcedony geodes and veinlets and as spherulites on analcime crystal faces, in the effusive rocks from Karadag, Eastern Crimea (Suprychev 1968): crystallization temperature was estimated to be $150°C - 230°C$.

India: as fine needles, on which octahedra of quartz pseudomorphs after cristobalite are perched, in the basalt of Ellora Cave, Hyderabad State (Van Valkenburg and Buie 1945); as radial aggregates in the Deccan Traps (Sukheswala et al. 1974).

Canada: in the basalts of Morden (type locality), Nova Scotia, with apophyllite and barite (How 1864).

U.S.A.: on bluish chalcedony in cavities of augite andesite fragments of a conglomerate of Green and Table Mts., Jefferson Co., Colorado ("ptilolite" of Cross and Eakins 1886); as a material ranging from finely fibrous to cottony, through radially fibrous to compact porcellaneous material intimately mixed with granules of quartz in the amygdales of the andesite at Challis, Custer Co., Idaho (Koch 1917; Ross and Shannon 1924); as pink needles ("ptilolite") embedded in calcite at Coyote, Garfield Co., Utah (Schaller 1932a); as fibrous woolly aggregates building up the central filling of amygdales of the basalt of Lanakai Hills, Hawaii (Dunham 1933).

Brazil: in the basalts of southern Brazil (Franco 1952).

Japan: as fibrous or cottony white or pink material, in vesicles of a rhyolite in the Yoshida area, Kagoshima Pref. (Tomita et al. 1970): mordenite and host rock share nearly the same composition; as aggregates of fine or cottony fibres with dachiardite, clinoptilolite and silica minerals in a vein in a rhyolitic mass, at Tsugawa, northeastern Japan (Yoshimura and Wakabayashi 1977).

5.1.7 Applications and Uses

Only few applications of natural mordenites have been described, in spite of the manifold uses of synthetic mordenites as adsorbers and catalysts. It is probable that natural mordenite, like many other natural and synthetic zeolites, is used in drying and purification of acidic gases, in view of its particular resistance to thermal cycles in acidic environments, but the authors are not aware of any publication on the matter.

Minato and Tamura (1978) devised an oxygen generator based on the differential adsorption of N_2 and O_2 on natural mordenite from the tuffs of Itado, Akita Pref., Japan. After a pre-treatment column for the removal of H_2O, CO_2 and SO_2, the air enters a zeolite-column, which adsorbs N_2, so that a gas with up to 93% of O_2 is obtained; generators are available in sizes ranging in oxygen outputs from 15 l/h to 300 m^3/h.

Ohtani et al. (1972) patented a hydrocarbon-conversion catalyst for the disproportionation of toluene to benzene, based on a hydrogen-exchanged natural mordenite. Also Lu et al. (1981) prepared a catalyst by supporting the natural mordenite of Dai Shi Kou Mine, China, with Ni and Bi; this catalyst gives results comparable with those of synthetic mordenites in the disproportionation of toluene.

Kalló et al. (1982) reported on catalysts based on modified (H- and Ag-form) natural mordenite, for hydrodealkylation of C_9 alkyl-aromatics, toluene disproportionation and xylene isomerization.

5.2 Dachiardite

$(Na, K, Ca_{0.5})_4 (Al_4 Si_{20} O_{48}) \cdot 18 H_2O$
C2/m
$a = 18.69$
$b = 7.50$
$c = 10.26\,(\text{Å})$
$\beta = 107°54'$

5.2.1 History and Nomenclature

The mineral was described as a new species by Giovanni D'Achiardi (1906), and named dachiardite (without apostrophe) in honour of his father Antonio D'Achiardi, a distinguished mineralogist. The spelling "d'achiardite" has also been used, incorrectly, as it does not correspond to the original intention of the proposer. Maleev (1976) proposed the new name svetlozarite for a mineral which was demonstrated later by Gellens et al. (1982) to be dachiardite with some kind of stacking disorder.

5.2.2 Crystallography

The crystal structure was determined by Gottardi and Meier (1963) and refined by Vezzalini (1983). To emphasize similarities with mordenite, Gottardi and Meier used a unit cell different from that given in the title page, and related to it by a matrix 100/001/010.

Fig. 5.2 A. Dachiardite from Elba, Italy (×4)

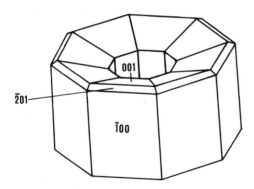

Fig. 5.2 B. Eightling of dachiardite twinned on {110}

   There is a tendency towards Si/Al order in the framework: schematically, the sheet of 6-rings is Al-free, and the "handles" which decorate the sheet and form 4-rings, are Al-rich. Cations are distributed between the 6-ring sheets, in part without coordinating framework oxygens; they are in part of type III and in part of type IV of Chap. 0.4.
   Alberti (1975b) found that all diffractions with k odd are diffuse in the Alpe di Siusi specimen, indicating disorder in the lattice; the same is true for Tsugawa dachiardite (Yoshimura and Wakabayashi 1977). Gellens et al. (1982) found similar features in the diffractions of svetlozarite, with diffuse

streaks parallel to c^*. This is interpreted as due to $\{001\}$-twins and $\{100\}$-stacking faults, which occur every $40-100$ Å within the crystal (every 2 to 5 unit cells), as shown by TEM observations.

The morphology is quite simple; most samples are simple fibres along b; only the forms $\{100\}$ $\{001\}$ $\{110\}$ $\{201\}$ have been observed in the Elba specimens, which are nearly always twinned on $\{110\}$, forming characteristic polysynthetic twins of eight individuals, which are similar to an octagonal beaker (Fig. 5.2A and B). This is possible because the angle between two individuals is nearly $45°$, so that there is an error of $1°$ at the boundary between the first and the eighth individual (Berman 1925). Meier (1968) gave a structural explanation of this $\{110\}$ twinning. Also frequent is the normal twin on $\{001\}$ (Bonatti and Gottardi 1960); less frequent a parallel twin on $\{100\}$ (Bonatti 1942; twin axis not given).

5.2.3 Chemistry and Synthesis

Table 5.2I gives the chemical analyses of all dachiardites described so far. The Si/Al ratio is quite constant, Si occupying $\sim 83\%$ of the tetrahedra, a value very near to that proper for a mordenite, but which has more variable values. The content of exchangeable cations is subject to large oscillations; the ratio monovalent : divalent cations ranges from 1 to 114 and the ratio Na : K ranges from 0.43 to 2.31. These wide ranges for some important constituents in such a small number of analyses and the absence of any synthesis prohibit drawing any conclusion on the possible chemical milieu from which this zeolite, instead of mordenite, crystallizes. It is also noteworthy that there are two kinds of dachiardite: non-fibrous (Elba) with nearly 1% of Cs_2O, fibrous (all others) without Cs_2O.

5.2.4 Optical and Other Physical Properties

The optical data are listed in Table 5.2II. Most dachiardites seem to have $\alpha \to b$, 2V large with both positive and negative sign, and $c\hat{\gamma}$ (in the obtuse angle) small ($\sim 20°$).

Dachiardite is often remarkably transparent in comparison with other zeolites, the lustre is vitreous to greasy, but sometimes may be translucent with silky lustre especially on cleavage surfaces; the colour may be absent (Elba) or pinkish (Siusi, Tsugawa).

Perfect cleavages (100) and (001): this corresponds to the [010] fibrous character of most specimens; in Elba dachiardite, the cleavages are the same but the uneven glassy fracture is more easy than in the other specimens. Streak: white.

Mohs-hardness: 4 to 4.5. Vickers-microhardness $170-200$ kg/mm^2 parallel to the fibre has been measured in svetlozarite (Maleev 1976).

Table 5.2I. Chemical analyses and formulae on the basis of 48 oxygens and lattice constants in Å

	1	2	3	4	5	6	7	8
SiO_2	65.15	65.72	69.58	67.38	72.41	71.33	75.09	70.10
Al_2O_3	12.91	12.15	10.26	12.65	10.15	12.15	10.98	10.44
Fe_2O_3		0.50	0.67	0.27	0.08	0.02	0.01	
MgO		0.08	0.30	0.03				
CaO	3.88	1.65	3.91	0.51	1.78	3.77	0.70	0.02
Na_2O	2.57	4.47	0.85	5.15	3.04	1.06	4.01	5.18
K_2O	0.95	1.87	3.00	0.97	1.30	2.55	2.69	0.98
H_2O	14.42	13.50	10.94	12.91				12.76
Total	99.88	100.01	99.51	100.00				99.55
Si	19.50	19.61	20.19	19.76	20.62	20.03	20.48	20.47
Al	4.55	4.27	3.51	4.37	3.40	4.02	3.53	3.59
Fe		0.11	0.15	0.06	0.02			
Mg		0.04	0.13	0.01				
Ca	1.24	0.53	1.22	0.16	0.55	1.14	0.21	0.01
Sr	0.02							0.01
Ba		0.01		0.02				
Na	1.49	2.59	0.48	2.93	1.67	0.58	2.12	2.93
K	0.36	0.71	1.11	0.36	0.47	0.91	0.93	0.36
Cs	0.12							
H_2O	14.39	13.43	10.59	12.63				12.43
E%	+1.3	−1.8	−14.7	+20.7	+5.5	+6.6	−1.1	+7.8
a	18.676	18.647	18.7	18.641	18.62	18.62		18.67
b	7.518	7.506	7.5	7.512	7.50	7.50		7.488
c	10.246	10.296	10.3	10.299	10.24	10.24		10.282
β	107°52'	108°22'	108°	108°29'	108°16'	108°16'		108°44'
D	2.14	2.16	2.17	2.16	2.16	2.16		2.14

1. Elba, Italy (Gottardi, 1960); also SrO 0.13% and Cs_2O 0.96% (Bonardi, 1979); lattice constants by Vezzalini (1983)

2. Alpe di Siusi, Italy; also SrO 0.01% and BaO 0.06% (Alberti, 1975b)

3. Svetlozarite, Rhodopes, Bulgaria (Maleev, 1976); lattice constants by Gellens et al.(1982)

4. Tsugawa, Japan (Yoshimura and Wakabayashi, 1977), plus BaO 0.13%

5. Altoona, Washington (average of two analyses of Wise and Ischernich, 1978a)

6. Cape Lookout, Oregon (do)

7. Agate Beach, Oregon (do)

8. Quebec (Bonardi et al., 1981), plus SrO 0.07%

Table 5.2 II. Optical data

	1	2	3	4	5 - 6	8
α	1.494		1.481	1.471	1.480	1.471
β	1.496	1.480	1.482		1.481	1.475
γ	1.499		1.485	1.477	1.485	1.476
2 V	+73°		+23°	-78°	+57°	-52°
$\underline{c}\hat{}\gamma$	38°			8°	2°	18°
	$\alpha \to \underline{b}$	$\alpha \to \underline{b}$		$\alpha \to \underline{b}$	$\gamma \to \underline{b}$	$\alpha \to \underline{b}$

1. Elba (Bonatti and Gottardi 1960).
2. Alpe di Siusi (Alberti 1975b).
3. Svetlozarite, Rhodopes (Maleev 1976).
4. Tsugawa (Yoshimura and Wakabayashi 1977).
5-6. Altoona and Cape Lookout (Wise and Tschernich 1978a).
8. Quebec (Bonardi et al. 1981).

5.2.5 Thermal and Other Physico-Chemical Properties

The thermal curves (Fig. 5.2 C, see also Yoshimura and Wakabayashi 1977) show a main water loss at 100°C, which continues with a decreasing rate and a shoulder at 200°C, up to 450°C. Beyond this limit, the lattice undergoes a sharp contraction.

Wise and Tschernich (1978a) report that Altoona dachiardite rehydrates after heating at 600°C, without any apparent damage of the framework.

No other physico-chemical data have been published on dachiardite, as far as we know.

5.2.6 Occurrences and Genesis

Only hydrothermal genesis has been ascertained for dachiardite. The holotype was collected (D'Achiardi 1906) in the "Filone (dyke) della Speranza", about 200 m south of the cemetery near the village of San Piero in Campo, island of Elba, Italy. The western part of the island is built by a Miocene granodiorite, crossed by aplitic to pegmatitic veins with cavities where the latest hydrothermal deposition consists of many zeolites: dachiardite as well developed twinned crystals, mordenite, epistilbite, stilbite, heulandite.

The second description was given for a sodium-rich variety by Alberti (1975b) who found fibrous crystals as aggregates with mordenite and clinoptilolite in a basalt at Alpe di Siusi, Bolzano, Italy. Demartin and Stolcis (1979) found a similar occurrence in the same formation in the nearby Val di Fassa. Maleev (1976) described svetlozarite (calcium-rich) with chalcedony in veinlets

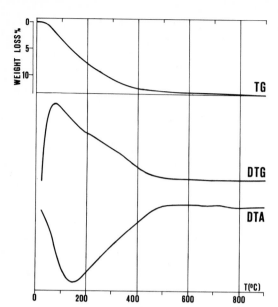

Fig. 5.2C. Thermal curves of dachiardite from Alpe di Siusi, Italy; in air, heating rate 20°C/min (by the authors)

cementing brecciated andesites to the west of Zvezdel, eastern Rhodopes, Bulgaria. The mode of occurrence of the dachiardite from Onoyama gold mine, Kagoshima Pref., Japan, was described by Kato (1973) without any detailed description of the zeolite, which occurs as subparallel aggregates of slender prismatic crystals, with heulandite. Yoshimura and Wakabayashi (1977) found a sodium-rich dachiardite as fibrous crystals with mordenite, clinoptilolite, heulandite in veins in a Miocene rhyolite in Tsugawa district, northeastern Japan. Nishido et al. (1979) found dachiardite as radiated aggregates of prismatic fibres, with mordenite and heulandite, in amygdales in the altered andesite pillow lava from Susaki, on the west coast of Chichijima, the Ogasawara islands, Japan; on the east coast of the same island, at Hatsuneura, Nishido and Otsuka (1981) found dachiardite as radiated aggregates of platy prismatic crystals, with mordenite, in chalcedony veins in an altered boninite pillow lava. Wise and Tschernich (1978a) reported on fibrous dachiardite with mordenite and ferrierite in the basalt east of Altoona, Wahkiahum Co., Washington, with erionite, phillipsite, mordenite in vesicular cavities in the basalt of Cape Lookout, Tillamook Co., Oregon, and with erionite and clinoptilolite at Agate Beach, Oregon. Bonardi et al. (1981) found a sodium-rich dachiardite as acicular crystals with mordenite and analcime in a pocket of a silico-carbonatite sill at Francon quarry, St. Michel, Montreal island, Quebec.

Two occurrences in geothermal fields have been described in the U.S.A.: the first (Sheridan and Maisano 1976) is the Tertiary basin near Hassayampa, Arizona, with andesite overlain by a basaltic breccia; dachiardite occurs both in amygdales and in the matrix from the surface down to the lower part of the basin; the second (Bargar et al. 1981) is the hot spring and geyser area of Yellowstone National Park, where dachiardite occurs at low and great (not intermediate) depth; its morphology is similar to that of Alpe di Siusi.

It is worth noting that the Elba dachiardite is the only one to feature fine crystals which give only sharp X-ray diffractions; all other dachiardites are fibrous and probably all give diffuse diffractions with k odd (but this property has been experimentally ascertained only for the samples from Alpe di Siusi, Rhodopes, Tsugawa). Elba dachiardite is unique also for its high Cs_2O content, which could be related to the perfection of the crystals.

5.3 Epistilbite

$Ca_3(Al_6Si_{18}O_{48}) \cdot 16H_2O$
C2/m
$a = 9.10$
$b = 17.77$
$c = 10.22 (\text{Å})$
$\beta = 124°36'$

5.3.1 History and Nomenclature

Rose (1826) proposed the name epistilbite for a zeolite found in Iceland and the Faröe Islands, so there are two type localities for this mineral. Other obsolete synonyms for epistilbite are monophane, parastilbite, reissite and orizite (see Galli and Rinaldi 1974). Epistilbite was thus named because of its similarity with stilbite (= desmine, but epidesmine = stellerite has nothing to do with epistilbite).

5.3.2 Crystallography

A structural model for the framework was first described by Kerr (1964) who found that the calculated X-ray intensities coincided well with the powder data. Merlino (1965) checked the validity of the model with single crystal photographic data obtaining a residual of 0.23. Perrotta (1967) refined the structure in C2/m down to a weighted R = 0.16. There is some Al-enrichment in the tetrahedra of the 4-rings joining the sheets which are parallel to (010). The only cation site lies on the mirror plane, coordinating three oxygens of the above mentioned 4-rings, plus six water molecules (type III of Chap. 0.4).

Slaughter and Kane (1969) refined the structure of an epistilbite, which was considered as disordered because of streaking in the single crystal photographs and the presence of sausage-like zones of high electron density in the channels. They adopted different axes, with transformation matrix 100/010/101, and also a different space group, C2. Their framework is the same, but cation and water sites are a little different.

The morphology (see Fig. 5.3C) is always characterized by (100) twins on c-elongated prismatic crystals, with the forms {110}, {001}, {010}, sometimes also with {011}; less common is {101}. Twins on {110} are also possible.

5.3.3 Chemistry and Synthesis

The chemistry was studied by Galli and Rinaldi (1974), who found a small variability around the mean formula

$$Ca_{2.5}Na_{0.8}K_{0.2}(Al_6Si_{18}O_{48}) \cdot 16 H_2O$$

which is very near to the schematic one of the title page. The composition of the framework is nearly constant, whereas the ratio $Ca:(Na+K)$ ranges from 9 to 2 (see Table 5.3I). The lattice constants too are nearly unchanging.

The hydrothermal synthesis was performed by Barrer and Denny (1961), in the system $CaO - Al_2O_3 - nSiO_2$ using powdered SiO_2 glass; the best results were obtained with $n = 7$ at 250°C. Similar results were obtained by Lo (1981). Wirsching (1981) synthesized the zeolite with a hydrothermal experiment simulating an "open system", viz. with frequent renewal of the solution $(0.1 N$ $CaCl_2)$ reacting with a rhyolitic glass; no epistilbite was obtained using a basaltic glass, or nepheline, or oligoclase.

5.3.4 Optical and Other Physical Properties

Optical data are scanty; they should be rather constant with $\alpha = 1.500 - 1.505$, $\beta = 1.510 - 1.515$, $\gamma = 1.512 - 1.519$, optical sign negative, 2V large $\beta \rightarrow b$, $\hat{c\gamma} \sim 10°$ in the obtuse angle. The mineral is transparent, usually colourless but sometimes reddish, with vitreous lustre.

Perfect (010) cleavage; irregular fracture; Mohs-hardness = 4.

Epistilbite is piezo-electric (Bond 1943; Ventriglia 1953) in spite of its centric structure.

5.3.5 Thermal and Other Physico-Chemical Properties

Figure 5.3D represents the thermal curves (DTA, TG, DTG) of three samples of Table 5.3I. Although their chemical composition is very similar the curves are different. In the DTG's, there is always a first peak at nearly 100°C, and so there is a first peak at nearly 130°C in the DTA's. Between 200°C and 400°C there are three peaks at 250°C, 300°C, 345°C in the DTG's, which can be distinguished in the third DTA only, because the first two show one broad peak instead. Galli and Rinaldi (1974) attribute the different dehydration of the samples to possible stacking faults.

Fig. 5.3A. Epistilbite from Pula, Sardinia, Italy (×10)

Fig. 5.3B. Epistilbite from Pula, Sardinia, Italy (SEM Philips, ×60)

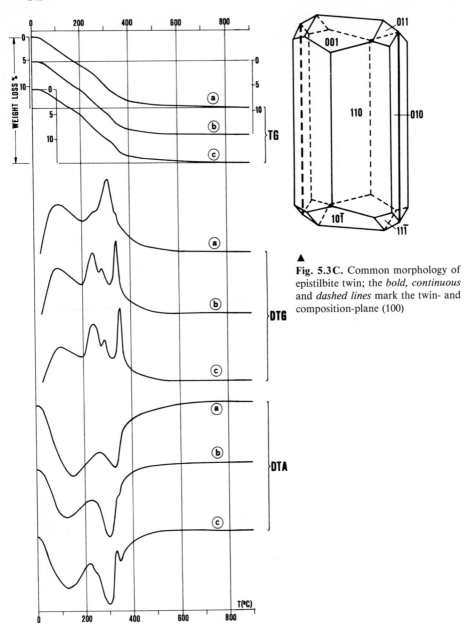

Fig. 5.3 C. Common morphology of epistilbite twin; the *bold, continuous* and *dashed lines* mark the twin- and composition-plane (100)

Fig. 5.3 D. Thermal curves of epistilbite from **a** Elba, Italy, **b** Berufjord, Iceland, and **c** Yugawara, Japan; in air, heating rate 20°C/min for TG (and DTG), 10°C/min for DTA (by the authors)

Table 5.3I. Chemical analyses and formulae on
the basis of 48 oxygens and lattice constants in Å

	1	2	3	4
SiO_2	61.9	60.5	57.4	59.7
Al_2O_3	15.5	16.9	17.7	16.3
CaO	6.8	7.2	8.4	8.1
Na_2O	1.3	1.9	0.6	0.5
K_2O	0.8	0.3	0.1	
H_2O	14.5	15.2	15.3	16.0
Total	100.8	102.1	99.5	100.6
Si	18.6	18.1	17.7	18.2
Al	5.5	6.0	6.4	5.9
Ca	2.2	2.3	2.8	2.7
Na	0.7	1.1	0.4	0.3
K	0.3	0.1	0.1	
H_2O	14.5	15.2	15.7	16.3
E%	+1.8	+2.1	+7.3	+5.2
a	9.088	9.088	9.096	9.101
b	17.741	17.756	17.761	17.766
c	10.225	10.236	10.236	10.242
β	124°41'	124°36'	124°35'	124°37'
D	2.22	2.23	2.25	2.25

1. Elba, Italy (Galli and Rinaldi
1974).
2. Berufjord, Iceland (do) with 0.1%
BaO.
3. Yugawara, Japan (do).
4. Kumoni, Japan (do).

5.3.6 Occurrences and Genesis

Epistilbite occurs always in vugs of massive rocks, and hence probably has
hydrothermal genesis. Galli and Rinaldi (1974) give a list of 39 occurrences
which is nearly complete.

Occurrences in basic volcanic rocks are:

Iceland: Berufjord, after Hintze (1897) the most probable location of the occurrence is near the observatory Djupivogur, at the entrance of Berufjord
(see also Betz 1981).
Faröe Islands: (Hintze 1897; Betz 1981).
Scotland: Skye (Dana 1914).

England: Hartlepool, Durham Co., in the cobblestones of a road (Hintze 1897).

Germany: Finkenhubel, Silesia (Hintze 1897).

Hungary: Nadap, Com. Fejer (Erdélyi 1940) at the contact of granite and andesite, with heulandite, chabazite, stilbite; Satoros, Com. Nograd (Erdélyi 1942) in an amphibole andesite with chabazite, heulandite, laumontite, stilbite.

Italy: Capo Santa Vittoria, Pula, Sardinia (Pongiluppi 1974) with chabazite, heulandite, stilbite, laumontite.

Greece: Thera, Santorini (Hintze 1897)

U.S.S.R.: Azerbaydzhan, in zeolitized andesite and in amygdales of albitized rocks, with mesolite (Amirov 1979).

India: Poona (Dana 1914) and Igatpuri, Bombay (Hintze 1897).

Japan: Kuroiwa, Niigata Pref. (Takeshita et al. 1975) in an andesite with scolecite and laumontite.

Canada: Margaretville, Annapolis Co., Nova Scotia (Dana 1914) with stilbite.

U.S.A.: Bergen Hill, New Jersey, with stilbite (Dana 1914); Lanakai Hills, Hawaii (Dunham 1933) with laumontite, heulandite, mordenite.

Argentina: Cerro China Muerta, with wairakite and levyne (Cortelezzi 1973).

Occurrences in rocks other than volcanic are:

Switzerland: Giebelsbach, Fiesch, Valais, with heulandite in gneiss (Hintze 1897).

Italy: as late hydrothermal deposition in aplitic-pegmatitic veins in granodiorite with many other zeolites (see dachiardite) at Fonte del Prete, Elba (Merlino 1972); in a calc-silicate rock with scolecite, laumontite, heulandite at Novate Mezzola, Sondrio (Cerio 1980).

U.S.A.: Baylis Quarry, Bedford, New York (Pough 1936) as late deposition in a pegmatite, on bertrandite which in turn lies on beryl.

5.4 Ferrierite

$(Na, K) Mg_2 Ca_{0.5} (Al_6 Si_{30} O_{72}) \cdot 20 H_2O$
Immm
$a = 19.18$
$b = 14.14$
$c = 7.50 (\text{Å})$

5.4.1 History and Nomenclature

Graham (1918) proposed the name for a mineral found at Kamloops Lake, British Columbia, Canada, in honour of W. F. Ferrier, who discovered and collected the first samples.

Fig. 5.4 A. Ferrierite from Weitendorf, Austria (SEM Philips, ×120)

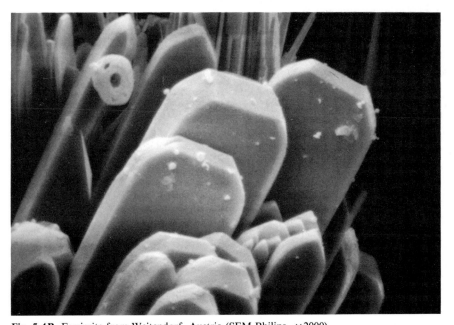

Fig. 5.4 B. Ferrierite from Weitendorf, Austria (SEM Philips, ×2000)

Fig. 5.4C. Ferrierite from Markleeville, California (SEM Philips, ×240)

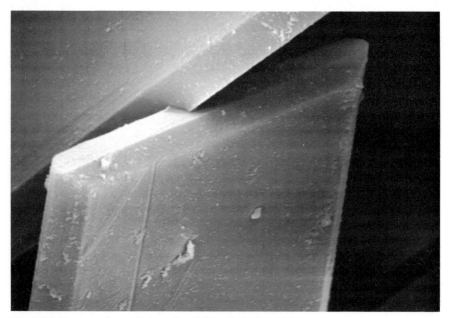

Fig. 5.4D. Ferrierite from Markleeville, California (SEM Philips, ×2000)

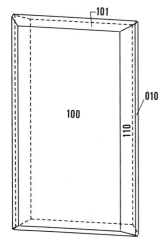

Fig. 5.4E. Common morphology of ferrierite

Fig. 5.4F. Sedimentary ferrierite from Lovelock, Nevada (SEM Philips, ×500)

5.4.2 Crystallography

Vaughn (1966) determined and refined the structure (see Chap. 0.5). There is a slight Al-enrichment in the tetrahedra, M(1) and M(2), of the double handle which form a six-ring between the hexagonal sheets parallel to (100). Two Mg are at the origin of the unit cell, in a twofold position between two of the above mentioned 6-rings, where Al is concentrated. Mg is surrounded by

6 H$_2$O only, to form a rather regular octahedron. Sodium was not located with certainty, but is probably placed in the main channel surrounded by 10-rings and parallel to c.

The morphology is characterized by (100)-lamellae. Graham (1918) found the forms {010} and {101} too. [001] prismatic needles have also been observed.

5.4.3 Chemistry and Synthesis

The chemical formula is not always near to the schematic one on the title page. Characteristic features are a rather constant Al/Si ratio, near to 1/5 (80% of the tetrahedra occupied by Si), and a frequent high content of Mg, with lesser amount of Ca, Na, K. This magnesian character is not general (see Table 5.4 I), some samples, for instance no. 6, are Mg-poor and alkali-rich. Only a few other zeolites may be Mg-rich: faujasite, offretite, mazzite and, to a lesser extent, erionite. More chemical analyses may be found in Wise and Tschernich (1976a).

The synthesis of ferrierite has been performed by Barrer and Marshall (1964a, 1965) from aqueous gels of composition SrO · Al$_2$O$_3$ · nSiO$_2$ at 340°C – 400°C with 7 $\leqslant$ n $\leqslant$ 9. Hawkins (1967a) did the synthesis not only from a strontian gel (n = 7), but also from a calcian one, again with n = 7, nearly at 350°C. Cormier and Sand (1976) were able to synthesize the zeolite from alkaline batches 3(Na$_2$O + K$_2$O) · Al$_2$O$_3$ · 13 SiO$_2$ plus some CO$_2$ at ~300°C; ferrierite crystallized in a certain time interval (1 to 2 days) then disappeared in favour of feldspar or mordenite. A synthesis from sodian batches was performed also by Malashevich et al. (1981). Nanne et al. (1980) synthesized a ferrierite-type phase heating at 100°C – 200°C an aqueous mixture of piperidine and/or alkyl-substituted piperidine (NR), Na-silicate, Al-sulphate, silica in the ratios mNR · nAl$_2$O$_3$ · SiO$_2$ · hH$_2$O with 0.05 < m < 1.2, n 0.07, h < 500 with a yield up to 100%.

5.4.4 Optical and Other Physical Properties

The mineral is biaxial positive with $\gamma \to c$, $\beta \to b$ and $\alpha \to a$, and 2V ~ 50°. Refractive indices for five natural samples are given in Table 5.4 II; the lowest values were measured in the alkali-rich samples. Similar values ($\alpha = 1.473$, $\gamma = 1.488$) are given also for the synthetic Sr-ferrierite by Barrer and Marshall (1965). Colour may be absent (Hokkaido) or reddish (Kamloops Lake) or definitely red (Vicenza). Crystals are sometime transparent or more frequently milky or cloudy.

Cleavage (100) perfect. Lustre bright vitreous, but pearly on (100). Mohs-hardness = 3.5. Streak: white.

5.4.5 Thermal and Other Physico-Chemical Properties

Data on the thermal behaviour are scanty. Kirov and Filizova (1966) state that the DTA curve is characterized by three broad endothermic peaks at 90°C,

Table 5.4I. Chemical analyses and formulae on the basis of 72 oxygens and lattice constants in Å

	1	2	3	4	5	6
SiO_2	56.80	66.33	62.62	70.14	71.21	71.50
Al_2O_3	12.71	13.18	9.94	11.45	9.84	10.18
Fe_2O_3	3.29	0.27	0.58	0.13	0.05	0.01
MgO	4.12	3.41	2.61	2.56	1.70	0.59
CaO	5.52	1.39	5.78	0.15		0.11
SrO		0.34	0.18	0.18		0.13
BaO		0.41		0.38		0.13
Na_2O	0.27	0.50	0.13	1.86	1.59	3.64
K_2O	0.82	1.17	1.13	0.63	2.85	1.73
H_2O	14.32	13.00	13.22	13.05	12.88	11.98
Total	100.79	100.00	100.16	100.53	100.12	100.00
Si	27.50	29.05	30.22	30.22	30.93	30.86
Al	7.25	6.80	5.65	5.81	5.04	5.15
Fe	1.20	0.09	0.21	0.04	0.02	
Mg	2.98	2.23	1.87	1.64	1.10	0.38
Ca	0.99	0.65	0.41	0.07		0.05
Sr		0.09	0.05	0.04		0.03
Ba		0.07		0.07		0.02
Na	0.25	0.43	0.12	1.55	1.34	3.06
K	0.51	0.65	0.70	0.35	1.58	0.97
H_2O	23.12	18.99	21.28	21.01	18.66	17.25
E%	+2.9	-3.8	+5.8	+5.6	-1.3	+3.0
a	19.227	19.222	19.187	19.164	19.18	18.90
b	14.146	14.147	14.161	14.151	14.14	14.141
c	7.503	7.54	7.498	7.496	7.51	7.487
D	2.21				2.06	

1. Vicenza, Italy (Alietti et al. 1967), with TiO_2 0.10%, CO_2 2.84%; formula calculated after deduction of 6.46% of $CaCO_3$; lattice constants by Passaglia (personal communication).
2. Monte Lake, British Columbia, Canada (Wise and Tschernich 1976a); H_2O by difference to 100.
3. Weitendorf, Styria, Austria (Passaglia 1978b), with CO_2 3.91%; formula calculated after deduction of 8.89% of $CaCO_3$.
4. Kamloops Lake, British Columbia, Canada (Wise and Tschernich 1976a); H_2O from Graham (1918).
5. Itomuka Mine, Hokkaido, Japan (Yajima et al. 1971); lattice constants deduced from their powder data.
6. Altoona, Washington, U.S.A. (no.1 of Wise and Tschernich 1976a); H_2O by difference to 100.

Table 5.4 II. Refractive indices

	1	2	4	5	6
α	1.488		1.479	1.483	1.473
β	1.489	1.482	1.480	1.484	1.474
γ	1.491	1.485	1.483	1.486	1.477

Numbers refer to the same smples of Table 5.4 I.

220°C and 380°C and that cordierite is the only phase present after heating at 1100°C. Yajima et al. (1971) published DTG and DTA curves, which however are quite flat because the heating rate (5°C/min) is too low, so that only one very broad peak at ~280°C is visible; their TG shows that dehydration is complete at 400°C. Figure 5.4G shows new thermal data obtained by the present authors, which besides the first two peaks at ~80°C and ~240°C, features a shoulder at ~160°C, whereas the third peak of Kirov and Filizova at 380°C is absent; a final small water loss at 680°C is also present.

Hawkins (1967b) tested the Ca – Sr exchange capacity for ferrierite synthesized from Ca and Sr batches; the value is nearly the same (0.6 meq/g) for both phases, but they strongly differ in the separation factor $\alpha^{Sr}_{Ca} = \dfrac{Sr_z Ca_s}{Ca_z Sr_s}$

where Sr_z is the equivalent concentration in the zeolite and Sr_s is the eq. concentration in the solution, and so on. This factor gives, for instance, the ratio Sr:Ca in the zeolite when the same ratio is unity in the solution; now this factor is 2.6 in the ferrierite synthesized in the Sr medium, and only 1.5 in the ferrierite synthesized in the Ca medium: the selectivity for Sr is high only if the synthesis was done in the presence of the same element.

5.4.6 Occurrences and Genesis

There are few sedimentary occurrences. At Lovelock, Nevada, pyroclastics are altered to ferrierite and mordenite (Regis 1970). At Tadami-Machi, Fukushima Pref., Japan (Hayakawa and Suzuki 1969) and at Wayaomono, Iwate Pref., Japan (Utada et al. 1974) ferrierite occurs as alteration (low-hydrothermal to diagenetic) of tuffs, in the clinoptilolite-mordenite zone.

Most known occurrences are in vugs of massive rocks and probably have hydrothermal genesis.

Graham (1918) described a decomposed olivine-basalt occurring near the north shore of Kamloops Lake, British Columbia; the mineral, associated with calcite, is embedded in chalcedony veins cutting this rock.

Baríc (1965) found ferrierite as a thin layer around calcite crystals in a spilitic basalt at Gotalovec, Croatia. Similar (Kirov and Filizova 1966) is the occurrence at Chernichino, Bulgaria in Tertiary volcanics, where spherical aggregates of the mineral are associated with chalcedony; this sample is calcium-rich and alkali-poor. Alietti et al. (1967) found the sample richest in magnesium at Albero Bassi, Vicenza, Italy, in a green, weathered porphyrite,

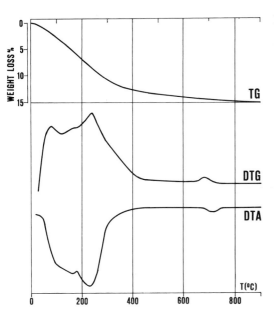

Fig. 5.4G. Thermal curves of ferrierite from Albero Bassi, Italy; in air, heating rate 20°C/min (by the authors)

cut by several fractures filled with calcite; the red lamellar mineral occurs at the border of these calcite veins.

Wise et al. (1969) found it as tiny colourless needles in cavities of an andesite breccia at Agoura, California, with clinoptilolite; both crystals are grown on a cavity lining which may be montmorillonite + chalcedony, or calcite. A similar occurrence exists at Sonora Pass, California.

Yajima et al. (1971) found it as interstitial filling in a propylite and as colourless spherical aggregates of radiating blades in druses, at the Itomuka Mine, Hokkaido, Japan. Two other Japanese occurrences are at Futatsumori, Akita Pref., and at Kobayashi, Fukushima Pref. (Nambu 1970, p. 53).

Small colourless needles of ferrierite were found in a shoshonite at Weitendorf, Styria, Austria, by Zirkl (1973).

Klitchenko and Suprychev (1974) described radial aggregates in monchiquites and camptonites in the Donets Basin, U.S.S.R.

Wise and Tschernich (1976a) gave the first complete description of the following occurrences: at Monte Lake, British Columbia, it occurs as thin blades, colourless to salmon-coloured, lining vesicles in a fine grained, greenish black basalt; at Pinaus Lake, British Columbia, as red to white blades which line cavities filled with calcite; at Francois Lake, British Columbia, as minute needles on collinsite and carbonate-fluorapatite; at Silver Mountain, southwest of Markleeville, Alpine Co., California, as very thin white to orange blades in an altered andesite. Wise and Tschernich (1978a) mention the presence of ferrierite in paragenesis with dachiardite and mordenite in a basalt breccia at Altoona, Washington.

England and Ostwald (1978) described spherical colourless clusters in the amygdales of a latite at Unanderra, New South Wales, Australia, with calcite, chalcedony, laumontite, heulandite. Birch and Morvell (1978) described flesh-coloured or white crusts, spherical aggregates or tufts in a weathered basalt with calcite and heulandite at Phillip Island, Victoria, Australia.

Schorr (1980) found fine radial aggregates in agate or druses in a pyroxene dacite at Reichweiler, Germany.

5.4.7 Uses and Applications

The sedimentary ferrierite of Nevada was shown by Chang et al. (1978) to be a valuable Claus tail-gas catalyst, namely a material capable of competing with alumina as a catalyst for the reaction

$$2H_2S + SO_2 \rightarrow 3S + 2H_2O .$$

H-ferrierites and mixed (Na, H)-ferrierites are acid and thermal resistant and surpass from this point of view NaX (synthetic Na-faujasite) which should have the same activity.

5.5 Bikitaite

$Li_2(Al_2Si_4O_{12}) \cdot 2H_2O$
$P2_1$
$a = 8.61$
$b = 4.96$
$c = 7.60\,(\text{Å})$
$\beta = 114°27'$

5.5.1 History and Nomenclature

Hurlbut (1957) proposed the name for a mineral from Bikita, Zimbabwe (Southern Rhodesia, at the time of the discovery). It is not certain that biki-taite is a typical zeolite, because its ion exchange capacity is uncertain, even though it dehydrates and re-hydrates over a wide temperature range.

5.5.2 Crystallography

The topology of the bikitaite framework (see Chap. 0.2.5) was described by Appleman (1960) but he was unable to finish the refinement because of the poor quality of his experimental data. A full refinement was published by Kocman et al. (1974).

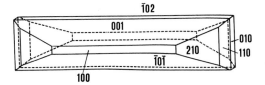

Fig. 5.5 A. Common morphology of bikitaite

The hexagonal sheets are parallel to (001) and their tetrahedra contain 50% Al; the pyroxene-chains, midway between the hexagonal sheets, are occupied by Si only. Li cations are placed on the walls of the main channels parallel to b, so as to be near to the hexagonal sheets, where Al is present. Each Li coordinates three framework oxygens and one water molecule (type III of Chap. 0.4).

The morphology (Fig. 5.5 A, Hurlbut 1958; Leavens et al. 1968) is characterized by a prismatic b-elongation; in this zone, common forms are {001} and {$\bar{1}$01} and less frequent {100}, {$\bar{1}$02}, {$\bar{2}$01}, and the termination is given by {010} {110} {210} {011} {121}.

5.5.3 Chemistry and Synthesis

Table 5.5 I contains the analyses of the only two known samples of bikitaite: both correspond very closely to the stoichiometric formula in the title page. Some Na and K, given by Hurlbut (1957) on the basis of wet chemical methods, were not found by Kocman et al. (1974).

A hydrothermal synthesis was performed by Drysdale (1971) from gels ranging from $Li_2O \cdot Al_2O_3 \cdot 4SiO_2$ to $2Li_2O \cdot Al_2O_3 \cdot 4SiO_2$, at 2 Kbar and 300°C, in the presence of CO_2.

5.5.4 Optical and Other Physical Properties

The zeolite is optically biaxial negative (see Table 5.5 II), with $\gamma \rightarrow b$ and $\widehat{\alpha c} = 28°$ in the obtuse angle.

Cleavages: (001) good, and (100) poor. Mohs-hardness 6.

5.5.5 Thermal and Other Physico-Chemical Properties

Figure 5.5 B shows the DTA and TG curves of Kocman et al. (1974); a curve of the static dehydration is given by Phinney and Stewart (1961); taking due account of the obvious difference between a static and a dynamic measurement, the two TG-curves are similar, showing a continuous water loss between 150°C and 400°C, but the static curve begins and ends more sharply. The DTA shows a sharp peak at 260°C where the water loss is at a maximum. Rehydration is complete up to 240°C, nearly complete after heating between 240°C and 360°C, scarce between 360°C and 600°C. Rapid heating in air above 750°C decomposes the zeolite to form β-spodumene or β-eucryptite

Table 5.5 I. Chemical compositions and formulae on the basis of 12 oxygens and lattice constants in Å

	1	2
SiO_2	58.7	58.63
Al_2O_3	25.1	25.09
Li_2O	7.27	7.71
H_2O	8.98	8.62
Total	100.05	100.05
Si	3.99	3.97
Al	2.01	2.00
Li	2.00	2.10
H_2O	2.04	1.95
E%	-14.6	+5.0
a	8.613	8.614
b	4.962	4.957
c	7.600	7.603
β	114°27'	114°19'
D	2.28	2.30

1. Bikita (Kocman et al., 1974).
2. King's Mt. (Leavens et al., 1968).

Table 5.5 II. Optical data

	1	2
	1.510	1.509
	1.521	1.520
	1.523	1.522
2V	-45°	-45°
$c\widehat{\ }\alpha$	28°	29°
in the obtuse angle		
	$\gamma \rightarrow \underline{b}$	

1. Bikita (Leavens et al., 1968).
2. King's Mt. (do).

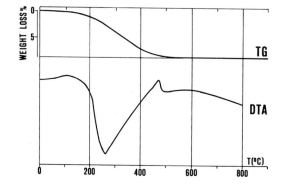

Fig. 5.5 B. Thermal curves of bikitaite from Bikita, Zimbabwe; in air, heating rate 5°C/min (Kocman et al. 1974)

(Phinney and Stewart 1961). After the same authors, bikitaite does not feature any ion exchange with ammonium, and 5% of the Li may be replaced by H_3O^+ if leached with 1 N acetic acid; no other ion exchange was tried.

Bikitaite is decomposed (Phinney and Stewart 1961) at 400°C under hydrothermal conditions from 0.5 to 4 Kbars, following the equation

$$3\,LiAlSi_2O_6 \cdot H_2O \rightarrow 2\,LiAlSiO_4 + LiAlSi_4O_{10} + 3\,H_2O$$
bikitaite   eucryptite   petalite

5.5.6 Occurrences and Genesis

Only two occurrences are known, both in pegmatites.

Bikita, 40 miles east of Fort Victoria, southern Rhodesia (now Zimbabwe): a lithium-rich pegmatite contains spodumene, amblygonite, petalite, lepidolite, eucryptite: bikitaite is the latest mineral and fills small fractures within eucryptite and interstices between quartz and eucryptite (Hurlbut 1957).

King's Mountain, North Carolina: a uniform pegmatite, intruded in amphibolite and mica-schist, is composed of microcline, quartz, spodumene, muscovite, and contains quartz, albite, rhodocroisite, several phosphates, eucryptite and bikitaite in the fractures (Leavens et al. 1968).

6 Zeolites of the Heulandite Group

6.1 Heulandite, Clinoptilolite

Heulandite	Clinoptilolite
$(Na, K)Ca_4(Al_9Si_{27}O_{72}) \cdot 24 H_2O$	$(Na, K)_6(Al_6Si_{30}O_{72}) \cdot 20 H_2O$
C2/m	C2/m
$a = 17.70$	$a = 17.62$
$b = 17.94 (\text{Å})$	$b = 17.91$
$c = 7.42 (\text{Å})$	$c = 7.39 (\text{Å})$
$\beta = 116°24'$	$\beta = 116°16'$
or quasi-orthogonal cell I2/m	
$a_0 = 15.87$	
$b_0 = 17.94$	
$c_0 = 7.42 (\text{Å})$	
$\beta_0 = 91°38'$	
$\vec{a}_0 = \vec{a} + \vec{c} \quad \vec{b}_0 = \vec{b} \quad \vec{c}_0 = \vec{c}$	
or second monoclinic cell C2/m	
$a' = 17.33$	
$b' = 17.94$	
$c' = 7.42 (\text{Å})$	
$\beta' = 113°43'$	
$\vec{a}' = \vec{a} + 2\vec{c} \quad \vec{b}' = \vec{b} \quad \vec{c}' = -\vec{c}$	

6.1.1 History and Nomenclature

The name heulandite was proposed by Brooke (1822) for minerals which were previously known but considered the same species as stilbite; he attributed the name (in honour of H. Heuland, an English mineralogical collector) to those samples which were distinctly monoclinic (as a matter of fact, stilbite too is monoclinic, but nearer to orthorhombicity than heulandite is). There is no type locality. It may be useful to remember that for a long time many authors, especially the German-speaking ones, continued to call "stilbite" the minerals we now name heulandites, using the name "desmine" for our stilbite.

Clinoptilolite is a name proposed by Schaller (1932b) for a mineral from the Hoodoo Mts., Wyoming, considered as "ptilolite" (= mordenite) by Pirsson (1890); the introduction of the new name was justified by the mono-clinicity of the mineral, which even Pirsson gave evidence for, in contrast to the orthorhombicity of mordenite; Hey and Bannister (1934) emphasized the strict similarity of clinoptilolite with heulandite. Mason and Sand (1960) and Mumpton (1960) contemporaneously and in the same issue of a magazine, proposed two redefinitions of clinoptilolite: after Mason and Sand, clinoptilo-lite is a zeolite of the heulandite group with $(Na + K) > Ca$; after Mumpton, each zeolite of the heulandite group must be named clinoptilolite if its crystal structure survives an overnight heating at $450\,°C$, otherwise it is a heulandite. Although the first definition conforms to the praxis of mineralogical nomenclature and the second not, nowadays most authors accept the second one. Alietti (1972) and Boles (1972) investigated the relationships between chemical composition and thermal behaviour, so clarifying the differences of the two definitions. Boles (1972) also proposed naming these zeolites clinoptilolite if $Si/Al > 4$, heulandite if $Si/Al < 4$.

The nomenclature of this genus is very uncertain indeed; adhering to common rules of mineralogical nomenclature, the present authors adopt Mason and Sand's nomenclature, naming heulandite those minerals with $(Ca + Sr + Ba) > (Na + K)$, and clinoptilolite the others, also in view of the practical impossibility of finding chemical parameters which give limits in accordance with the three possible responses to heating (see Chap. 6.1.5).

6.1.2 Crystallography

First of all, it is useful to emphasize the possibility of choosing between the three unit cells of the title page; the second one is near to orthorhombic but only from the metrical point of view, because the highest possible (topological) sym-metry for this structure (see Chap. 0.2.6) is monoclinic C2/m only; this situa-tion is remarkably different from stilbite, whose pseudo-orthorhombic cell cor-responds to a truly orthorhombic topological symmetry of the framework.

The structure of heulandite was resolved and refined in space group Cm by Merkle and Slaughter (1968) and later on refined also by Alberti (1972) in space groups C2/m, Cm and C2 and, with neutron diffraction, by Bartl (1973) in space group I2/m. Alberti (1972), who also published a table with Merkle and Slaughter's coordinates referred to a new origin on a pseudocentre of symmetry, showed that the evidence in favour of acentricity was rather feeble, so that it is quite reasonable to assume the centric C2/m symmetry until a con-vincing proof of the contrary is given.

There are two main channels parallel to c (Fig. 6.1 I), A and B, formed by ten tetrahedra-rings; there are also channels C, formed by eight member rings, parallel to a or to [102] (codes by Koyama and Takéuchi 1977).

There is some Si, Al-order in the framework: T2, a tetrahedron with one oxygen-vertex on the mirror plane, contains 50% of Al, hence 4 Al per unit

Fig. 6.1 A. Heulandite from Val di Fassa, Italy (×2)

Fig. 6.1 B. Clinoptilolite from Agoura, California (SEM Philips, ×120)

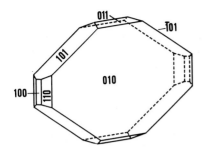

Fig. 6.1 C. Common morphology of heulandite with the indices of the quasi-orthogonal cell

cell; the other tetrahedra contain lesser fractions of Al, a little more in T3 and T5, nearly 0 in T1 and T4.

There are two cation sites (later on coded C1 and C2 by Alberti 1975a, and CS1 CS2 by Alberti and Vezzalini 1983a) each in one of the two main channels parallel to c and both coordinate water molecules and framework oxygens on one side only (type III of Chap. 0.4). The presence of these two cation sites only in heulandites was confirmed by Bresciani-Pahor et al. (1980) and by Alberti and Vezzalini (1983a); the first authors give evidence for CS2 being filled with Ca only, whereas all monovalent cations are present, with all remaining divalent cations, in CS1.

For clinoptilolite, four refinements are available: Alberti (1975a) studied samples from Agoura, California, and Alpe di Siusi (= Seiseralp), Italy; Koyama and Takéuchi (1977) those from Kuruma Pass, Japan, and Agoura, California. Both papers give the highest Al-fraction (~30% for Agoura, ~40% for the other samples) in T2, as in heulandite. Both papers accept as major cation sites the same two of heulandite, but Alberti (1975a) found a third site coded C3 at the origin of the unit cell, at the crossing of the two channel systems, with a 22% occupancy in Agoura crystals, 10% in Alpe di Siusi crystals. Koyama and Takéuchi (1977) neglect this site, although they find there some electron density, especially in their Agoura sample, and introduce two new cation sites: M3, which is nearly superposed on W1, a water site found also by Alberti (1975a); the M3 – W1 distance is 0.7 Å only, so that simultaneous occupancy of the two sites is impossible; M3 is in the minor channel C and coordinates water molecules and framework oxygens on two sides (type II of Chap. 0.4); the site M4 is on a symmetry centre in the main channel A, and is completely surrounded by water only (type IV of Chap. 0.4).

Galli et al. (1983) refined a K-exchanged heulandite; the large number of monovalent cations was accommodated not only in sites C2 and C3 of Alberti (1975a) and M3 + W1 of Koyama and Takéuchi (1977) but also in site $H_2O(6)$ or W6 previously attributed to water molecules; by contrast, site C1 is here filled with water. Refinements on heulandites partially or fully Ag-exchanged were published by Bresciani-Pahor et al. (1980, 1981).

The morphology of heulandite is somewhat confused due to the different crystallographic axes which may be chosen as reference. Hintze (1894) and

Fig. 6.1D. Sedimentary clinoptilolite from Hector, California (SEM Philips, ×3800)

Fig. 6.1E. Sedimentary clinoptilolite from Creede, Mineral Co., Colorado (SEM Philips, ×3800)

Dana (1914) use the nearly orthogonal I2/m reference, with a and c interchanged, and their c (our a) halved; Winchell and Winchell (1951) use the same I2/m reference interchanging a and c without halving anything (and so do Merkle and Slaughter 1968). Using the indices of the quasi-orthogonal I2/m cell, the crystals are always 010 lamellae, bounded by the forms $\{100\}$ $\{101\}$ $\{\bar{1}01\}$ $\{110\}$ $\{011\}$ (Fig. 6.1 C). (100) may be twin and contact plane, but twinning is not so frequent.

Clinoptilolite occurs in the large majority of cases as microcrystalline sedimentary masses; SEM pictures show that the crystals are always (010) lamellae; only few occurrences of larger crystals are known (see Chap. 6.1.6), and some of these again feature the (010) lamellae but evidence on the bounding faces is scanty, with one exception, the holotype clinoptilolite from Hoodoo Mts., which shows the same forms as heulandite.

6.1.3 Chemistry and Synthesis

There is a continuous solid solution series along the join between the two stoichiometric formulae given in the title page, and also on both sites of the join, so that altogether this zeolite genus has a very wide chemical range. Large collections of chemical analyses may be found in Alietti (1972), Boles (1972), Alietti et al. (1977), Stonecipher (1978), some of these being listed in Tables 6.1 I and II. Careful examination of these analyses allows one to summarize the chemical range as follows: the Si ranges from 26.8 to 30.2 per unit cell (from 74% to 84% of the tetrahedra) and Al correspondingly from 9.2 to 5.8, Fe''', always considered as tetrahedral, is a little higher than in other zeolites (up to 1.5% of the tetrahedra); Ca ranges from 4 to 0 per unit cell, but also Sr and Mg may be as high as 1.25, whilst Ba is very low in most samples, but is 1.06 in one occurrence; Na and K both oscillate between 0 and 4.4 per unit cell. On the whole, there is a certain correlation between Si and (Na + K); clinoptilolites are richer in silicon than heulandites on the average, and only few clinoptilolites with Si < 28 have been found.

The first synthesis of a phase with the heulandite framework is due to Ames (1963), who treated hydrothermally at $250°C - 300°C$ gel mixtures of composition $Li_2O : Al_2O_3 : SiO_2 : H_2O = 1 : 1 : 8 : 8.5$ or nearly so, the time needed being 2 or 3 days; final pH in the solution at $25°C$ was $7.1 - 8.7$. Hawkins (1967a) obtained hydrothermally Ca-, Sr- and Ba-based heulandites (he called them "clinoptilolites" because of the similarity of their powder pattern with that of clinoptilolite, as he did not know that the powder pattern of heulandite is undistinguishable from the first), starting from mixtures $CaO \cdot Al_2O_3 \cdot nSiO_2$ with $5 \leqslant n \leqslant 7$ at $340°C - 380°C$, $SrO \cdot Al_2O_3 \cdot nSiO_2$ with $4 \leqslant n \leqslant 10$ at $300°C - 360°C$, $BaO \cdot Al_2O_3 \cdot nSiO_2$ with $6 \leqslant n \leqslant 8$ at $250°C - 300°C$; the synthesis was possible only in the presence of clinoptilolite seeds; without seeding, mordenite was obtained (but mordenite was present also in some successful runs with the seeds).

Table 6.11. Heulandites: chemical analyses and formulae on the basis of 72 oxygens and lattice constants in Å

	1	2	3	4	5	6	7	8	9
SiO_2	57.38	57.17	56.8	58.57	55.77	56.1	59.61	67.45	58.90
Al_2O_3	16.91	17.03	16.6	16.95	15.88	15.5	14.52	13.27	14.17
Fe_2O_3		0.04				0.1	0.25	0.31	0.71
MgO	0.01	0.02		0.03	0.08		0.60	1.03	1.39
CaO	7.00	7.13	5.8	3.78	3.04	4.4	5.60	3.86	3.77
SrO	1.55	0.19	2.0	7.89	4.44	3.3	0.12		0.79
BaO		0.31		0.77	2.44	5.6	0.18		0.69
Na_2O	0.10	1.39	1.6	0.45	1.54	0.7	1.40	0.26	0.76
K_2O	1.38	0.73	0.8	0.38	1.34	0.6	0.55	2.26	1.82
H_2O	16.73	16.71	15.75		15.52		17.30		17.00
Total	101.06	100.72	99.35		100.05		100.13		100.00
Si	26.80	26.70	26.84	26.94	26.97	27.01	27.88	29.24	27.83
Al	9.28	9.37	9.25	9.19	9.05	8.79	8.00	6.78	7.89
Fe		0.01				0.04	0.08	0.10	0.25
Mg				0.02	0.06		0.40	0.67	0.98
Ca	3.48	3.57	2.94	1.76	1.57	2.27	2.80	1.79	1.91
Sr	0.40	0.05	0.55	2.10	1.25	0.92	0.04		0.22
Ba		0.06		0.14	0.46	1.06	0.04		0.13
Na	0.08	1.26	1.46	0.40	1.44	0.65	1.28	0.23	0.70
K	0.88	0.43	0.48	0.22	0.83	0.37	0.32	1.25	1.09
H_2O	26.08	26.02	24.82		25.02		27.00		26.79
E%	+6.4	-3.5	+3.7	+6.1	+1.1	-7.2	-3.2	+7.5	+1.5

	1	2	3	4	5	6	7	8	9
a	17.73	17.718	17.725	17.655	17.722	17.717	17.691	17.689	17.72
b	17.82	17.897	17.864	17.877	17.856	17.986	17.973	18.044	17.91
c	7.43	7.428	7.427	7.396	7.458	7.424	7.407	7.408	7.43
β	116°20'	116°25'	116°24'	116°27'	116°22'	116°24'	116°37'	116°21'	116°19'
D	2.20			2.27	2.28				

1. Hydr., TB1, Giebelbach, Switzerland (Merkle and Slaughter, 1968)
2. Hydr., TB1, Faroer (Alberti, 1972)
3. Hydr., TB1, Cape Blomidon, Nova Scotia, Canada (no.1 of Boles, 1972)
4. Hydr., TB1, Campegli, Liguria, Italy (Lucchetti et al., 1982)
5. Hydr., TB1, Kozakov, Czekoslovakia (Cerny and Povondra, 1969)
6. Hydr., Blairmore Group, Alberta, Canada (RS'-SC-1 of Miller and Ghent, 1973)
7. Hydr., TB2, Val di Fassa, Italy (no.1 of Alietti, 1967, 1972)
8. Hydr., TB2, Taringatura Hills, New Zealand (no.11 of Boles, 1972)
9. Sed., TB1, Garbagna, Alessandria, Italy (Passaglia and Vezzalini, 1985)

Note: Hydr. = hydrothermal, Sed. = sedimentary,
TB=heat behaviour of type 1 or 2 after Alietti (1972)

Table 6.1II. Clinoptilolites: chemical analyses and formulae on the basis of 72 oxygens and lattice constants in Å

	10	11	12	13	14	15	16	17	18	19
SiO_2	54.58	65.17	61.14	61.77	64.70	72.40	69.93	62.29	64.00	66.40
Al_2O_3	15.86	13.38	14.52	13.26	12.43	13.17	11.89	13.81	12.50	11.17
Fe_2O_3	1.55	1.06		0.59	0.44	0.20	0.02	0.75		0.57
MgO	0.90	0.53	0.28	0.51	0.34	0.64	0.47	0.14	0.25	0.17
CaO	3.16	3.22	3.60	0.89	1.26	0.24	1.07	2.65	3.89	1.94
SrO	2.00			0.03				0.50		
BaO	0.60			0.10	0.53			0.33		
Na_2O	1.01	1.62	4.48	4.19	4.32	2.92	2.96	3.58	1.99	2.27
K_2O	3.90	2.82	0.94	4.04	2.28	4.93	3.47	1.68	1.81	3.58
H_2O	15.99	11.43	14.16	14.78	13.56			14.20	15.56	13.31
Total	99.55	99.39	99.12	100.57	99.86			99.93	100.00	99.41
Si	26.48	28.78	28.00	28.50	29.19	29.71	30.02	28.42	29.20	29.88
Al	9.08	6.96	7.84	7.21	6.61	6.37	6.01	7.43	6.72	5.93
Fe	0.56	0.		0.20	0.15	0.06	0.30	0.26		0.19
Mg	0.64	0.35	0.19	0.35	0.23	0.39		0.09	0.17	0.11
Ca	1.64	1.52	1.77	0.44	0.61	0.11	0.49	1.29	1.90	0.94
Sr	0.56			0.01				0.13		
Ba	0.12			0.02	0.09			0.06		
Na	0.96	1.39	3.98	3.75	3.78	2.32	2.45	3.17	1.76	1.98
K	2.40	1.60	0.55	2.38	1.31	2.58	1.88	0.98	1.05	2.06
H_2O	25.84	16.84	21.74	22.72	20.40			21.60	23.7	19.98
E%	+3.9	-8.6	-7.2	+4.4	-2.7	+8.9	+1.7	-5.1	-3.3	+0.3

a	17.498	17.683	17.670	17.69	17.627	17.67	17.67	17.637	17.660
b	17.816	18.025	17.982	17.92	17.955	17.89	17.89	18.024	17.963
c	7.529	7.408	7.404	7.40	7.399	7.408	7.41	7.399	7.400
β	116°04'	116°24'	116°24'	116°18'	116°17'	116°09'	116°16'	116°23'	116°28'

10. Hydr., TB1, Albero Bassi, Vicenza, Italy (Passaglia, 1972)
11. Hydr., TB2, Shizuma, Japan, with TiO$_2$ 0.16% (Minato and Utada, 1971; no.10 of Boles, 1972)
12. Hydr., TB2, Challis, Idaho (Ross and Shannon, 1924; no.6 of Boles, 1972)
13. Sed., TB3, Waitemata Group, New Zealand (Sameshima, 1978; plus 0.41% of TiO$_2$)
14. Sed., TB3, San Bernardino Co., California (no.1 of Sheppard and Gude, 1969a)
15. Sed., TB3, Deep sea sample (A of Boles and Wise, 1978)
16. Hydr., TB3, Agoura, California (no.2 of Wise et al., 1969)
17. Hydr., TB3, Alpe di Siusi, Italy (Alberti, 1975a)
18. Hydr., TB3, Kuruma Pass, Japan (Koyama and Takeuchi, 1977)
19. Hydr., holotype of Hoodoo Mts., Wyoming (Pirsson, 1890; Schaller, 1932b)

Note: Hydr.=hydrothermal; Sed.=sedimentary.
TB=heat behaviour of type 1, 2 or 3 after Alietti (1972)

Goto (1977) synthesized clinoptilolites hydrothermally starting from gel mixtures of composition $(Na_2O + K_2O): Al_2O_3 \cdot 7\,SiO_2$ $(Na_2O: K_2O = 1)$ at 200°C in 25 days, both if the added solution was pure water or 0.1 N HCl solution (the product with distilled water included some mordenite); if the runs were prolonged to 65 days, mixtures of mordenite and clinoptilolite were obtained in both cases; the final pH in the residual liquid at room temperature was 7.9; the synthesized clinoptilolite has a thermal behaviour of type 3 after Alietti (1972). Hawkins et al. (1978) and Hawkins (1981) described a synthesis from natural materials, namely a rhyolitic (Mg – Fe-poor, Na – K-rich) glass treated at 150°C for 15 days with a 2M solution of alkali-carbonates (Na : K = 1), mordenite being present after longer runs at higher T, phillipsite after shorter runs at lower T (the second quoted paper gives more complex information which cannot be summarized here). Similar results were obtained by Wirsching (1981), who operated with Na – Ca-solutions; if the natural material was not a rhyolitic glass, but a basaltic glass or nepheline or oligoclase, other zeolites were obtained.

6.1.4 Optical and Other Physical Properties

The optical data on heulandite are rather confused because of the different choices of the crystallographic axes made by the authors. However, the reference is unequivocal if the $c = c_0$ of the title page is choosen; this corresponds to a of Hintze (1897), Dana (1914), Winchell and Winchell (1951) and also to a_0 of Merkle and Slaughter (1968), Wise et al. (1969). The most reliable data on the optical orientation of heulandite are those of Nawaz (1980a), which are as follows: $\gamma \rightarrow b$, $\hat{c a} = 15°$ in the obtuse angle. Other data on the refractive indices, 2V and orientation are collected in Table 6.1 III; the data of Černý and Povondra (1969) correspond to those of Nawaz, but Merkle and Slaughter (1968) found $\beta \rightarrow b$, which is doubtful.

The data on the optical orientation in clinoptilolite are rather uncertain (Table 6.1 III nos. 15 and 18); probably the data of Pirsson (1890) are wrong, and those of Wise et al. (1969) are right if properly re-interpreted. As a matter of fact, these authors present in their Fig. 4 a morphology which is based on a sketch quoted with angles calculated from the unit cell data; by analogy with Fig. 1 of Nawaz (1980a), and because the face $(20\bar{1})$ was never found before in heulandites, we believe that their (101) is really the (001) with their axes, or the (100) with our quasi orthogonal axes, so that the orientation should be $\hat{c a} = $ small, $\beta \rightarrow b$. In other words the optical orientation in heulandites and clinoptilolites could be similar, with β and γ interchanged. The refractive indices of clinoptilolites are, on the average, smaller than those of heulandites.

Heulandites may be transparent colourless, but also reddish or brick red; (010) cleavage perfect; Mohs-hardness 3.5 to 4.

Clinoptilolites also may be transparent colourless (Hoodoo Mts.) or brick red (Alpe di Siusi); (010) cleavage perfect; Mohs-hardness probably 3.5.

Table 6.1 III. Optical data

	1	4	5	7	16	19
α	1.496	1.501	1.504	1.492	1.478	1.476
β	1.498	1.502	1.507	1.494	1.479	
γ	1.504	1.510	1.515	1.501	1.481	1.479
2V	$+35°$	$+30-70°$	$+10-40°$	$+62°$	$+32-48°$	negative
	$\beta \to \underline{b}(?)$	$\gamma \to \underline{b}$	$\gamma \to \underline{b}$		$\beta \to \underline{b}$	$\alpha \to \underline{b}$
	$\widehat{\alpha\ c} = 20°$	$\widehat{\alpha\ c} = 43°$			$\widehat{\alpha\ c} = small$	$\widehat{\gamma\ c} = 15°$
		all in the obtuse angle				

Note: numbers refer to samples of Tables 6.1 I and II.

Heulandite is piezoelectric (Ventriglia 1953), in spite of the good refinements performed in the centric space group.

6.1.5 Thermal and Other Physico-Chemical Properties

The thermal curves of a typical calcian heulandite are shown in Fig. 6.1 F; there is a first water loss in a broad interval from 50°C to 200°C corresponding to nearly 11 of the total 24 H_2O, whereas the second water loss of 8 H_2O corresponds to a sharp DTG peak at 280°C; the remaining 5 H_2O are lost gradually at higher T, up to 900°C. Figure 6.1 G shows the thermal curves of a clinoptilolite; there is only one continuous water loss from 50°C which also goes on at a low rate also over 500°C.

Rinne (1923) was the first to point out that heulandite, if *stark erhitzt*, undergoes a transition to "metaheulandite II" with a sharp 8% contraction of *b*. Slawson (1925) studied the same inversion and called the heated phase heulandite B and this name was retained also by Mumpton (1960), who established that heulandite contracts to heulandite B at 230° ± 10°C (if the heating is long enough); this phase B is stable for a long time when brought back to room conditions, but its lattice is destroyed by an overnight heating at 450°C, whereas clinoptilolite does not suffer any contraction, nor is the lattice destroyed below 750°C; indeed he redefined clinoptilolite as a mineral which is undestroyed and stable after the above-mentioned overnight heating at 450°C, leaving the name heulandite to all other zeolites of the same genus. Alietti (1972) and Boles (1972) studied many more zeolites and completed the scheme of the thermal behaviour of this genus which may be summarized as follows, also considering the data of Simonot-Grange et al. (1968) and Simonot-Grange and Cointot (1969):

Thermal behaviour of type 1: if heated up to 200°C, the zeolite loses up to 12 H_2O with a small contraction of the lattice and returns to the original cell

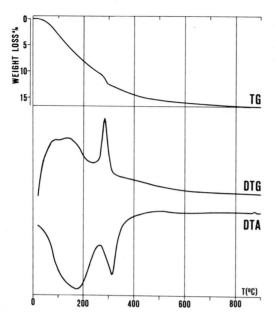

Fig. 6.1F. Thermal curves of heulandite from Mossyrock Dam, Washington; in air, heating rate 20°C/min (by the authors)

dimensions readsorbing its water when cooled to room conditions and it is called phase A in this temperature interval; increasing the temperature over 200°C, there is a sharp contraction (15% of the cell volume, 8% of b) at about 230°C, but definitely complete at 260°C: at this point, nearly $5\,H_2O$ remain in the structure (now called phase B) which, if brought back to room conditions, maintains its shorter dimensions and does not re-hydrate immediately; after months the crystals invert to heulandite I (for Intermediate), very close both in water content and cell dimensions to the original phase A; after heating up to a temperature in the range 200°C – 260°C, both A, B and I phases may be present in different ratios, depending on the heating history; d_{020} ranges from 8.97 to 9.05 Å for phase A, from 8.25 to 8.35 for phase B, from 8.73 to 8.87 for phase I; very prolonged heating at 450°C destroys the lattice.

Thermal behaviour of type 2: the crystals show the reversible dehydration with a corresponding very small contraction up to a temperature higher than 200°C, but at 280°C – 400°C there is a contraction of part of the sample, so that after cooling the three phases A, B and I are all present, even if the original sample is chemically quite homogeneous; the lattice resists without destruction up to 550°C and over.

Thermal behaviour of type 3: the sample undergoes continuous reversible dehydration with only a very small lattice contraction and the lattice is not destroyed by heating if not over 750°C.

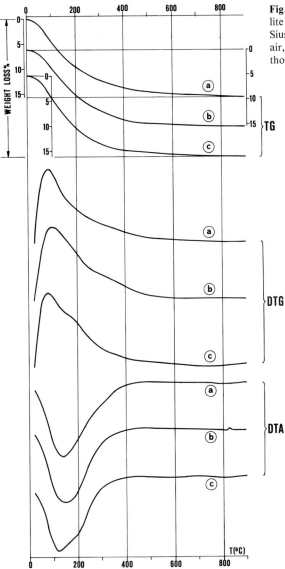

Fig. 6.1 G. Thermal curves of clinoptilolite from **a** Agoura, California, **b** Alpe di Siusi, Italy, and **c** Castle Creek, Idaho; in air, heating rate 20°C/min (by the authors)

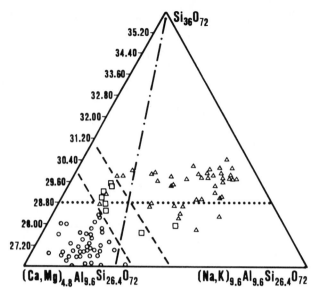

Fig. 6.1 H. Mole plot of heulandites and clinoptilolites (Alietti et al. 1977). Samples with thermal behaviour 1 are *circles*, thermal behaviour 2 *squares*, thermal behavior 3 *triangles*; the *dashed* lines give reasonable boundaries between the three kinds of samples; the *dashed-dotted* line is the boundary where $(Na+K) = (Ca+Mg)$, hence the boundary heulandite-clinoptilolite after the authors of this book; at the *dotted* line $Si:Al = 4$, which is the boundary between heulandite and clinoptilolite after Boles (1972)

A relation between chemical composition and thermal behaviour was presented, on the basis of a large set of data, by Alietti et al. (1977); these results are summarized in a simpler form in Fig. 6.1 H, which is the upper part of the triangular plot $Si_{36}O_{72} - (Mg + Ca + Sr + Ba)_9Al_{18}Si_{18}O_{72} - (Na + K)_{18}Al_{18}Si_{18}O_{72}$; in this triangle one may draw two lines which nearly separate the fields of zeolites with the three types of thermal behaviour; uncertainties may be given by analyses of insufficiently purified sedimentary clinoptilolites and/or by the influence of the ratio Na/K. The diagram clearly shows that a high-silica high-alkali content favours thermal behaviour 3.

Shepard and Starkey (1966) and Alietti et al. (1974) investigated the influence of the exchangeable cations and $Si:(Si+Al)$ ratios on the thermal behaviour. These researches clearly demonstrate that the exchangeable cation content is the main influencing factor; if a low-silica heulandite $(Si:(Si+Al)) = 0.76)$ and a high-silica clinoptilolite $(Si:(Si+Al)) = 0.81)$ are exchanged with Rb, K, Na, and Ca, both Rb-forms do not contract at all, both K-forms do not contract after a short heating (but prolonged heating allows a small and continuous contraction), both Na-forms contract, but to a different extent, both Ca-forms contract. At 800°C all diffractions are still observable for all Rb- and K-forms, and also for the Na-form of the high-

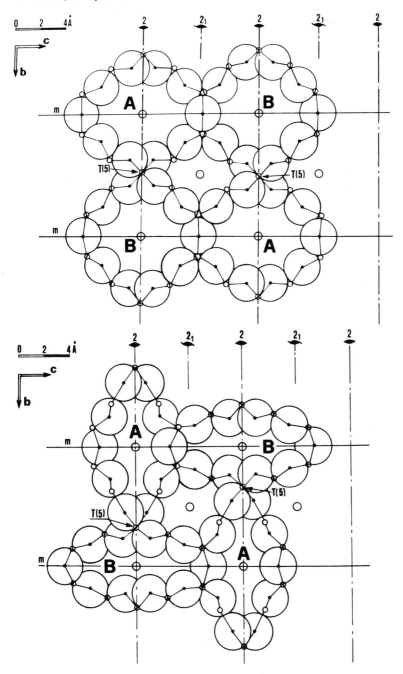

Fig. 6.1I. *Top* projection along *c* of the structure of hydrated heulandite: the channels are blown up; *bottom* projection along *c* of the structure of dehydrated heulandite. The channels are flattened

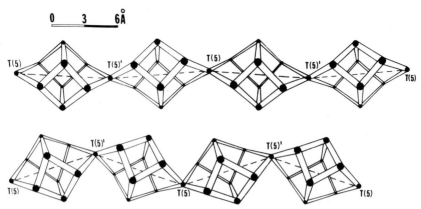

Fig. 6.1J. The chain of 4–4–1 units along [102]. *Top* in hydrated heulandite the chain is straightened; *bottom* in dehydrated heulandite the chain has a marked zig-zag pattern

silica clinoptilolite, whereas the Na-form of the low-silica heulandite loses its diffracting power at 450 °C; the Ca-forms are destroyed at 450 °C or 500 °C, the higher value being that of the high-silica clinoptilolite. On the whole, there is also a minor influence of the silica content on the thermal behaviour.

The structural explanation of the thermal behaviour of the zeolites of this genus has been given by Alberti and Vezzalini (1983a, 1984) also on the basis of the previous work of Alberti (1973) and Galli et al. (1983). First of all, the dehydration causes a deformation of the framework with a flattening of the A and B channels parallel to c (see Fig. 6.1I) and a shortening of the zig-zag chain of the heulandite-units in the [102] direction (see Fig. 6.1J); a measure of this shortening is given by the T5 – T5′ – T5 angles, which are 162° – 165° in natural heulandites and clinoptilolites, 157° in dehydrated K-heulandite, 147° to 144° in dehydrated calcium-rich heulandite. This deformation is proportional to the number of cations in the channels and to their charge. Considering the problem in more detail, it is just one of the cations, namely C2 (or CS2 or M2) in channel B, which is responsible for this deformation; the shortest distance from a cation to a framework oxygen in these structures is always C2 – O1 (~2.55 Å), and during dehydration, this cation moves towards the coordinates 0, 1/2, 1/2 and it pulls O1 by 0.5 Å in K-heulandite, but as much as 1 Å in Ca-heulandite. If that were all, one should always observe a reversible deformation, viz. under room conditions the crystals should re-adsorb the lost water and return to the original configuration of the framework; this is true if C2 is sparsely occupied and/or occupied by mono-valent cations so that the O1 shift is small, but this is not true if C2 is highly occupied by divalent cations so that the O1 shift is large: in this case the pre-existing framework is broken and a new one is formed. The interruption occurs in the T1 – O9 – T3 bridge; T1 and O9 do not change their position but the atom in T3 jumps into a new site T3D on the other side of the tetrahedral face O2,O3,O7 and a new oxygen site O9D is created, so as to form a new

on the basis of the power of the different cations to compete with Cs. Ames (1962) studied also the caesium loading kinetics of Li-, Na-, K-, Ca-, Ba-based clinoptilolites and also of a H-based (acid treated) form; the caesium loading on the Li-form is the most rapid, the slowest is on the H-form. Ames (1964a, b) published four exchange isotherms (see Fig. 6.1L) for the couples Na$\leftrightarrow$K, Na$\leftrightarrow$Sr, Ca$\leftrightarrow$Na and Ca$\leftrightarrow$Sr, and several thermodynamic constants relative to these exchanges. Barrer et al. (1967) studied the exchange of Na-clinoptilolite with NH_4^+, $CH_3NH_3^+$, $(CH_3)_2NH_2^+$, $C_2H_5NH_2^+$, n- and iso-$C_3H_7NH_3^+$, n-$C_4H_9NH_3^+$, $(CH_3)_3NH^+$. The isotherms show a clear preference for the first four cations, a nearly equal acceptance for Na and n-$C_3H_7NH_3^+$, and a preference for Na over the remaining three cations, which allow an incomplete exchange only. Barrer and Townsend (1976) studied the exchange of clinoptilolite with ammines of Co, Cu, and Zn.

Pioneering work on gas adsorption of heulandite has been done by Sameshima (1929), Rabinowitsch (1932) and Barrer (1938), who also measured rather low values for gases with small molecules. This result is quite predictable in view of the contraction of its channels during the activation (= dehydration) which must preceed any gas adsorption. Therefore, more interesting data of this kind may be obtained on clinoptilolite. Breck (1974) gives an adsorption of 0.08 g CO_2 per 1 g Hector clinoptilolite at nearly S.T.P., of 0.019 g O_2 (at T = $-183\,°C$, P = 700 torr), 0.014 g N_2 (at T = $-196\,°C$, P = 700 torr). Barrer and Coughlan (1968) found 0.098 g CO_2 with the same clinoptilolite in nearly the same conditions; 20% larger adsorption capacities were obtained by acid treatment (= decationization) of the zeolite.

Klopp et al. (1980) determined for samples from Ratka, Hungary, an adsorption of 0.05 g methanol and 0.01 g ethanol per 1 g zeolite at nearly S.T.P., but the value for ethanol could be raised to 0.05 g with the K-exchanged clinoptilolite.

Kalló et al. (1982) determined for samples from the Tokaj Hills, Hungary, an adsorption of 0.07 g NH_3, 0.11 g SO_2 and 0.066 g CO_2 per 1 g zeolite at S.T.P. Torii et al. (1977) investigated gas separations on modified clinoptilolite: O_2 and N_2 can be separated on a K-clinoptilolite column; the following elution sequence was observed at 30°C on columns of Na-, K- and Ca-clinoptilolite: methane, ethane, propane, iso-butane, n-butane.

The catalytic properties of seven clinoptilolite samples from different occurrences in California and Nevada were examined by Chen et al. (1978) with a test involving the conversion of a mixture of hexane isomers; n-hexane can enter the eight-tetrahedra rings, the other isomers cannot, so the test gives informations on the catalytic sites which can be reached only through these ports. The seven clinoptilolites, in the H-form, show different activities and rates of deactivation, which may be related with some features of their powder patterns: samples with sharp odd k-lines convert n-hexane preferentially, samples with weak and/or broad odd k-lines convert all hexanes at the same rate; there are some samples with intermediate activity and odd k-lines. The different catalytic rates of these samples should be related to different structures in a broad sense, viz. including a different frequency of stacking faults. Kalló et al. (1982) studied several catalytic reactions on the H-clinoptilolite from Hungary: xylene isomerization, n-butene isomerization, toluene hydrodemethylation, methanol dehydration to dimethylether. The isomer rate of n-butane (at 200°C, 133 kPa) to cis-2-butene r_{12} was found to be 3.5×10^{-6}, and to trans-2-butene r_{13} 5.6×10^{-6} (both mol $g_{cat}^{-1} s^{-1}$), with a selectivity r_{12}/r_{13} 0.62 nearly equal to that of Hector H-clinoptilolite, which allows 80% higher rates.

6.1.6 Occurrences and Genesis

There are only a few sedimentary occurrences of heulandite with proper evidence (at least the thermal test of Mumpton 1960) that the mineral was not clinoptilolite, which much more frequently has this kind of genesis.

Surface genesis has never been attributed to heulandite.

Heulandite with genesis in a hydrologic closed system (saline lake) has been found only once, in the Big Sandy formation, Arizona (Sheppard and Gude 1973); these authors name it clinoptilolite because of its high silica content, although divalent cations prevail over the monovalent ones; the thermal behaviour of this sample is not known.

Deep sea sediments commonly contain some clinoptilolite, but if very high in $CaCO_3$, a zeolite with $(Mg + Ca) > (Na + K)$, hence a heulandite, may be present (see for instance the analyses of the "clinoptilolites" from samples $163 - 19 - 5$ and $163 - 27 - 1$ in Stonecipher 1978); marine geologists usually name these zeolites clinoptilolites, because their structures survive an overnight heating at 450°C (test of Mumpton 1960). For more details on this genesis, see the description of deep sea clinoptilolite, later in this chapter.

Heulandite has been described a few times in marine sediments now on land. Chumakov and Shumenko (1977) identified the zeolite chemically and thermally in a Lower Pliocene fluvio-marine sediment of the Aswan region, Nile Valley, Egypt. Passaglia and Vezzalini (pers. comm.) chemically analyzed a heulandite of the uppermost level of an Oligocene-Miocene tuffite from Garbagna, Alessandria, Italy which is phillipsite-rich on the whole; the same tuffite is not zeolitized in most occurrences of the Northern Apennines.

Barrows (1980) chemically and thermally determined both heulandite and clinoptilolite in the Miocene volcaniclastic sequence from the southern Desatoya Mts., Nevada, 2500 m thick, together with mordenite, analcime, erionite: this is probably a case of burial diagenesis although no zoning was observed and an early partial crystallization at close to surface conditions (open system) cannot be excluded. Miller and Ghent (1973) and Ghent and Miller (1974) found a Sr – Ba-rich heulandite building up with laumontite the cement of a non-marine Cretaceous sandstone from the Blairmore group, Alberta, Canada; they give the upper P/T limit (T = 280°C, P = 2 kbars) for this crystallization but not the lower one, so that the framing of this occurrence into a genesis scheme is uncertain. Read and Eisbacher (1974) proved the existence of heulandite in the fluviatile sediments from the Sustut Group, British Columbia, from Late Cretaceous to Eocene in age, up to 2300 m thick; the zeolite is present in the tuffaceous layers, no zoning is observed.

Heulandite is mentioned by Coombs et al. (1959) in their proposal of the "zeolite facies", but the name used is sensu lato for the whole genus. In his general scheme of zeolitic crystallization in diagenesis and very low metamorphism, Utada (1970) attributes heulandite to the "zone III" (clinoptilolite is mainly in the "zone II"); there are two subzones, IIIa − regional with analcime or heulandite and IIIb − layer-like with analcime and other zeolites (clinoptilolite, mordenite or laumontite, without heulandite). Typical occurrences of subzone IIIa are the Dewa Hills Lands formation, Japan; in the volcaniclastic sequence analcime occurs as cementing material, heulandite as replacement of plagioclase; the coexistence of the two is rare. In agreement with Utada's scheme, Yoshitani (1965) found that the Kurosawa-Zawa tuff members in the Tanzawa Mts., Japan, can be divided into three zones in ascending order: laumontite-, heulandite- and mordenite-zones. The Taveyanne formation, Switzerland and France (Sawatzki and Vuagnat 1971; Coombs et al. 1976) is made up of volcanic greywackes and intercalated shales; most formations can be attributed to the laumontite facies (zone IV of Utada 1970), but in a restricted area of the Thones syncline, Savoie, France, heulandite appears as groundmass mineral, indicating conditions of lower grade; a higher grade zone (prehnite-pumpellyite zone 6 of Utada 1973) appears sparingly in other parts of the formation. In general, heulandite genesis seems to be possible from shallow to moderate depth, up to the first zone of very low metamorphism (analcime-heulandite is the beginning stage of the zeolite metamorphic facies for Coombs et al. 1959, or the highest diagenetic zone for Utada 1970: this is not a big difference). The development towards either heulandite or clinoptilolite could be driven more by the chemistry than by depth or P/T conditions; high calcium activity is possible only if pH ~ 7 and this allows minor silica activity only: this situation would favour heulandite; a high alkali concentration is a good prerequisite for a high pH, which permits a high silica activity: this situation seems to be more probable in the presence of volcanic glasses and would favour clinoptilolite, which is much more frequent as a sedimentary mineral. The attribution of clinoptilolite and heulandite to different zones (II and III of Utada 1970) probably has more a statistical basis than a physico-chemical one; moreover ion exchange after crystallization cannot be excluded.

The most common type of occurrence for heulandite is in veins and geodes of various types of rocks with fine crystals of probable hydrothermal genesis. Already Hintze (1897) and Dana (1914) list so many occurrences of this kind, that only a brief summary of these classical localities can be given here:

Iceland and Faröes: fine crystals can be collected in very many occurrences in Tertiary basalts; Betz (1981) gives a long list of localities.

Zeolites of the Heulandite Group

Scotland: red crystals occur in the Campsie Hills, Stirlingshire; white crystals at Storr, Skye.

Germany: at Andreasberg, Harz, with stilbite.

Italy: fine red crystals in the porphyrites of Val di Fassa, Trento, with thermal behaviour of type 2 (see Alietti 1967, 1972); clear crystals in the basalts of Montecchio Maggiore, Vicenza.

Switzerland: clear crystals at Giebelsbach, Wallis (see also Merkle and Slaughter 1968).

Canada: in several localities of Nova Scotia (see also Walker and Parsons 1922; Aumento 1964; Aumento and Friedlander 1966).

Newer descriptions in basalts and other basic volcanic rocks are:

Northern Ireland: in the Tertiary basalts (Walker 1960a).

Poland: red crystals in a melaphyre at Rudno (Piekarska and Gawel 1954).

Czechoslovakia: Sr- and Ba-rich crystals in the melaphyres of Kozakov, North Bohemia, with chabazite (Černý and Povondra 1969).

Hungary: in the andesites of Matra Mts., with chabazite, stilbite and laumontite (Mezösi 1965); in the andesites of Nadap (Alberti et al. 1983a; Alberti and Vezzalini 1983a).

Italy: in the ophiolites of Campegli, Liguria (Lucchetti et al. 1982) with the highest Sr-content (8.70%) ever reported.

Bulgaria: in the basalts of Nanovitza, Bulgaria; this zeolite was named by Kirov (1967) "calcium-rich clinoptilolite" (thermal behaviour unknown).

U.S.S.R.: in an igneous rock of the Crimea, with gmelinite, wellsite and analcime (Shkabara 1940); in basaltic breccias and tuffs of Lower Tunguska region with mordenite, thomsonite, stilbite, analcime, chabazite, fibrous zeolites (Shkabara and Shturm 1940).

India: in the Deccan Traps, with many other zeolites (Sukheswala et al. 1974).

Japan: in an altered basalt from Mazé, Niigata Pref., with phillipsite and natrolite (Harada et al. 1967); in an altered basalt from Hudonotaki, Kanagawa Pref. (Harada et al. 1969b).

U.S.A.: in the basalts of Lanakai Hills, Hawaii, with epistilbite, laumontite, mordenite (Dunham 1933).

Brazil: in the basalts of southern Brazil (Franco 1952; Mason and Greenberg 1954).

The following heulandites were found in rocks other than basic volcanics:

Norway: in a quartz-rich gneiss at Kragerø with natrolite, stilbite and laumontite (Saebø and Reitan 1959); in a granite near Drammen, with stilbite, harmotome and chabazite (Raade 1969).

Austria: in a pseudotachylyte in Jam Valley, Vorarlberg, with stilbite, stellerite and chabazite (Koritnig 1940).

Italy: in a schist at Prascorsano, Piemonte, with stilbite (Resegotti 1929); in the granite of Baveno, with laumontite, stilbite and chabazite (Pagliani 1948).

U.S.A.: as filling of fossils at Eugene, Oregon, with analcime and stilbite (Staples 1965).

Clinoptilolite, a rare hydrothermal mineral similar to heulandite in earlier times, has now been shown to be the main constituent of many sediments and hence much more abundant in the earth's crust than heulandite (but words like "heulandite, a silica-poor variety of clinoptilolite" of Sands and Drever 1978 p 271, are somewhat exaggerated...).

Evidence for a surface genesis of clinoptilolite has never been published.

The genesis of clinoptilolite in a hydrologically closed system like a saline, alkaline lake is very common and has been investigated in detail many times (Hay 1978 and quoted references). In desert areas, the water of a basin without effluent streams may reach high salinity and alkalinity, mainly if some volcanic glassy components are present in the sediment; this brine may react easily with several minerals (glass, clay minerals, opal, feldspar, quartz) to form zeolites. A typical feature of these deposits is the lateral concentric zonation: at the centre, which is or was a saline lake and where the pH is highest, alkali feldspar is formed; an analcime zone follows with a subsequent zeolite zone often including clinoptilolite; at larger distances from the lake, only clay minerals (montmorillonite) and glass are present and the pore water is nearly fresh. Sheppard (1971) lists nearly fifty clinoptilolite occurrences of this kind in the western United States; good examples with fine descriptions are Lake Tekopa, California (Sheppard and Gude 1968) without the analcime zone and with a wide zeolite zone including clinoptilolite, alone or with chabazite, phillipsite, erionite; the Barstow Formation, California (Sheppard and Gude 1969a) with clinoptilolite (up to 100%) alone or with erionite, mordenite, phillipsite, the zonation being only lateral here, not concentric; the Green River formation, Wyoming (Hay 1966; Surdam and Parker 1972) with a well developed zeolite zone (clinoptilolite and mordenite) around an analcime zone, which in turn encircles a feldspar zone. Similar is certainly the genesis of clinoptilolite from the Lake Magadi region, Kenya (Surdam and Eugster 1976) but no clear zoning was detected there between the areas where glass, analcime and other zeolites (mainly erionite, with minor clinoptilolite, mordenite, chabazite) are present.

Zeolite genesis in a hydrologic open system is given by the slow percolation of meteoric water in tuffs and tuffaceous sediments; the pore water becomes more and more saline and alkaline during the descent to the water table and the solution thus formed is suitable for zeolite crystallization at the expense of the dissolving volcanic glass (Sheppard 1971; Hay 1978); this kind of genesis is obviously possible only over the continents and above the water table and hence obviously in not very thick sedimentary layers (but the water table changes a lot in geological time). The increasing alkalinity of the descending solution allows the crystallization of different mineral assemblages in the following zones in descending order: I, fresh glass + montmorillonite; II, zeolites; III, analcime; IV, feldspar. This is nearly the same sequence which is found in burial diagenesis or metamorphism, so it may be difficult to distinguish between the two. The different thicknesses of the layers affected by the zonation could give a key for the distinction; moreover the transition between zones is sharp in open systems, gradational in burial diagnesis. After Hay (1978), a typical example for clinoptilolite genesis in an open system is the Vieja Group, Texas, studied by Walton (1975): several formations of the sequence, mainly two reworked tuffs nearly 1000 m thick, show a fine zonation with glass and montmorillonite at the top, clinoptilolite in the middle and analcime at the bottom. Another typical "open system" is the John Day formation, Oregon, U.S.A., studied by Hay (1963a) who proposed this model for the presence of clinoptilolite at the bottom and of glass plus montmorillonite at the top of a series of tuffs and tuffaceous claystones with rhyolitic and dacitic glass.

A similar genesis may be attributed also to the few known occurrences where clinoptilolite grew at the expense of impactite glasses, as in Noerdlinger Ries, Germany (Stoeffler et al. 1977; Fuechtbauer 1977), and in Boltyshsk, U.S.S.R. (Bobonich 1979).

A third type of near-to-surface genesis was proposed independently by Lenzi and Passaglia (1974) and by Aleksiev and Djourova (1975) and by this last author named "geoautoclave". The model may be considered only for ash flow deposits, also called ignimbrites, which are rocks which somewhat resemble tuffs and were often so called in the past and then coded with this name in geological maps. These rocks were actually deposited at high temperature (800°C – 900°C); the above mentioned authors think that this temperature was retained in part also when surface or meteoric or marine water entered the still hot deposit, so generating a particular hydrothermal homogeneous environment favouring a speedy complete zeolitization: the particular feature of this genesis should be just the homogeneous zeolitization of the ash flow deposit. Lenzi and Passaglia (1974) proposed this model for the zeolitization to chabazite of the Quaternary *"tufo rosso a pomici nere"* occurring near Bracciano Lake, Italy; Aleksiev and Djourova (1975) proposed it for the clinoptilolite rich Oligocene marine deposit of the north-eastern Rhodopes, Bulgaria. The same genesis is probably true for the ash flow deposit, completely zeolitized to clinoptilolite, of Tokaj Hills, Hungary (Varjú 1966).

The deep sea occurrence of clinoptilolite is quite common: Stonecipher (1976) and Kastner and Stonecipher (1978) give a good and complete review of the ample literature on this subject; we have widely used this article for the following description; many chemical analyses are in Stonecipher (1978). The crystals are never larger than 0.045 mm, usually are in the range 0.020 mm – 0.010 mm. The chemistry is always characterized by a high silica content, normally the ratio $Si : (Si + Al)$ is $83\% \pm 1\%$; usually the alkalis are by far more abundant than alkali earths but, obviously enough, Ca content is not so low in clinoptilolites from carbonate bearing sediments and chalks: in some cases the percentage of $Ca(+Mg)$ may be so high (for instance in the zeolites of samples $163 - 19 - 5$ and $163 - 27 - 1$ of Stonecipher 1978) that they must be named heulandites.

The presence of clinoptilolite in deep sea sediments is, from several points of view, different from and in contrast with the presence of phillipsite: this dichotomy has already been illustrated in Chap. 3.2.6. Clinoptilolites are more frequent in carbonate bearing sediments (with high deposition rate), than in the clayey ones (with low deposition rate), but may be also associated with siliceous sediments with opal and/or quartz. There is a general agreement that most deep sea clinoptilolite crystallizes as an alteration product of acid (mostly rhyolitic) volcanic glass, spread over the oceans via atmosphere by the explosive volcanic activity at the continental margins, but there is no obvious relation between the presence of glass and clinoptilolite. As a matter of fact, it takes a long time for a glass shard in the sea to be transformed into clinoptilolite, which has been found only in sediments older than Eocene (and not older than Cretaceous); moreover clinoptilolite is quite uncommon in the first 100 m depth from the sea bottom, being rather abundant from 700 m to 1200 m depth. In young sediments acid glass is not yet transformed into clinoptilolite (but a basic glass has begun its alteration to phillipsite): in medium aged (Neogene) sediments both glass and clinoptilolite are present; in older sediments (Eocene to Cretaceous) clinoptilolite is rather abundant and glass absent, because it has all turned into the zeolite. It may be worth noting that it is not certain which of the two, age or depth from sea bed, is more important in controlling the rate of glass alteration. The general trend does not exclude the possibility of a genesis of clinoptilolite from basic glass in the presence of a high silica activity, or even in the absence of glass, but in this last case it is not clear what could be the source of the necessary alkalinity. A genesis by transformation of pre-existing zeolites is also probable, because "other" zeolites, mainly phillipsite, disappear as sediments of increasing age are considered, probably in favour of clinoptilolite. Clinoptilolite is frequent in all oceans, whereas phillipsite is rare in the Atlantic, and at high latitudes; this fact is related (see Chap. 3.2.6) to the presence of acid explosive activity at the border of all oceans; some prevalence of clinoptilolite in the Atlantic ocean may be associated with the volcanic debris transported by rivers, whose total water flow into the Atlantic is much larger than into the other oceans.

Clinoptilolite is also common as a major constituent of submarine volcanic sediments, or tuffites, now on land. Sameshima (1978), in his study of the Early Miocene submarine deposits of the Waitemata group, New Zealand, found almost monomineralic clinoptilolite tuffite beds in the Tom Bowling formation. Typical marine clinoptilolites have been found not only in the Cretaceous chalks of the English Channel (Estéoule and Estéoule-Choux 1972) but also in the marls and limestones of the same age from Anjou, France (Estéoule et al. 1971). Brown et al. (1969) found relevant percentages of clinoptilolites in several Cretaceous marine sediments of the Upper Greensand formation, England. These marine sediments with zeolites cannot be distinguished clearly from those inserted in the following type, burial diagenesis, including both continental and marine sediments, in which there is a more or less evident depth-zoning.

Burial diagenesis of sediments containing volcanic debris very often favours the crystallization of clinoptilolite. Utada (1970), as mentioned during the discussion of heulandite, regards clinoptilolite and mordenite as the most common and typical zeolites of his zone II, which is under zone I (glass + clay minerals) and above zone III (analcime or heulandite). Examples of this kind can be found with both marine and continental sediments. Utada (1970) reviews the Japanese literature and gives many fine examples of zoning, which may be concordant with stratification as in the Shinjo basin and in the Akita oil field, or oblique to the stratification as at Itaya; another good example, the Niigata oil field with marine sediments, was described by Iijma and Utada (1971) who measured also the actual temperature at the top ($\sim 50\,°C$) and the bottom ($\sim 90\,°C$) of the clinoptilolite-mordenite zone. The best example outside Japan was described by Moncure et al. (1981) in Pahute Mesa, Nevada test site, where Tertiary volcanic silicic rocks are altered differently in three zones, the intermediate one containing clinoptilolite and mordenite.

There are many other clinoptilolite occurrence probably due to burial diagenesis, which are part of a sedimentary sequence where zeolites appear at one level only, and the zonal character of the diagenesis does not emerge from available data. There are many such occurrences in U.S.S.R. (see Gottardi and Obradovic 1978, for a short description and references): in Oxfordian and Albian beds of the Penza region, west Kazakhstan, in Miocene and Oligocene siltstones of Georgia, in marine Cretaceous formations of the Southern Russian Platform, in Cretaceous tuffites of Azerbaydzhan, and many others. For Yugoslavia, Gottardi and Obradovic (1978) report on the presence of clinoptilolite with analcime in the "pietra verde" (see Chap. 2.1.6) of Montenegro, Bosnia, Croatia and Slovenia, which probably underwent a diagenesis intermediate between Utada's zones II and III; the crystallization of clinoptilolite with mordenite in tuff and tuffites of the Northern part of Backa, Serbia, probably correspond to a diagenesis of zone II. To an undefined burial diagenesis is perhaps due the clinoptilolite of the Oligocene Sarmiento formation, Patagonia, Argentina, with interbedded tuffs, described by Mason and Sand (1960).

The diffuse hydrothermal deposition of clinoptilolite in geothermal fields has been observed for instance by Honda and Muffler (1970) in rhyolitic detritus and rhyolite of Yellowstone National Park, Wyoming, where it is associated with mordenite from 10 m to 50 m depth (temperature from $90\,°C$ to $150\,°C$), and by Sheridan and Maisano (1976) in a volcaniclastic formation near Hassayampa, Arizona, where it is the main zeolite with stilbite in the rock matrix at nearly 200 m depth; the estimated temperature of deposition is $\sim 150\,°C$.

Hydrothermal deposition of clinoptilolite in veins and vugs of basic volcanic rocks is known, in contrast to heulandite, only for few localities:

- in a decomposed amygdaloid basalt of Hoodoo Mts., Wyoming (the type locality, Pirsson 1890; thermal behaviour unknown);

- in the porphyrites of Valle dei Zuccanti, Vicenza, Italy (Alietti 1967; thermal behaviour 2);
- in the porphyrite of Albero Bassi, Vicenza, Italy (Passaglia 1972) as Al-rich red lamellae (thermal behaviour 1);
- in the andesite from Challis, Idaho (Ross and Shannon 1924; thermal behaviour 2) with mordenite and analcime;
- in the brecciated porphyritic andesite from Agoura, California (Wise et al. 1969; thermal behaviour 3) with ferrierite;
- in the porphyrites from Alpe di Siusi (= Seiseralp), Bolzano, Italy (Alberti 1975a; Alberti and Vezzalini 1975; thermal behaviour 3) with dachiardite, mordenite and other zeolites;
- in the rocks from Kuruma Pass, Japan (Koyama and Takéuchi 1977; thermal behaviour 3).

6.1.7 Applications and Uses

As sedimentary clinoptilolite is available as large masses which can be excavated at low cost in quarries, many applications have been widely studied and deployed: building stones, soil conditioner, paper filling, dietary supplement for pigs and roosters, waste water purification, dehydration, gas separation (more information may be found in Mumpton 1975).

We are not able to quote a paper on the uses of clinoptilolite tuff for cutting building stones, but this use does certainly exist, as zeolitic materials are always suitable for this application, no matter which zeolite is present.

Torii (1978) reports on the increased agricultural crops obtained in Japan when clinoptilolite ($1 \, \text{ton}/1000 \, \text{m}^2$) was added; retention of ammonia and potassium could be the cause of this favourable effect. Clinoptilolite is also used in agriculture as a chemical carrier, both for herbicides and fertilizers.

Torii (1978) reports that $3300-4000$ tons per month are used in Japan as paper filler; organic and iron impurities must be avoided or dissociated by oxidation and bleaching.

The literature on the use of clinoptilolite as dietary supplement is so wide that we quote only two papers where many other references can be found. Torii (1978) presents interesting data on the advantages in feeding pigs with a diet including 6% of this zeolite, especially in the first month. Hayhurst and Willard (1980) showed that roosters raised with a diet including 7.5% clinoptilolite showed increased weight gain over a seven weeks period; no toxic effects were observed. Olver (1983) reported on the influence of this dietary supplement on laying hens; there is no effect on the live mass nor on the eggmass, but significant effects were noticed in the days to first egg, numbers of eggs laid per hen, egg size percentages, feed consumed per hen and feed efficiency (kg eggs/kg feed).

Mercer and Ames (1978) reported the use of clinoptilolite beds for recovering Cs from radioactive wastewater, and ammonia from municipal wastewater; for this last application see also Murphy et al. (1978), Torii (1978) and

Sheppard et al. (1983). Similar is the application proposed by Bower and Turner (1982) to add some clinoptilolite to the plastic shipping bags containing goldfish, to reduce ammonia concentration during transport.

The most common industrial use of clinoptilolite is probably to fill drying beds for the dehydration of liquids and gases, for instance before a catalytic reaction. Klopp et al. (1980) report widely on this kind of use of Hungarian clinoptilolite, and Torii (1978) for the zeolite from Japan; both authors mention the possibility of filling drying cartridges in compression-type refrigerators.

A particular way of exploiting the water adsorption power of zeolites is to build a refrigerator based on solar energy (Tchernev 1980): the zeolite is heated and dehydrated in the sun during the day in a special cell, the outgoing water being collected in a reservoir; during the night or when the solar heating is interrupted, the water is re-adsorbed in the zeolite cell and evaporation from the liquid state secures the wanted refrigeration.

Galabova et al. (1978) studied an oxygen enrichment of air using clinoptilolite, but Minato and Tamura (1978) obtained better results with mordenite.

We are not aware of any deployed industrial process which exploits the catalytic properties of clinoptilolite.

6.2 Stilbite, Stellerite, Barrerite

Stilbite	Stellerite	Barrerite
$NaCa_4(Al_9Si_{27}O_{72})$ $\cdot 30\,H_2O$	$Ca_4(Al_8Si_{28}O_{72})$ $\cdot 28\,H_2O$	$Na_8(Al_8Si_{28}O_{72})$ $\cdot 26\,H_2O$
C2/m	Fmmm Z = 2	Amma Z = 2
$a = 13.61$ $b = 18.24$ $c = 11.27\,(\text{Å})$ $\beta = 127°51'$	$a = 13.60$ $b = 18.22$ $c = 17.84\,(\text{Å})$	$a = 13.64$ $b = 18.20$ $c = 17.84\,(\text{Å})$
or pseudo-orthorhombic cell		
F2/m Z = 2		
$a_0 = 13.61$ $b_0 = 18.24$ $c_0 = 17.80\,(\text{Å})$ $\beta_0 = 90°45'$ $\vec{a}_0 = \vec{a} \quad \vec{b}_0 = \vec{b}$ $\vec{c}_0 = 2\vec{c} + \vec{a}$		

6.2.1 History and Nomenclature

Haüy (1801) introduced the name stilbite from the Greek $\sigma\tau\iota\lambda\beta\eta$ = lustre, for orthorhombic lamellar zeolites, including the present species heulandite; this last name was introduced only in 1822 by Brooke, who recognized the distinct monoclinicity of the second mineral, and hence distinguished it from pseudo-orthorhombic stilbite (for more historical details see Hintze 1894 or Dana 1914, who also record many synonyms now fully obsolete). For the same mineral Breithaupt (1818) introduced the name "desmine" which was long used, especially by German-speaking authors; now it has been nearly completely abandoned. It may be useful to remember that some of those authors, who preferred the name "desmine", used the name "stilbite" for the species which is now called heulandite. "Foresite" is a name introduced by Pullé and Capacci (1874) for a mineral which was shown by Cocco and Garavelli (1958) to be stilbite.

Morozewicz (1909) introduced the name stellerite for a Ca-rich orthorhombic mineral similar to stilbite, which, meanwhile, had been recognized as truly monoclinic; stellerite was found in the Komandor Islands, Bering Sea, and so named in honour of G. W. Steller, the discoverer of the islands. Epidesmine, a synonym of stellerite, was introduced by Rosický and Thugutt (1913) for a mineral occurring at Schwarzenberg, Saxony; the name is now obsolete (and has nothing to do with epistilbite!).

The name barrerite was proposed by Passaglia and Pongiluppi (1975) for a Na-rich orthorhombic zeolite from Capo Pula, Sardinia, and so named in honour of R. M. Barrer, the eminent zeolitologist.

6.2.2 Crystallography

The structure of stilbite was determined by Galli and Gottardi (1966) and refined by Galli (1971) and by Slaughter (1970) (see Chap. 0.2.6). The topological symmetry of this framework is Fmmm, but the real symmetry is C2/m only, so that a monoclinic reference with $\beta \sim 128°$ is commonly chosen, in order to avoid the face centred cell, which has $\beta_0 \sim 90°$; obviously, there is a second possible monoclinic reference C2/m with a′ in the direction [102] and nearly the same β. There is no Si/Al-order in the framework. There are two cation sites; Ca fully occupies a site in the centre of the main channel parallel to a, on the mirror plane near to a screw diad, and is completely surrounded by 8 water molecules without any contact with framework oxygens (site type IV of Chap. 0.4); Na is in a site with low occupancy, again in the main channel but nearer to the framework, so as to coordinate not only water molecules but also some oxygens (site type III of Chap. 0.4).

Stellerite (Galli and Alberti 1975a) has the same framework and again no Si/Al-order, but a higher symmetry F 2/m 2/m 2/m, which coincides with the topological symmetry for the framework. This fact can be explained as from Fig. 6.2A: in stilbite electrostatic repulsive forces between extraframework

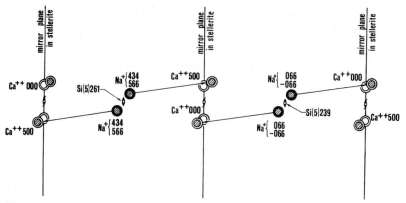

Fig. 6.2A. Projection along b of some atoms in the stellerite-stilbite structures; three digit numbers give $(y/b) \times 1000$; in stellerite there is no Na^+ and Ca^{2+} lies on the (010) mirror planes; in stilbite, the repulsion between Na^+ and Ca^{2+} pushes this last atom out of the (010) planes lowering the symmetry from Fmm to C2/m

cations are strong enough to push the calcium out of the mirror planes, so allowing only symmetry F2/m (or C2/m); in stellerite there is no sodium, and its site is partially occupied by water, so calcium can peacefully remain on the mirror plane, and symmetry is F 2/m 2/m 2/m.

In barrerite (Galli and Alberti 1975b), the larger number of extraframework cations (~ 8) is distributed mainly in C1 and C1P sites corresponding to the Ca site of stilbite and stellerite, in part in sites C2 and C2P broadly corresponding to the Na site of stilbite, and also in site C3 without counterpart in the other two species. The lower charge of Na in comparison with Ca and the distribution with low occupancy over the two sites C2 and C2P with opposite repulsive electrostatic forces against C1 and C1P, allow the mirror plane normal to b to be maintained. The symmetry is only A 2_1/m 2/m 2/a, because the site C3, slightly shifted on one site of the 001-mirror plane, forces the framework to rotate around a screw diad parallel to a. Passaglia and Sacerdoti (1982) refined an Na-exchanged stellerite and found that in this phase the C3 sites are statistically distributed on both sides of the 001-mirror plane so that, on the average, the symmetry remains at its highest value Fmmm. If, by contrast, a barrerite is Ca-exchanged, the resulting phase has symmetry Fmmm as expected (Sacerdoti and Gomedi 1984).

Some information on the structure of modified stilbites is contained in Chap. 6.2.5.

Stilbite, like phillipsite, has a morphology characterized by the absence of untwinned crystals, and this means, in our opinion, that the crystals nucleate on the contact plane of the twin. These stilbite twins are definitely pseudomerohedral in view of the pseudo-orthorhombicity of the structure, but unlike phillipsite, there is not always a definite contact plane. The frequent appearance of these fourlings or eightlings is given in Fig. 6.2F, and Fig. 6.2G

Fig. 6.2B. Stilbite from Poona, India ($\times 4$)

Fig. 6.2C. Stilbite from Poona, India ($\times 8$)

Fig. 6.2 D. Stilbite from Varallo Sesia, Italy (SEM Philips, ×60)

Fig. 6.2 E. Stilbite from Varallo Sesia, Italy (SEM Philips, ×120)

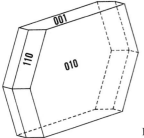

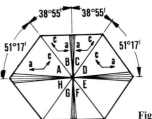

Fig. 6.2F

Fig. 6.2G

Fig. 6.2 F. Common appearance of a stilbite eightling

Fig. 6.2 G. View on (010) of a stilbite eightling, with *a* and *c* in each individual; microdomains with uncertain orientation fill the *striped zones*

Fig. 6.2 H. Stellerite from Villanova Monteleone, Sardinia, Italy (×8)

shows how a 010-section may be divided into 8 individuals (A to H) with 4 orientations: A and H have a_0 or c_0 (both pseudo-diads) parallel, the contact plane $(001)_0$ being replaced by a zone with a mosaic of microindividuals of uncertain orientation; the same is true for B and C, the contact zone being here nearly $(100)_0$; the twin axis between A and B is the same, but the contact plane is $(10\bar{1})_0$ for A and $(101)_0$ for B. Sheaf-like aggregates of *a*-elongated crystals are common and typical.

Fig. 6.2I. Stellerite from Villanova Monteleone, Sardinia, Italy (SEM Philips, ×120)

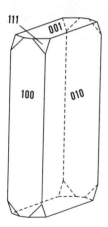

Fig. 6.2J. Common morphology of stellerite

The morphology of orthorhombic stellerite is quite simple, with lamellar (010) crystals terminated with {100} {001} {111} and hence with rectangular outline (see Fig. 6.2J); no pseudo-merohedral twins are possible, neither are other kinds of twins known.

Barrerite usually occurs as (010) lamellae, up to 5 cm in diameters, without a definite outline; (010) platy crystals, with {001} and {111} are rare (see Fig. 6.2L).

Fig. 6.2 K. Barrerite from Capo Pula, Sardinia, Italy (×8)

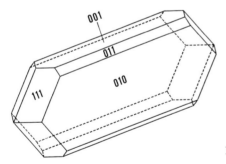

Fig. 6.2 L. The morphology of barrerite

6.2.3 Chemistry and Synthesis

A wide spectrum of chemical analyses has been published by Passaglia et al. (1978b) and some of these are listed in Table 6.2I. The composition corresponding to the stoichiometric formula of the title page (no. 4) is quite common; more aluminous formulae are possible (nos. 1 and 2), with constant Ca (~4) and up to 2.58 Na-atoms. Ca rarely, as in no. 3, is lower than 4, and in these cases Na may be over 4; only three of these sodian stilbites are known.

Table 6.21. Chemical analyses and formulae on the basis of 72 oxygens and lattice constants in Å

	1	2	3	4	5	6
SiO_2	51.97	53.77	58.34	56.85	59.10	58.82
Al_2O_3	18.34	16.77	15.32	15.58	14.62	14.75
Fe_2O_3	0.03	0.08	0.18	0.08	0.08	0.04
MgO	0.03	0.03	0.16	0.06	0.02	0.24
CaO	7.88	7.75	3.31	7.89	7.96	1.66
Na_2O	2.73	1.81	4.78	1.00	0.24	5.97
K_2O	0.22	0.07	0.88	0.03	0.02	1.76
H_2O	18.85	19.75	17.00	18.55	17.98	16.40
Total	100.05	100.03	99.97	100.04	100.02	99.64
Si	25.34	26.26	27.47	27.14	27.83	27.72
Al	10.54	9.66	8.50	8.77	8.11	8.19
Fe	0.01	0.03	0.06	0.03	0.03	0.01
Mg	0.02	0.02	0.11	0.04	0.01	0.17
Ca	4.12	4.06	1.67	4.04	4.01	0.84
Na	2.58	1.71	4.36	0.93	0.22	5.45
K	0.14	0.04	0.53	0.02	0.01	1.06
H_2O	30.65	32.17	26.69	29.53	28.23	25.78
$E\%$	-4.0	-2.3	+1.3	-3.4	-1.8	+3.9
a_o	13.657	13.621	13.596	13.601	13.583	13.643
b_o	18.252	18.263	18.309	18.237	18.237	18.200
c_o	17.783	17.786	17.834	17.799	17.845	17.842
β_o	90°55'	90°51'	90°20'	90°42'		
D	2.18	2.19	2.14	2.16	2.13	2.13

1. Stilbite, Montresta, Nuoro, Sardinia, Italy (Passaglia et al., 1978b).
2. Stilbite, Flodigarry, Skye, Scotland (do).
3. Stilbite, Capo Pula, Cagliari, Sardinia (do).
4. Stilbite, Poona, Maharastra State, India (do).
5. Stellerite, Shimohutsakari, Ohtsuki, Yamanashi Pref., Japan (do)
6. Barrerite, Capo Pula, Sardinia (Passaglia and Pongiluppi, 1974).

Stellerites are usually very near to the stoichiometric formula of the title page (no. 5), and the same is true for the only known barrerite (no. 6) if the substitution $K \rightarrow Na$ is neglected. SrO and BaO are always very low: Passaglia et al. (1978b) found 0.13% BaO in only one case, and nearly 0.06% SrO in two cases, and less than 0.01% in all their remaining 62 samples.

There is no report available on the synthesis of phases with the stilbite framework, as far as we know.

Liou (1971a) studied the reaction

stilbite = laumontite + quartz + water

which may be true if the chemical formulae are written as follows

$$Ca_4(Al_8Si_{28}O_{72}) \cdot 30\,H_2O = Ca_4(Al_8Si_{16}O_{48}) \cdot 16\,H_2O + 12\,SiO_2 + 14\,H_2O$$

neglecting the sodium content of stilbite. In this research, Liou obtained the growth of stilbite crystals, but no nucleation, as he started from mixtures 9:1 and 1:9 of (natural stilbite) + (natural laumontite + quartz): a nearly constant limit at $T = 175\,°C$ from 2 to 6 Kbars of P(fluid) was found for stilbite which is stable below this temperature; at higher T, laumontite is converted to wairakite (see Chap. 2.2.3). Juan and Lo (1973) studied the reaction

stilbite = Ca – analcime + quartz + water

which again is possible if the chemical formulae are written as follows

$$2\,Ca_4(Al_8Si_{28}O_{72}) \cdot 30\,H_2O = Ca_8(Al_{16}Si_{32}O_{96}) \cdot 16\,H_2O + 24\,SiO_2 + 14\,H_2O \ .$$

The starting material is a natural stilbite, which decomposes when the temperature goes over $222\,°C$, P(fluid) ranging from 0.7 to 2.5 Kbars. The existence of stilbite from $175\,°C$ to $222\,°C$ in the absence of laumontite nuclei is probably metastable in view of the above mentioned results of Liou (1971a).

6.2.4 Optical and Other Physical Properties

Stellerite is biaxial negative with $\alpha = 1.484$, $\beta = 1.492 - 1.496$, $\gamma = 1.495 - 1.498$ and $\alpha \rightarrow a$, $\beta \rightarrow b$, $\gamma \rightarrow c$, $2V \sim -45°$ (Mattinen 1952; his a and b must be interchanged).

Stilbite is also biaxial negative with nearly the same orientation, namely $\alpha \hat{a}_0 \sim 5°$ in the obtuse angle β_0 or β_m, $\beta \rightarrow b$, $\gamma \hat{c}_0 \sim 5°$ in the acute angle β_0, as correctly given by Dana (1914) and by Ramdohr and Strunz (1978) (Winchell and Winchell 1951, make some confusion using the monoclinic reference in the text and the orthorhombic one in Fig. 234, which is nearly correct if a and c are interchanged, as obviously necessary in view of the position of the 129° angle of the outline).

$\alpha = 1.489$ $\beta = 1.498$ $\gamma = 1.501$

$2V$ negative and variable around $40°$

Barrerite is biaxial negative with the same orientation

$\alpha \to a$ $\beta \to b$ $\gamma \to c$ $2V \sim -78°$

$\alpha = 1.479$ $\beta = 1.485$ $\gamma = 1.489$.

All three minerals are transparent colourless; brick red stilbite exists but is rare. Streak: white. (010) cleavage is always perfect; in stilbite also a poor (001) cleavage is observable. Mohs-hardness $3.5 - 4$. Stilbite is piezoelectric (Ventriglia 1953) in spite of its centric structure.

6.2.5 Thermal and Other Physico-Chemical Properties

The thermal behaviours of stilbite (Fig. 6.2M) and of stellerite (Fig. 6.2N) are similar, with three main water losses, the smallest one at 70°C, the largest at 175°C, and a medium one at 250°C; there is a fourth water-. or more probably hydroxyl-loss at ~500°C in stilbite, and at ~680°C in stellerite. Out of the 30 water molecules present in six sites, nearly fully occupied at 20°C (Galli 1971), nearly 15 are lost between 20°C and 200°C, and nearly other 13 are lost between 200°C and 400°C, the weight loss at the highest temperature being very small, corresponding to all the residual 1.5 H_2O.

Barrerite behaves in a slightly different way (Fig. 6.2O), because the three main water losses are shifted down to lower temperatures, which are now 50°C, 150°C and 200°C; the fourth weight (hydroxyl) loss occurs in a wide temperature interval (from 500°C to 900°C) with no DTG or DTA peak. This four step water loss is associated with a corresponding lattice shrinkage. From the data of Simonot-Grange et al. (1970) and Alberti et al. (1978), the variation in lattice dimensions of all three minerals may be summarized as follows:

1. In the interval 20°C – 180°C the shrinkage is continuous and reversible, as is the rehydration (phase A of Alberti et al. 1978).
2. At 200°C there is a sharp reduction in cell volume (12% for stilbite and stellerite, 16% for barrerite) (phase B of Alberti et al. 1978).
3. If phase B is cooled down to room temperature, it rehydrates immediately but not completely, only to 70% of the original water content, and so its unit cell remains 7% smaller than the original value for stilbite and stellerite, 12% smaller for barrerite (phase C of Alberti et al. 1978). After a year, the phases C of stilbite and stellerite have water contents and cell dimensions similar to those of the original phase A.
4. If the heating is pushed up to 350°C – 400°C, the zeolites are nearly dehydrated, their powder pattern is diffuse, but b is certainly near to 16.50 Å, and this means a 10% reduction in this parameter only (and hence a much larger reduction in cell volume). It rehydrates to a reduced extent when cooled down to room temperature (phase D of Alberti et al. 1978).

Passaglia (1980) prepared ion-exchanged phases for one stellerite, two stilbites and one barrerite, with Li, Na, K, Rb, Mg, Ca, Sr, Ba, Gd, La, and tested the influence of temperature on their 020 powder peaks. His results

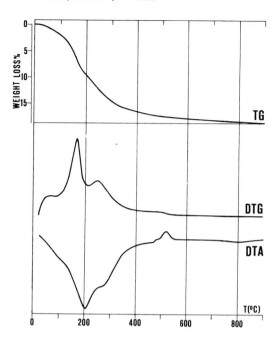

Fig. 6.2M. Thermal curves of stilbite from Poona, India; in air, heating rate 20°C/min (by the authors)

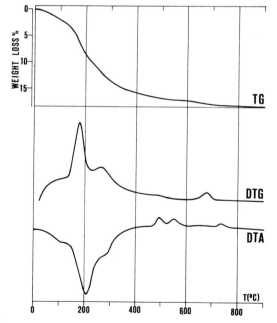

Fig. 6.2N. Thermal curves of stellerite from Villanova Monteleone, Sardinia, Italy; in air, heating rate 20°C/min (by the authors)

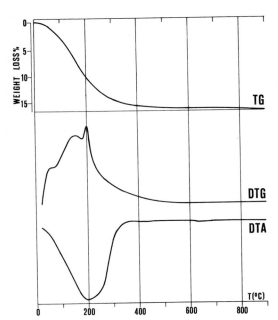

Fig. 6.20. Thermal curves of barrerite
from Capo Pula, Sardinia, Italy; in air,
heating rate 20°C/min (by the authors)

show a sharp contraction at 150°C – 200°C and destruction of lattice at
~500°C for Li and nearly all divalent cations, a sharp contraction at
150°C – 200°C and destruction of the lattice at ~850°C for Na, and a small
continuous contraction of the lattice with its destruction only at 1000°C for K
and Rb. Also the hydrogen form, prepared by ammonium exchange and
heating, has a lattice which remains undestroyed at this high temperature
(Jacobs et al. 1979).

These different thermal behaviours have been interpreted by Alberti and
Vezzalini (1978), Alberti et al. (1978), Alberti et al. (1983b), Alberti and
Vezzalini (1984) as due to the breaking of T – O – T bridges of the framework
in the following ways:

1. In all minerals, nearly up to 180°C, the reversible dehydration is associated
   with a distortion of the framework, with a rotation of the 4 – 4 – 1 units
   which characterize the structure.
2. In stellerite (and probably also in stilbite) the heating at ~200°C causes
   such a distortion of the framework as to cause the breaking of some
   T – O – T bridges and the transition to phase B, whose interrupted framework has an oxygen sharing coefficient (Zoltai 1960) slightly lower than 2.
   Where the framework is interrupted, a hydroxyl is present instead of an
   oxygen, in accordance with the presence of a fourth high temperature loss
   in the thermal curves. This phase B cannot invert back to phase A at room
   conditions, because of its different framework.

3. In barrerite the heating at 200°C causes again a strong distortion of the framework and breaking of the same $T - O - T$ bridges, but in such a way as to form new $T - O - T$ bridges, at least in part; an interrupted framework is again formed, but different from that of stellerite B. Barrerite B also does not invert to barrerite A when brought back to room conditions.

4. The other monionic modifications behave more or less like stellerite, with the exception of the Rb- and K-forms, whose framework undergoes only a minor distortion when dehydrated by heating, probably in view of the small charge and large radius of these two ions; the same is true for the H-form, because of its low charge and its different site in the structure.

Simonot-Grange (1979) studied the dehydration of stilbite controlling both T and $P(H_2O)$ and found a breaking of linearity for instance in the adsorption isotherm, when stilbite A inverts to stilbite B.

Ion exchange is certainly easy in these zeolites, as demonstrated by the large number of forms obtained by Passaglia (1980), and known much earlier through the pioneering work of Zoch (1915) and Beutell and Blaschke (1915). Ames (1966) published stilbite exchange isotherms for the couples Cs – Na, K – Na, Cs – K, Sr – Cs; the selectivity quotients for 1:1 concentration in the solution, total normality = 1, at 23 °C, are in the same order 2.36 ($Cs_z > Na_z$), 4.37 ($K_z > Na_z$), 1.11 ($Cs_z > K_z$), 3.46 ($Sr_z > Cs_z$).

Gas adsorption data on these zeolites are scanty: after the pioneering work of Rabinowitsch (1932) and Sameshima and Hemmi (1934), some data on the adsorption of CO_2, O_2, Ar, Kr, may be found in Barrer and Vaughan (1969) and on the adsorption of methanol and Ar in the Mg-form in Zyla and Zabinski (1978). Jacobs et al. (1979) remarked on the thermal stability of the H-form, which shows catalytic properties similar to those of H-clinoptilolite in the isopropanol dehydration to propylene.

6.2.6 Occurrences and Genesis

Stilbite is quite a common zeolite, but in some way exceptional from the genetical point of view, as it has never been synthetized, and hydrothermal genesis can be attributed to the majority of the natural crystals, although a low grade metamorphism may be responsible for some occurrences; nobody has described a stilbite whose crystallization may be attributed to surface or near surface conditions.

Utada (1970, 1971) does not include stilbite amongst the zeolites which crystallize in his five zones of diagenesis and low metamorphism, but reports on its presence in some zones of sediments affected by contact metamorphism around intrusive masses; stilbite, sometimes with heulandite, may occur in the outermost zone, the following nearer to the centre being the laumontite zone, which in turn is followed by the wairakite zone. Seki at al. (1969b) studied a thick pile of submarine volcanic materials subjected to burial diagenesis/metamorphism, influenced by a nearby quartz-diorite intrusion in the

Tanzawa Mts., Japan; in the zone of lowest grade, stilbite is associated with clinoptilolite or heulandite; laumontite and wairakite are present in the subsequent zone of more intense conditions. Murata and Whiteley (1973) reported on the zeolites occurring in the tuffaceous Miocene Briones sandstones and related formations of California; they find a widespread crystallization of clinoptilolite, and development of laumontite in places where the sandstones appear to have been faulted to greater depths; stilbite and heulandite appear only in fractures in a few places. Metamorphic (or deep diagenetic) stilbites have been found twice in U.S.S.R., by Zaporozhtseva et al. (1961) and by Amirov et al. (1976), but the descriptions are inadequate for a framing in the current schemes. Thus, on the whole, Utada's view seems acceptable, as stilbite occurs in thick piles of sediments only if they are affected by some contact metamorphism or hydrothermalism (for this last one see Utada 1980).

The typical hydrothermal genesis with formation of fine crystals in veins and geodes, is quite common for stilbite so that only a restricted number of occurrences may be listed here. The host rock is commonly a basic volcanic one, but may be also different, like a granite or a gneiss. Different zeolite zones may be found in this situation also; Kostov (1969) found in basic volcaniclastic formations of Bulgaria that stilbite is associated with other lamellar silica-rich zeolites in an "outer" zone, preceeded by a laumontite zone and a fibrous zeolite zone; Kristmanndóttir and Tómasson (1978) found the following zones in the basalts of Iceland, in order of increasing temperature of deposition: (1) chabazite; (2) fibrous zeolite; (3) stilbite; (4) laumontite.

Classical occurrences, already listed by Hintze (1894) and Dana (1914), are:

Scotland: in the basalts at Long Craig, Dunbartonshire, as fine red crystals; at Storr, Skye;

Northern Ireland: in the Tertiary basalts of Co. Antrim (see also Walker 1960a);

Norway: at Kongsberg and Arendal.

Faröe Islands: in the Tertiary basalt (see also Betz 1981);

Germany: at Andreasberg, Harz;

Italy: in the basalts of Forra di Bulla (= Puflerloch), South Tyrol (see also Passaglia and Moratelli 1972);

Iceland: in the Tertiary basalts (see also Betz 1981);

India: in the basalts of Poona (see also Sukheswala et al. 1974);

Canada: in the basalts of Nova Scotia (see also Walker and Parsons 1922; Aumento 1964; Aumento and Friedlander 1966; Gardiner 1966);

U.S.A.: at Bergen Hill, New Jersey (see also Manchester 1919).

Newly described occurrences in basic volcanic rocks are the following:

Hungary (see Koch 1978, for a general review): at Dunabogdany, in the andesite (Reichert and Erdélyi 1935) with chabazite, analcime; at Balaton Lake, in the basalts with natrolite (Mauritz 1938); at Matra Mts., in an andesite with heulandite, chabazite and laumontite (Mezösi 1965).

Yugoslavia: at Bor, East-Serbia, in an andesite (Majer 1953) with chabazite.
Italy: in the andesites of Sardinia (Pongiluppi et al. 1974; Pongiluppi 1974).
Japan: at Chojabaru, Nagasaki Pref., in an altered basalt with levyne, erionite, chabazite (Shimazu and Mizota 1972).
U.S.A.: in the basalts of Table Mt., Colorado (Johnson and Waldschmidt 1925); in the basalts of Ritter Hot Spring, Grant Co., Oregon (Hewett et al. 1928); in the basalts of Alexander Dam, Kauai, Hawaian Islands (Dunham 1933).
Brazil: in the basalts of southern Brazil (Franco 1952; Mason and Greenberg 1954) with heulandite, chabazite and mordenite.
Australia: in the basalts of Tasmania with chabazite, gonnardite, natrolite and phillipsite (Sutherland 1965).

Occurrences of stilbite in other kinds of rocks are:

Norway: at Østland, Kragerø, in a gneiss, with heulandite and natrolite (Saebø and Reitan 1959); at Honningvåg, Magerø, in a gneiss with stellerite (Saebø et al. 1959); at Elveneset, Innhavet, in a pegmatite (Saebø and Sverdrup 1959); this is a late hydrothermal deposition in the pegmatite, not a magmatic deposition; at Drammen, at the contact of granite with Cambro-Silurian, with heulandite, harmotome, chabazite (Raade 1969).
Switzerland: in the granite-gneisses of the Aar Massif (Parker 1922); in a gneiss from Pedemonte, Tessin (Gschwind and Brandenberger 1932).
Italy: at Val Nambrone, Trento, in a tonalite (Sanero 1938; Alberti 1971); in the granite of Baveno, near the Lago Maggiore (Pagliani 1948) with chabazite and laumontite; at Val Malenco, Lombardia, in an amphibole-chlorite schist (Scaini and Nardelli 1952).
U.S.A.: in the gneiss of Jones Falls, Maryland (Shannon 1925); as replacement of fossils at Eugene, Oregon, with analcime and heulandite (Staples 1965).
Brazil: in the northeastern region, in a micaschist with chabazite and heulandite (Rao and Cunha e Silva 1963).
Taiwan: in a basalt with natrolite, thomsonite and heulandite (Juan et al. 1964).
Japan: at Onigajo, Mié Pref., in a granite porphyry with heulandite and laumontite; this is a rare sodium-rich stilbite (Harada and Katsutoshi 1967).

Stellerite is known only from depositions in veins and geodes.
In volcanic basic rocks the occurrences are: in the diabasic tuff of the Komandor Islands, U.S.S.R., the type locality, with analcime (Morozewicz 1909, 1925). At Gelbe Birke, Fürstenberge bei Schwarzenberg im Erzgebirge, Germany; this is the type locality for "epidesmine" (Rosický and Thugutt 1913). In an andesite at Szob, Malomvölgy, Hungary, with stilbite and chabazite (Erdélyi 1943) and in the basalt of Nadap, Hungary (Erdélyi 1955). In the andesites of Villanova Monteleone, Sardinia, Italy (Galli and Passaglia

1973) and of the near locality of Seremida (Alberti et al. 1978). In the basalts of Franca, Sao Paulo, Brazil (Franco 1952).

The following occurrences of stellerite in other types of rocks are known: in a micaschist at Juneau, Alaska (Wheeler 1927). In a magnetite-pyroxene skarn at Stillböle, Finland (Mattinen 1952). In a gneiss from Mt. Medvezhyi, Siberia, with chabazite and heulandite (Shkabara 1941); in fissures of a granodiorite at Bukuka, Siberia (Barabanov 1955); in calcite veins in the Pb – Zn-deposit at Savinskoe, Siberia (Taldykina 1958); in the geothermal field at Paratumskoe, Kamchatka, where the deposition in vugs with laumontite, stilbite, heulandite and mordenite occurs at $190\,°C - 220\,°C$ (Petrova and Trukhin 1975). In the calc-schist of Val Varenna, Liguria, Italy, with chabazite (Pelloux 1949). In a pseudotachylyte in the Jam Valley, Vorarlberg, Austria, with heulandite, stilbite, chabazite (Koritnig 1940). In narrow veins in dioritic and granitic gneisses at Kongsberg, Norway (Neumann 1944). In Eocambrian schists on the island of Magerø, Norway, with stilbite (Saebø et al. 1959). In an oligoclase granite porphyry in China (Young 1965).

Barrerite is known to occur only in the type locality as deposition in a nearly monomineralic vein in a lava (andesitic to rhyolitic) at Capo Pula, Sardinia, Italy (Passaglia and Pongiluppi 1974).

6.3 Brewsterite

$Sr_2(Al_4Si_{12}O_{32}) \cdot 10\,H_2O$
$P2_1/m$
$a = 6.77$
$b = 17.51$
$c = 7.74\,(\text{Å})$
$\beta = 94\,°18'$

6.3.1 History and Nomenclature

Brooke (1822) described the new mineral from Strontian, Argyllshire, Scotland, and named it in honour of Sir David Brewster.

6.3.2 Crystallography

Perrotta and Smith (1964) determined the structure (see Chap. 0.2.6) and refined it down to R = 11% on a sample from the type locality; Schlenker et al. (1977b) reached a lower value (5%) using the same sample. There is some (Si, Al)-order: T(D), which is Al-free, is one of the two tetrahedral nodes shared by two successive heulandite units, forming chains along a. T(D) is the

Fig. 6.3 A. Brewsterite from Strontian, Scotland (×5)

centre of the only tetrahedron which does not contain O(1) and O(2), the two framework oxygens coordinated by the extraframework cation; all other tetrahedra T(A), T(B) and T(C) have O(1) and/or O(2) at their vertices. There is only one extraframework cation site, occupied mainly by Sr with some Ba, in the middle of the channels along a, coordinating four framework oxygens (two on each side) plus five water molecules (type II of Chap. 0.4).

The morphology is characterized by prismatic crystals, sometimes flattened (010), an elongation being possible in both a or c directions, with the forms {100}, {010}, {001}, {110} and {016}, less frequent {120}.

6.3.3 Chemistry and Synthesis

The brewsterites from only two occurrences have been analyzed (see Table 6.3 I). The composition is similar to that of the stoichiometric formula, apart some substitution of Sr by Ca and Ba.

No brewsterite-like phase has yet been synthesized.

6.3.4 Optical and Other Physical Properties

Some data were published by Larsen and Berman (1934) without giving the sample occurrence, although it should be Strontian, which is by far the most

Fig. 6.3B. Brewsterite from Strontian, Scotland (SEM Philips, ×30)

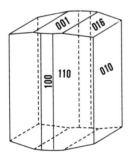

Fig. 6.3C. Common morphology of brewsterite

probable source of the mineral; other data are available for the Siberian sample (see Table 6.3 II). The two sets are similar, with one disagreement: Larsen and Berman (1934) (and also Hintze 1897; Dana 1914) give $\overset{\wedge}{\alpha c} = 22°$, whereas Khomyakov et al. (1970) give $\overset{\wedge}{\alpha a} = 28°$, both in the obtuse angle. The second value is probably true, because the old authors assumed the elongation to be always c, but Khomyakov et al. (1970) showed that crystals may also be a-elongated.

Transparent colourless to white or yellowish, streak white, lustre vitreous, also nacreous on the (010)-cleavage planes.

Table 6.31. Chemical analyses and formulae on the basis of 32 oxygens and lattice constants in Å

	1	2	3
SiO_2	54.42		54.02
Al_2O_3	15.25		15.86
Fe_2O_3	0.08		0.11
CaO	1.19		0.80
SrO	8.99		11.80
BaO	6.80		3.01
Na_2O			0.21
K_2O			0.14
H_2O	13.22		13.72
Total	99.95		99.67
Si	12.01	11.9	11.89
Al	3.97	4.1	4.12
Fe	0.01		0.02
Ca	0.28		0.19
Sr	1.15	1.42	1.51
Ba	0.59	0.48	0.26
Na			0.09
K		0.02	0.04
H_2O	9.73	10.0	10.07
E%	-1.2	+7.3	+2.4
a		6.793	6.82
b		17.573	17.51
c		7.759	7.75
β		94°32'	94°16'
D	2.45		2.32

1. Strontian, Scotland (Mallet 1859).
2. Strontian, Scotland (Schlenker et al. 1977b); H_2O from structure refinement.
3. Burpala pluton, Siberia (Khomyakov et al. 1970).

Table 6.3 II. Optical data

	1	3
α	1.510	1.506
β	1.512	1.510
γ	1.523	1.522
2 V	+50°	+60°

$$\gamma \rightarrow \underline{b} \qquad \gamma \rightarrow \underline{b}$$
$$\alpha \widehat{} \underline{c} = 22° \qquad \alpha \widehat{} \underline{a} = 28°$$
in the obtuse angle

1. Probably from Strontian, Scotland (Larsen and Berman 1934).
3. Burpala pluton, Siberia (Khomyakov et al. 1970).

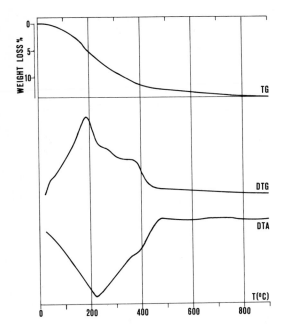

Fig. 6.3D. Thermal curves of brewsterite from Strontian, Scotland; in air, heating rate 20°C/min (by the authors)

Mohs-hardness 4 – 4.5, microhardness from 172 to 330 kg/mm^2 (Khomyakov et al. 1970).

Brewsterite is piezoelectric (Ventriglia 1953) in spite of its centricity.

6.3.5 Thermal and Other Physico-Chemical Properties

The DTG curve (Fig. 6.3 D) shows a shoulder at 50°C, and three peaks at 190°C, 260°C and 380°C, well in accordance with the presence of four water sites in the structure (Schlenker et al. 1977b); the water loss is nearly complete at 500°C, but continues at a slow rate up to 800°C.

6.3.6 Occurrences and Genesis

Brewsterite always occurs in vugs of massive rocks, the hydrothermal genesis being probable. The only Sr-zeolite seems to be quite rare. The type locality is Strontian, Argyllshire, Scotland, where it occurs with calcite and galena in veins in a gneiss.

Hintze (1897) and Dana (1914) list several other occurrences, which seem to be rather uncertain because of the absence of reliable chemical evidence for the presence of strontium in these samples. These doubtful occurrences are:

Scotland: Kilpatrick (perhaps the edingtonite locality).
Northern Ireland: Giant's Causeway, Antrim.
France: Departement de l'Isère; Mont Caperan, Barèges, Hautes-Pyrénées (see also Candel Vila 1949); Col de Bonhomme, south west of Mont Blanc.
Germany: Freiburg im Breisgau; St. Turpet, Münsterthal.

There are only a few more recent finds:

U.S.S.R.: in the Burpala pluton, North Baikal region, Siberia, in a joint cutting aplite and albitized syenite, with stilbite, heulandite, natrolite and pyrite (Khomyakov et al. 1970).
Canada: in the flattened vesicles of the porphyritic trachyte flows from Yellow Lake near Olalla, southern British Columbia, with thomsonite, mesolite, calcite, analcime (Wise and Tschernich 1978b).
U.S.A.: in a contact metamorphic zone in Danbury, Connecticut, with epistilbite (Pawloski 1965).

7 Zeolites with Unknown Structure

7.1 Cowlesite

$Ca_6(Al_{12}Si_{18}O_{60}) \cdot 36 H_2O$
orthorhombic
$a = 11.27$
$b = 15.25$
$c = 12.61$ (Å)

7.1.1 History and Nomenclature

Wise and Tschernich (1975) found and described the new mineral from 7 different occurrences; the first on the list, Goble, Oregon, is assumed to be the type locality; the mineral is named for James Cowles of Rainer, Oregon, amateur mineralogist and enthusiastic zeolite collector.

7.1.2 Crystallography

Cowlesite crystals are blade-shaped and very small, not more than 0.1 mm in width and only 2 μm thick, so that complete single crystal studies have been unsuccessful until now. Cell dimensions and symmetry were derived from powder data and from very faint precession photographs and hence are reliable within the limits of the methods applied; the orthorhombicity is also supported by the optical data.

The crystals are {010} lamellae with very narrow {100} and {101} faces.

7.1.3 Chemistry and Synthesis

The chemistry of cowlesite is very near to the stoichiometric formula of the title page in all seven analyses given by Wise and Tschernich (1975): two are listed in Table 7.1I together with one by Rinaldi (written communication). Mg, Sr, Ba are always absent, K is very low. Apart from the higher water content, cowlesite is a polymorph of scolecite. No other analysis has been published yet, nor is any synthesis of a cowlesite-like phase known.

Lattice constants are $a = 11.29$, $b = 15.24$, $c = 12.68$ (Å) for a sample from Kuniga, Japan (Matsubara et al. 1978b).

Fig. 7.1 A. Cowlesite from Spray, Oregon (SEM Philips, ×120)

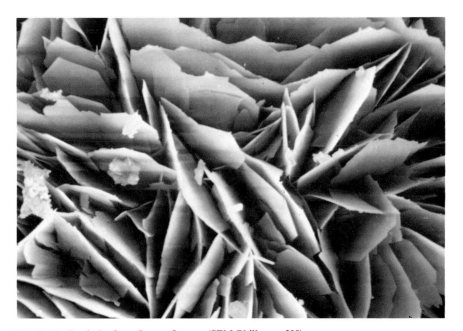

Fig. 7.1 B. Cowlesite from Spray, Oregon (SEM Philips, ×500)

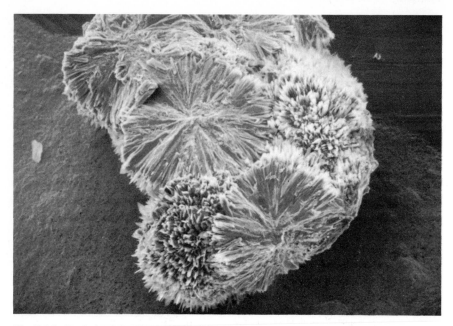

Fig. 7.1C. Cowlesite from Skye, Scotland (SEM Philips, ×60)

Fig. 7.1D. Cowlesite from Skye, Scotland (SEM Philips, ×120)

Fig. 7.1E. Cowlesite from Skye, Scotland (SEM Philips, ×950)

Fig. 7.1F. Cowlesite from Skye, Scotland (SEM Philips, ×3800)

Table 7.1I. Chemical analyses and formulae on the basis
of 60 oxygens and lattice constants in Å

	1	2	3
SiO_2	40.92	41.47	43.46
Al_2O_3	23.29	23.13	23.36
Fe_2O_3		0.07	
MgO			0.11
CaO	12.32	13.52	11.09
Na_2O	0.67	0.72	0.52
K_2O		0.09	0.36
H_2O	22.8	21.0	21.10
Total	100.00	100.00	100.00
Si	17.94	17.88	18.50
Al	12.04	11.76	11.72
Fe		0.02	
Mg			0.07
Ca	5.79	6.25	5.06
Na	0.57	0.60	0.43
K		0.05	0.19
H_2O	33.34	30.20	29.95
E%	-0.9	-10.4	+7.7
a	11.27		
b	15.25		
c	12.61		
D	2.14	2.12	

1. Goble, Oregon (Wise and Tschernich
1975) with anhydrous part normalized
to 77.2%; lattice constants probably
from this sample, although not clearly
stated by the authors.
2. Superior, Arizona (do) with anhydrous
part normalized to 79.0%.
3. Skye, Scotland (Rinaldi, written
communication).

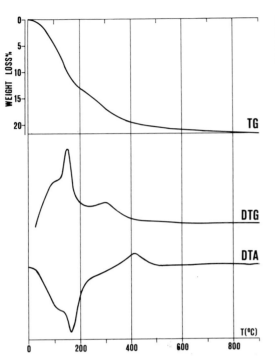

Fig. 7.1G. Thermal curves of cowlesite from Island Magee, Co. Antrim, Northern Ireland; in air, heating rate 20°C/min (by the authors)

7.1.4 Optical and Other Physical Properties

Cowlesite from the seven original occurrences is biaxial negative with $\alpha = 1.512$, $\beta = 1.515$, $\gamma = 1.517$, $2V = -44$ to $-53°$, $\alpha \rightarrow b$, $\beta \rightarrow a$, $\gamma \rightarrow c$ (Wise and Tschernich 1975). The sample from Kuniga, Japan (Matsubara et al. 1978b) has $\alpha = 1.505$, $\beta \simeq \gamma = 1.509$, $2V = -30°$, $\alpha \rightarrow a$, $\beta \rightarrow b$, $\gamma \rightarrow c$.

Transparent white, streak white. (010)-cleavage: perfect; Mohs-hardness probably lower than 4, the average value for zeolites.

7.1.5 Thermal and other Physico-Chemical Properties

Figure 7.1 G shows the TG, which has two main water losses at 150°C and 300°C and a shoulder at 100°C; the loss continues at a very low rate up to 600°C and perhaps over.

7.1.6 Occurrences and Genesis

All known occurrences are depositions in basic lavas.

Wise and Tschernich (1975) described the following occurrences:

U.S.A.: in small vugs in a late Eocene basalt, in a roadcut 0.4 miles northwest of Goble, Columbia Co., Oregon (type locality); in most vugs cowlesite,

as cavity lining of radially oriented blades, is the only zeolite, more rarely
analcime, garronite, phillipsite, levyne and thomsonite may be present; in
cavities of the basalt from the Beech Creek Quarry, Grant Co., Oregon,
with levyne (other zeolites may be present if cowlesite is absent); in the
basalt from a roadcut 10 miles northeast of Spray, Wheeler Co., Oregon,
as only zeolite present in some cavities, other zeolites being present if
cowlesite is absent; in tiny vesicles in the olivine basalt bombs in a Middle
Tertiary cinder cone 5.5 miles south of Superior, Pinal Co., Arizona,
closely associated with thomsonite, chabazite, analcime and calcite; in the
vugs of a basalt (?) from Capitol Peak, Thurston Co., Washington; in a
shoshonite from Table Mt., near Golden, Colorado.
Canada: in basalt scoria and breccias from roadcuts south of Monte Lake,
British Columbia.

Other occurences are:

Iceland: in small cavities of a levyne-rich basalt from Mjödalsá Canyon near
Hvammur (Betz 1981).
Faröes: in samples from Dalsnipa, Sandoy, and from Stremoy (Tschernich in
Betz 1981).
Scotland: in the basalt from Kingsburgh, Skye (Rinaldi, written communication).
Northern Ireland: in the basalt of Island Magee, Co. Antrim (H. Foy, written
communication).
Japan: as felt-like material on the walls of the amygdales of a trachybasalt
from Kuniga tunnel, Dōzen, Oki island (Matsubara et al. 1978b); the
locality is the same described by Tiba and Matsubara (1977) for levyne,
whose lamellae overlay the cowlesite felt; chabazite, thomsonite, phillipsite
and analcime may also be present.

7.2 Goosecreekite

$Ca_2(Al_4Si_{12}O_{32}) \cdot 10H_2O$
$P2_1/m$
$a = 7.52$
$b = 17.56$
$c = 7.35 (Å)$
$\beta = 105°43'$

7.2.1 History and Nomenclature

The mineral was found by Mr. George Brewer of Columbia, Maryland, when
collecting zeolites in the Goose Creek quarry, Loudoun Co., Virginia, and

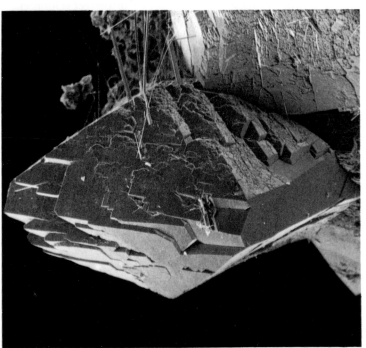

Fig. 7.2A. Goosecreekite from Goose Creek, Virginia (SEM, ×16) (Courtesy of P. J. Dunn, D. R. Peacor, N. Newberry, R. A. Ramik, and of "The Canadian Mineralogist")

Fig. 7.2B. Goosecreekite from Goose Creek, Virginia (SEM, ×32) (Courtesy of P. J. Dunn, D. R. Peacor, N. Newberry, R. A. Ramik, and of "The Canadian Mineralogist")

described by Dunn et al. (1980), who named it for the locality of the first and, so far, only find.

7.2.2 Crystallography

The mineral occurs both as polycrystalline aggregates and as imperfect single crystals up to 2 mm in size, the two pinacoids {100} {001} being present with other undetermined forms (see Figs. 7.2 A and B).

7.2.3 Chemistry and Synthesis

Dunn et al. (1980) give the following analysis: SiO_2 59.3, Al_2O_3 17.2, CaO 9.3, H_2O 15.0, total 100.8 (all %), corresponding to the formula:

$$Ca_{2.01}(Al_{4.08}Si_{11.94}O_{32}) \cdot 10.07\,H_2O$$

balance error E% = +1.7, D = 2.21; the lattice constants are under the title. No synthesis is known for this phase.

7.2.4 Optical and Other Physical Properties

Goosecreekite is biaxial negative with $\alpha = 1.495$, $\beta = 1.498$, $\gamma = 1.502$, $2V = -82°$, $\beta \rightarrow b$, $\overset{\wedge}{\gamma c} = 46°$ (in the obtuse angle?). Transparent colourless to white, streak white. Mohs-hardness 4+; cleavage (010) perfect (Dunn et al. 1980).

7.2.5 Thermal and Other Physico-Chemical Properties

The mineral loses (Dunn et al. 1980) 2.7% H_2O after 70 h in vacuum, and the remaining 12.3% in two steps, the first between 55 °C and 222 °C, the second at 328 °C.

7.2.6 Occurrences and Genesis

The first and only find of goosecreekite was as deposition in vugs of a Triassic diabase, where it is associated with quartz and chlorite; the rock is from the new Goose Creek quarry, Loudoun Co., Virginia.

7.3 Partheite

$Ca_8(Al_{16}Si_{16}O_{64}) \cdot 16\,H_2O$
C2/c
$a = 21.59$
$b = 8.78$
$c = 9.31\,(\text{Å})$
$\beta = 91°28'$

7.3.1 History and Nomenclature

Sarp et al. (1979) proposed the new name for a mineral from a rodingite from the Taurus Mts., Turkey, in honour of Erwin Parthé, professor of crystallography at the University of Geneva.

7.3.2 Crystallography

It occurs either as radial aggregates up to 0.3 mm in length including small (ca. 0.05 mm) grains, or as hypidiomorphic grains nearly 0.2×0.1 mm in size. It is not known which crystallographic axis the fibre axis corresponds to. No twins were observed.

7.3.3 Chemistry and Synthesis

The chemical composition is quite near to the chemical formula; the average values of the three analyses of Sarp et al. (1979) are: SiO_2 39.06, Al_2O_3 30.71, CaO 16.30, Na_2O 0.32, K_2O 0.23, H_2O 13.38, Total 100.00 (all %) corresponding to the formula

$$Ca_{7.14}Na_{0.26}K_{0.12}(Al_{15.41}Si_{16.63}O_{64}) \cdot 19.00\,H_2O$$

$D = 2.39$; the lattice constants are under the title. Composition and lattice constants of a sample from the Urals are similar (Ivanov and Mozzherin 1982), but with $\beta = 91°47'$ and $D = 2.44$.

Partheite is a polymorph of gismondine, if the water content is neglected. No synthesis is known for this phase.

7.3.4 Optical and Other Physical Properties

The mineral is biaxial positive with $\alpha = 1.547$, $\beta = 1.549$, $\gamma = 1.559$, $2V = +48°$ (Turkey) and $\alpha = 1.550$, $\beta = 1.552$, $\gamma = 1.565$, $2V = +45°30'$ (Urals). The orientation is not known, but there is a 23° to 30° extinction angle between the fibre axis and α.

Transparent white (Turkey) or dark blue (Urals), vitreous lustre, white streak. Mohs-hardness 4; cleavage planes {100} {110} {010}.

7.3.5 Thermal and Physico-Chemical Properties

The DTA curve (Ivanov and Mozzherin 1982) has endothermic peaks at 255 °C, 430 °C and 590 °C.

Sarp et al. (1979) took several Gandolfi patterns at room temperature after heating the mineral to higher temperatures: a heating at 150 °C does not change the normal pattern; after a heating at 300 °C, the pattern is still very similar to the original one, although some changes are clear; a heating at 350 °C inverts the mineral to a different unknown phase; no pattern is obtained after heating at 400 °C.

7.3.6 Occurrences and Genesis

Partheite has been found in the Taurus Mts., at 7 km southeast of Doganbaba, Burdur Province, Turkey (type locality), in a rodingitic vein of an ophiolitic formation, with prehnite, thomsonite, augite (Sarp et al. 1979).

The second find (Ivanov and Mozzherin 1982) is in a gabbro-pegmatite of Denezhkin Kamen, Urals, U.S.S.R.

Appendix 1

X-Ray Powder Patterns of Natural Zeolites

The first two digits of the table codes correspond to the chapter in which the zeolite is described.

Table 1.1 II. X-ray powder patterns of orthorhombic and tetragonal natrolites

Natrolite					Tetranatrolite		Natrolite					Tetranatrolite	
d(A)	I/Io	H	K	L	I/Io	d(A)	d(A)	I/Io	H	K	L	I/Io	d(A)
6.53	74	2	2	0	50	6.53	2.582	43	2	4	2	25	2.580
5.88	36	1	1	1	100	5.90	2.570	71	4	2	2		
4.66	35	0	4	0	25	4.61	2.552	16	6	4	0	–	–
4.59	30	4	0	0			2.448	88	1	7	1	25	2.425
4.39	58	1	3	1	50	4.38	2.410	86	7	1	1		
4.35	70	3	1	1			2.331	15	0	8	0	–	–
4.15	42	2	4	0	25	4.12	2.318	37	4	4	2	5	2.322
4.11	37	4	2	0					3	7	1	–	–
3.622	2	3	3	1	–	–	2.288	18					
3.261	12	4	4	0	–	–			8	0	0	–	–
3.192	42	1	5	1	50	3.171	2.260	36	0	6	2	5	2.250
3.151	52	5	1	1			2.239	7	6	0	2		
		0	2	2	25	3.114	2.222	7	8	2	0	–	–
3.098	29	2	0	2			2.194	58	2	6	2	15	2.187
		2	2	2	25	2.949			6	2	2		
2.939	36	2	6	0	5	2.914	2.177	100					
2.897	9	6	2	0					6	6	0	–	–
2.863	80	3	5	1	100	2.851	2.076	11	4	8	0	10	2.061
2.844	74	5	3	1			2.053	27	8	4	0		
							1.933	23	9	1	1	–	–
									1	5	3		
							1.883	21	3	9	1	15	1.881
									5	1	3		

Samples from Tierno, Trento, Italy and from Ilímaussaq, Greenland; natrolite data from Alberti et al. (1982a); tetranatrolite data from Krogh Andersen et al. (1969).

Table 1.1III. X-ray powder pattern of mesolite

d(A)	I/Io	H	K	L	d(A)	I/Io	H	K	L
6.59	100	2	6	0	2.472	8	1	21	1
6.12	4	1	1	1	2.443	1	6	14	0
5.86	38	1	3	1	2.421	9	7	3	1
5.41	4	1	5	1	2.361	2	0	24	0
4.72	49	0	12	0			5	17	1
4.60	29	4	0	0	2.312	4	3	21	1
4.46	22	3	1	1			8	0	0
4.40	3	1	9	1	2.270	3	0	18	2
4.35	24	3	3	1			6	0	2
4.20	32	2	12	0	2.237	4			
4.14	16	4	6	0			8	6	0
3.917	2	3	7	1	2.198	17	6	18	0
3.864	3	4	8	0	2.178	4	6	6	2
3.373	2	3	11	1			6	8	2
3.294	4	2	16	0			8	10	0
3.218	14	1	15	1			4	16	2
3.163	19	5	3	1	2.130	2			
3.091	7	0	6	2			7	13	1
3.084	7	2	0	2			1	5	3
2.975	10	2	18	0			1	25	1
2.931	16	2	6	2	2.081	1	6	20	0
2.885	55	3	15	1	2.069	1	8	12	0
2.857	56	5	9	1	2.044	2	3	3	3
2.598	2	4	18	0	2.022	1	3	5	3
2.582	4	2	12	2	2.000	1	8	14	0
2.574	5	6	12	0	1.986	1	1	27	1
2.568	5	4	6	2					

Sample from Harbabkhandi, Azerbaijan, Iran; data from Alberti et al. (1982a).

Table 1.1IV. X-ray powder pattern of scolecite

d(A)	I/Io	H	K	L	d(A)	I/Io	H	K	L
9.26	12	1	1	0	2.794	16	3	3	2
6.55	62	0	2	0	2.677	78	2	4	2
5.90	24	0	1	2			1	3	4
5.37	5	1	1	2	2.582	17			
4.64	98	2	2	0			3	1	4
4.38	34	2	1	2	2.564	10	5	1	0
		1	3	0	2.433	21	4	3	2
4.14	61				2.326	4	0	4	4
		3	1	0	2.313	6	4	4	0
3.953	6	1	3	1	2.284	6	2	5	2
3.789	5	2	2	2	2.255	21	3	3	4
		1	3	2			2	4	4
3.506	67				2.190	24			
		3	1	2			4	2	4
3.277	18	0	4	0	2.178	50	6	0	0
3.202	20	0	1	4			3	5	2
3.182	46	3	2	2	2.124	9			
3.083	6	3	3	0			5	3	2
		0	2	4	2.087	5	0	2	6
2.948	38				2.067	14	6	2	0
		2	0	4	2.025	2	5	1	4
2.926	40	4	2	0	1.976	3	2	6	2
2.859	100	4	1	2					

Sample from Deloraine, Tasmania; data from Alberti et al. (1982a).

Table 1.2 III. X-ray powder pattern of thomsonite

d(Å)	I/I₀	H	K	L	d(Å)	I/I₀	H	K	L
6.63	100	2	2	0	2.579	7	4	2	-2
5.87	59	1	1	-1	2.479	8	1	7	-1
4.74	74	0	4	0			1	7	1
4.62	49	4	0	0	2.448	2	5	5	1
4.40	35	1	3	1	2.440	3	7	1	-1
4.38	37	3	1	-1	2.421	8	7	1	1
4.34	8	3	1	1	2.371	1	0	8	0
4.22	31	2	4	0	2.336	1	4	4	-2
4.16	20	4	2	0	2.322	3	3	7	-1
3.643	10	3	3	1	2.316	4	4	4	2
3.308	4	4	4	0			3	7	1
3.222	14	1	5	1			8	0	0
3.187	10	5	1	-1	2.296	4	2	8	0
3.158	14	5	1	1	2.293	4	2	8	0
3.083	7	2	0	-2			7	3	-1
3.065	4	2	0	2	2.272	4	0	6	2
2.991	12	2	6	0	2.254	3	6	0	-2
2.934	31	2	2	-2	2.248	3	8	2	0
		6	2	0	2.208	29	2	6	-2
2.902	40	3	5	-1			6	6	0
2.889	54	3	5	1			2	6	2
2.882	48	3	5	1	2.170	3	6	2	2
		5	3	-1	2.143	1	1	1	3
2.858	42	5	3	1	2.111	3	4	8	0
2.607	3	2	4	-2	2.078	5	8	4	0
		6	4	0	2.063	1	7	5	-1
2.584	9	2	4	2	2.032	4	3	1	3
		4	2	-2	1.993	3	1	9	-1

Sample from Tiglieto, Savona, Italy. New data by the authors; Philips diffractometer, Ni filtered CuKα radiation ($\lambda = 1.54051$ Å), internal standard $Pb(NO_3)_2$ (cubic: $a = 7.8568$ Å), slits 1°, 0.1 mm, 1°, scanning speed 0.5°/min, indexing after Alberti's (1976) method taking into account the diffraction intensities from single crystal measurements; cell parameters: $a = 13.071(2)$, $b = 13.105(2)$, $c = 13.213(2)$ Å.

Table 1.3 III. X-ray powder patterns of orthorhombic and tetragonal edingtonite

Orthorhombic					Tetragonal		Orthorhombic					Tetragonal	
d(A)	I/I₀	H	K	L	I/I₀	d(A)	d(A)	I/I₀	H	K	L	I/I₀	d(A)
6.78	14	1	1	0	13	6.79	2.597	34	1	2	2		
6.51	40	0	0	1	39	6.54						36	2.594
5.40	33	0	1	1			2.592	47	2	1	2		
					37	5.40	2.469	11	2	3	1		
5.38	35	1	0	1			2.460	11	3	2	1	8	2.461
4.83	73	0	2	0			2.410	3	0	4	0	3	2.397
					75	4.81	2.340	4	1	4	0	–	–
4.77	64	2	0	0			2.289	27	0	3	2		
4.70	55	1	1	1	47	4.70						29	2.284
4.30	18	1	2	0			2.274	19	3	0	2		
					17	4.29	2.263	100	3	3	0	75	2.261
4.28	11	2	1	0			2.240	5	4	0	1	–	–
3.878	7	0	2	1	4	3.867	–	–	3	1	2	3	2.222
3.591	100	1	2	1			2.201	21	1	4	1		
					100	3.587						17	2.192
3.578	93	2	1	1			2.181	19	4	1	1		
3.393	65	2	2	0	46	3.392	2.153	22	2	4	0	–	–
3.256	7	0	0	2	6	3.262	2.137	35	4	2	0		
3.085	16	0	1	2	12	3.085			3	3	1	23	2.139
		1	0	2			2.116	7	0	1	3	–	–
3.050	65	1	3	0					1	0	3	–	–
					47	3.027	–	–	1	1	3	8	2.069
3.021	57	3	1	0			2.062	12	2	3	2		
3.010	53	2	2	1	33	3.011						9	2.063
2.936	24	1	1	2	17	2.939	2.057	11	3	2	2		
2.763	87	1	3	1			2.045	10	2	4	1		
					83	2.753						7	2.034
2.741	96	3	1	1			2.032	10	4	2	1		
2.668	12	2	3	0			1.976	4	2	0	3	4	1.981
					14	2.658							
2.657	17	3	2	0									

Orthorhombic sample from Böhlet Mine, Sweden; tetragonal sample from Ice River, British Columbia, Canada; data from Mazzi et al. (1984).

Table 1.4III. X-ray powder pattern of gonnardite

d(A)	I/Io	H	K	L		d(A)	I/Io	H	K	L
		2	0	0				5	1	0
6.64	76					2.608	11			
		0	0	1				3	1	2
		2	1	0				5	2	0
5.90	39					2.468	17	5	0	1
		1	0	1				4	3	1
		2	2	0				4	4	0
4.70	61					2.351	7			
		2	0	1				4	0	2
		3	0	0		2.306	5	4	1	2
4.41	37							5	3	0
		2	1	1		2.268	5			
4.20	24	3	1	0				3	3	2
		3	2	0				6	0	0
3.675	3					2.211	17	4	4	1
		3	0	1				4	2	2
3.580	4	3	1	1		2.178	3	1	0	3
		4	1	0				6	2	0
3.213	31	3	2	1		2.102	3			
		1	0	2				6	0	1
3.109	12	1	1	2				5	0	2
		4	0	1		2.067	4	4	3	2
2.954	27							2	1	3
		2	0	2				6	3	0
		4	1	1		1.981	4			
2.898	100							5	4	1
		2	1	2						

Sample from Gignat, France. New data by the authors; Philips diffractometer, Ni filtered CuKα radiation ($\lambda = 1.54051$ Å), internal standard $Pb(NO_3)_2$ (cubic: $a = 7.8568$ Å), slits 1°, 0.1 mm, 1°, scanning speed 0.5°/min, indexing after Alberti's (1976) method taking into account the diffraction intensities from single crystal measurements; tetragonal cell parameters: $a = 13.289(5)$, $c = 6.600(3)$ Å.

Table 2.1III. X-ray powder pattern of analcime

d(A)	I/Io	H	K	L		d(A)	I/Io	H	K	L
5.61	51	2	1	1				6	1	1
4.86	13	2	2	0		2.230	8			
3.804	<1	3	2	0*				5	3	2
3.669	7	3	2	1		2.172	1	6	2	0
3.436	100	4	0	0		2.120	1	5	4	1
		4	1	1*		2.025	1	6	3	1
3.242	<1							5	4	3
		3	3	0*		1.941	<1	7	1	0*
2.927	48	3	3	2				5	5	0*
2.804	5	4	2	2		1.905	12	6	4	0
		4	3	1		1.869	7	6	3	3
2.695	15					1.836	<1	6	4	2
		5	1	0*				6	5	1
2.508	13	5	2	1		1.745	15			
2.428	8	4	4	0				7	3	2

Sample from Alta Val Duron, Trento, Italy. New data by the authors; Philips diffractometer, Ni filtered CuKα radiation ($\lambda = 1.54051$ Å), internal standard $Pb(NO_3)_2$ (cubic: $a = 7.8568$ Å), slits 1°, 0.1 mm, 1°, scanning speed 0.5°/min, indexing after Alberti's (1976) method taking into account the diffraction intensities from single crystal measurements; cubic cell parameter: $a = 13.734(1)$ Å.

* Not in agreement with space group Ia3d.

322   Appendix 1

Table 2.1IV. X-ray powder pattern of wairakite

d(A)	I/I₀	H	K	L		d(A)	I/I₀	H	K	L
6.82	19	0	2	0				3	1	-4
		1	2	-1		2.677	14			
5.57	59	2	1	1				4	3	1
		1	2	1				5	1	2
4.83	21	2	2	0		2.488	20			
		1	3	2				1	5	2
3.636	19					2.417	9	4	4	0
		3	1	2		2.342	1	3	5	0
3.405	100	0	4	0		2.273	3	2	4	-4
3.212	3	1	4	1		2.262	2	2	4	4
3.062	4	4	0	-2				3	2	-5
		2	0	-4				1	6	-1
3.048	5	0	4	2		2.211	9	1	6	1
		4	0	2				5	3	2
		3	3	-2				3	5	2
2.914	43					2.146	2	0	2	6
		3	2	-3		2.109	2	5	1	-4
2.902	40	3	3	2		2.092	2	4	1	5
2.781	5	2	2	-4		2.053	1	2	2	-6
2.768	7	2	2	4		1.996	5	3	1	6

Sample from Wairakei, New Zealand. New data by the author; Philips diffractometer, Ni filtered CuKα radiation ($\lambda = 1.54051$ Å), internal standard $Pb(NO_3)_2$ (cubic: $a = 7.8568$ Å), slits 1°, 0.1 mm, 1°, scanning speed 0.5°/min, indexing after Alberti's (1976) method taking into account the diffraction intensities from single crystal measurements; cell parameters: $a = 13.685(7)$, $b = 13.642(5)$, $c = 13.557(5)$ Å, $\beta = 90°28'(2)$.

Table 2.1V. X-ray powder pattern of viseite

d(A)	I/I₀	H	K	L
5.68	40	2	1	1
4.98	10	2	2	0
3.46	50	4	0	0
2.92	100	3	3	2
		5	3	2
2.20	20			
		6	1	1
2.11	<10	5	4	1
2.014	<10	6	3	1
1.886	30	6	4	0
		6	5	1
1.740	60			
		7	3	2

Sample from Visé, Belgium; data from Mc Connell (1952).

Table 2.2 IV. X-ray powder pattern of laumontite

d(Å)	I/I₀	H	K	L	d(Å)	I/I₀	H	K	L
9.43	78	1	1	0	2.876	38	5	1	-1
6.83	56	2	0	0	2.876	38	4	2	-2
6.18	9	2	0	-1	2.644	2	5	1	-2
5.04	18	1	1	1	2.628	3	3	3	1
4.72	16	2	2	0	2.571	34	2	4	1
4.49	32	2	2	-1	2.537	3	1	3	2
4.15	100	1	3	0	2.517	5	2	2	2
3.762	8	1	3	-1	2.515	9	2	0	-3
3.657	42	4	0	-1	2.452	7	6	0	-1
3.506	94	0	0	2	2.437	43	4	4	-1
3.404	26	1	3	1	2.388	2	4	0	-3
3.358	34	3	1	-2	2.358	23	1	5	1
3.265	63	0	4	0	2.267	9	3	5	0
3.196	45	3	1	1	2.215	8	6	2	-2
3.091	4	4	0	-2	2.178	8	0	6	0
3.031	45	4	2	0	2.150	28	6	2	0
2.947	8	2	4	0	2.087	4	3	3	2
					1.990	5	5	3	-3

Sample from Tanzawa Mountains, Japan; data from Liou (1971b).

Table 2.3 III. X-ray powder pattern of yugawaralite

d(Å)	I/I₀	H	K	L	d(Å)	I/I₀	H	K	L
14.01	2	0	1	0	2.647	11	2	1	1
7.80	4	0	1	1	2.638	10	2	3	-2
7.01	26	0	2	0	2.602	2	2	3	0
6.28	4	1	0	0	2.575	7	2	2	-3
5.82	55	1	1	-1	2.559	4	1	5	0
5.61	2	0	2	1	2.515	5	1	0	-4
4.67	100	1	2	0 / 0	3	0	2.474	5	1	1	-4
4.44	9	0	1	2	2.427	7	1	1	3
4.41	12	1	1	-2	2.404	3	0	5	2
4.30	30	1	1	1	2.364	6	1	2	-4
4.18	16	0	3	1	2.360	7	2	4	-2
3.896	8	0	2	2	2.345	7	0	0	4
3.769	9	1	3	-1	2.336	8	2	3	1
3.307	8	0	3	2	2.292	1	2	1	-4
3.272	10	2	1	-1	2.227	3	2	1	2
3.237	36	1	0	2	2.205	3	3	1	-2
3.194	5	2	0	-2	2.191	5	1	6	-1
3.135	9	2	0	0	2.179	4	1	3	3
3.115	8	2	1	-2	2.148	5	2	2	2
3.049	93	1	4	0 / 0	1	3	2.136	6	2	4	1
2.997	5	1	2	-3	2.119	2	1	5	2
2.938	17	1	2	2	2.105	4	2	5	-2 / 3	1	-3
2.906	24	2	2	-2	2.089	9	3	0	0
2.863	5	2	2	0	2.042	3	1	4	-4
2.854	5	0	2	3	2.014	4	1	4	3
2.769	18	1	4	1	2.003	5	3	3	-2 / 3	3	-1
2.719	15	2	1	-3	1.974	5	0	7	0 / 1	0	4
2.685	13	0	5	1					
2.658	6	1	3	2					

Sample from Osilo, Sassari, Sardinia, Italy. New data by the authors; Philips diffractometer, Ni filtered CuKα radiation ($\lambda = 1.54051$ Å), internal standard Pb(NO₃)₂ (cubic: $a = 7.8568$ Å), slits 1°, 0.1 mm, 1°, scanning speed 0.5°/min, indexing after Alberti's (1976) method taking into account the diffraction intensities from single crystal measurements; cell parameters: $a = 6.728(1)$, $b = 14.007(2)$, $c = 10.056(2)$ Å, $\beta = 111°12'(1)$.

Table 2.4I. X-ray powder pattern of roggianite

d(A)	I/Io	H	K	L	d(A)	I/Io	H	K	L
13.08	100	1	1	0	2.67	20	4	4	2
9.27	90	2	0	0	2.62	10	6	3	1
6.13	90	2	1	1	2.57	10	5	5	0
5.81	35	3	1	0			7	1	0
		3	3	0			5	3	2
4.33	20				2.53	35	4	1	3
		1	1	2	2.45	35	6	2	2
4.12	35	2	0	2	2.37	10	4	3	3
3.74	20	2	2	2	2.31	35B	0	0	4
3.60	70	3	1	2	2.26	20	5	5	2
3.41	70	4	3	1			8	2	0
		4	0	2	2.19	10			
3.22	50B						2	2	4
		5	2	1	2.14	10	3	1	4
3.13	50	5	3	0	2.05	10	8	0	2
2.88	50B	2	1	3	2.02	10	9	1	0
2.82	50	5	1	2	1.93	35	9	3	0

Sample from Alpe Rosso, Val Vigezzo, Novara, Italy. The data from Passaglia (1969) have been re-indexed by the authors after Alberti's (1976) method taking into account the diffraction intensities from single crystal measurements; cell parameters: $a = 18.31(2)$, $c = 9.23(1)$ Å. B = broad.

Table 3.1III. Powder pattern of gismondine

d(A)	I/Io	H	K	L	d(A)	I/Io	H	K	L
10.00	3	1	0	0	2.744	76	3	2	-1
7.30	63	1	1	0	2.714	59	1	2	-3
5.94	7	1	1	-1	2.693	78	3	2	1
5.77	15	1	1	1	2.662	69	1	2	3
5.32	4	0	2	0	2.658	72	0	4	0
5.00	17	2	0	0	2.624	17	3	1	2
4.91	52	0	0	2	2.607	11	2	1	3
4.68	17	0	2	1	2.567	7	0	4	1
4 46	10	0	1	2	2.521	15	2	3	-2
4.33	6	1	0	2	2.492	7	1	4	-1
4.27	100	1	2	-1	2.475	8	2	2	-3
4.21	51	1	2	1	2.467	9	2	3	2
4.18	34	2	1	-1	2.458	8	0	0	4
4.05	30	2	1	1	2.407	13	0	3	3
4.02	6	1	1	2	2.389	7	4	1	-1
3.642	8	2	2	0	2.340	16	0	4	2
3.606	8	0	2	2	2.293	4	2	4	-1
3.587	5	2	0	-2	2.265	6	4	2	0
3.431	16	1	2	-2	2.242	5	2	0	-4
3.383	8	2	2	1	2.193	15	4	0	2
3.338	47	0	3	1	2.135	8	2	4	-2
3.186	90	3	1	0	2.102	6	2	4	2
3.132	71	0	1	3	2.080	6	1	5	0
3.065	4	3	1	-1			3	4	-1
3.022	5	1	1	-3	2.046	10			
2.993	16	3	1	1			4	3	0
2.955	1	1	1	3	2.033	10	1	4	-3
2.873	4	0	3	2	2.011	12	1	4	3
2.825	5	3	2	0			5	1	0
2.782	11	1	3	-2	1.967	8			
							1	3	4

Sample from Montalto di Castro, Viterbo, Italy. New data by the authors; Philips diffractometer, Ni filtered CuKα radiation ($\lambda = 1.54051$ Å), internal standard $Pb(NO_3)_2$ (cubic: $a = 7.8568$ Å), slits 1°, 0.1 mm, 1°, scanning speed 0.5°/min, indexed after Alberti's (1976) method taking into account the diffraction intensities from single crystal measurements; cell parameters: $a = 10.021(2)$, $b = 10.637(2)$, $c = 9.836(2)$ Å, $\beta = 92°28'(1)$.

Table 3.1IV. X-ray powder pattern of amicite

d(A)	I/Io	H	K	L	d(A)	I/Io	H	K	L
7.30	55	1	1	0	2.470	5	0	0	4
		1	0	1	2.424	8	4	1	1
7.20	10						4	1	-1
		0	1	1	2.390	7			
5.11	40	2	0	0			0	3	3
4.94	28	0	0	2	2.355	3	1	1	4
		2	1	1			1	1	-4
4.22	90				2.324	7			
		1	2	1			2	4	0
4.18	40	1	2	-1			0	4	2
		1	1	2	2.305	7	4	0	2
4.12	8						4	2	0
		2	1	-1	2.249	5	2	0	4
4.05	4	1	1	-2			4	0	-2
3.647	7	2	2	0	2.243	7			
3.585	5	0	2	2			0	2	4
		1	3	0			2	3	3
3.289	30				2.183	4			
		0	3	1			3	2	3
3.238	45	3	1	0	2.165	3	3	3	-2
3.141	80	0	1	3	2.112	10	2	4	2
2.965	10	2	2	2	2.090	10	2	4	-2
2.759	35	3	2	1			2	2	-4
		1	3	-2	2.025	5	4	3	1
2.722	100						3	4	-1
		3	2	-1			4	1	3
2.704	50	1	2	3	2.006	3	5	1	0
		3	1	-2			4	3	-1
2.674	20						1	4	-3
		1	2	-3	1.997	3	3	1	4
2.605	40	0	4	0			5	0	-1

Sample from Höwenegg in Hegau, Germany; data from Alberti et al. (1979).

Table 3.1V. X-ray powder pattern of garronite

d(A)	I/Io	H	K	L	d(A)	I/Io	H	K	L
7.15	75	1	0	1	2.674	72	3	1	2
4.95	59	2	0	0	2.573	11	0	0	4
4.15	78	1	1	2	2.415	2	1	1	4*
4.07	27	2	1	1	2.337	7	4	1	1
3.576	2	2	0	2	2.212	4	4	2	0
3.499	4	2	2	0	2.124	4	3	3	2
3.244	50	1	0	3	2.072	7	2	2	4
3.144	100	3	0	1			4	2	2
2.992	1	3	1	1*	2.031	3			
2.895	4	2	2	2*			3	0	4
2.744	3	3	2	0*	2.015	3	1	0	5
2.710	10	2	1	3	1.987	7	3	1	4

Sample from Goble, Oregon. New data by the authors; Philips diffractometer, Ni filtered CuKα radiation ($\lambda = 1.54051$ Å), internal standard $Pb(NO_3)_2$ (cubic: $a = 7.8568$ Å), slits 1°, 0.1 mm, 1°, scanning speed 0.5°/min, indexing after Alberti's (1976) method taking into account the diffraction intensities from single crystal measurements; cell parameters: $a = 9.898(2)$, $c = 10.295(3)$ Å.
* Not in agreement with space group I4/amd.

Table 3.1 VI. X-ray powder pattern of gobbinsite

d(A)	I/I_0	H	K	L	d(A)	I/I_0	H	K	L
		1	1	0	2.887	5	2	2	2
7.11	100B						3	0	2
		1	0	1	2.757	5			
5.78	20	1	1	1			2	0	3
5.06	50	2	0	0			3	2	1
4.89	30	0	0	2	2.699	80B			
4.41	25	1	0	2			3	1	2
4.12	100	2	1	1	2.651	40	2	1	3
4.03	20	1	1	2	2.539	25	4	0	0
3.515	10	2	0	2	2.435	20	4	0	2
3.326	30	2	1	2	2.379	20	1	0	4
		3	1	0	2.317	10	1	1	4
3.201	100				2.256	15	4	0	2
		3	0	1	2.206	25	2	0	4
3.106	80	1	0	3	2.153	25	2	1	4
3.040	10	3	1	1	2.057	20	4	4	2
2.968	10B	1	1	3	1.986	25	4	3	1

Sample from Gobbins, Co. Antrim, N. Ireland; data from Nawaz
and Malone (1982).
B = broad.

Table 3.2 II. X-ray powder patterns of three phillipsites

Casal Brunori, Italy		Montegalda, Italy		Indian Ocean		H	K	L*	H	K	L†
d(A)	I/I_0	d(A)	I/I_0	d(A)	I/I_0						
–	–			–	–	1	0	0	1	0	1
		8.15	5								
8.11	8			8.19	10	1	0	-1	1	0	-1
7.18	63	7.14	74	7.14	79	0	0	1	0	0	2
7.16	66	–	–	7.10	74	0	2	0	0	2	0
6.42	17	6.39	4	6.38	20	0	1	1	0	1	2
5.38	19					1	2	0	1	2	1
		5.36	24	5.36	25						
						1	2	-1	1	2	-1
5.07	23	5.04	15	5.03	24	0	2	1	0	2	2
4.94	27	4.95	15	5.00	16	2	0	-1	2	0,	0
4.67	4	–	–	–	–	2	1	-1	2	1	0
4.31	8					1	0	1	1	0	3
		4.29	5	4.29	11						
4.29	10					1	0	-2	1	0	-3
4.13	36					1	1	1	1	1	3
		4.11	32	4.11	47						
4.12	41					1	1	-2	1	1	-3
–	–	–	–			1	3	0	1	3	1
–	–	–	–	4.09	30	1	3	-1	1	3	-1
4.06	18					2	2	-1	2	2	0
		4.07	9	–	–						
–	–					2	0	-2	2	0	-2
3.971	3	3.950	6	3.943	11	0	3	1	0	3	2
3.922	6	–	–	–	–	2	1	0	2	1	2
–	–					1	2	1	1	2	3
		3.675	7	3.667	7						
3.688	3					1	2	-2	1	2	-3
3.481	4	3.466	3	3.461	8	0	1	2	0	1	4
3.433	3	–	–	–	–	2	3	-1	2	3	0
3.277	37	3.260	26	3.254	39	1	4	-1	1	4	-1
. –	–	3.215	20	3.245	29	3	0	-1	3	0	1
				–	–	0	2	2	0	2	4
3.206	100	3.191	100								
				3.177	100	0	4	1	0	4	2
3.136	35			–	–	3	1	-1	3	1	1
		3.139	21								

Table 3.2 II. (Continued)

Casal Brunori, Italy		Montegalda, Italy		Indian Ocean		H	K	L*	H	K	L†
d(A)	I/Io	d(A)	I/Io	d(A)	I/Io						
3.129	34	–		–		3	1	-2	3	1	-1
3.090	6	3.088	3	3.099	6	2	3	-2	2	3	-2
						3	2	-1	3	2	1
2.929	15	2.931	13	2.948	17						
						3	2	-2	3	2	-1
2.893	7					2	0	-3	2	0	-4
2.764	20			2.743	36	1	0	2	1	0	5
		2.744	17								
				–		1	0	-3	1	0	-5
2.753	36	–				1	4	1	1	4	3
				2.738	29						
		–				1	4	-2	1	4	-3
2.712	20	–		–		1	1	2	1	1	5
		2.687	25			1	5	0	1	5	1
2.702	36	–		–		1	1	-3	1	1	-5
2.683	21	–		2.690a	32	2	2	-3	2	2	-4
–				–		3	1	0	3	1	3
		2.665	13								
				2.683b	41	3	1	-3	3	1	-3
2.658	10										
		–		–		0	5	1	0	5	2
						1	2	2	1	2	5
2.572	5	2.561	4	2.559	8						
						1	2	-3	1	2	-5
2.539	8	2.535	2	2.548	7	3	2	0	3	2	3
2.534	9					3	2	-3	3	2	-3
2.468	1	–		–		4	0	-2	4	0	0
–		2.521	3	2.515	9	0	4	2	0	4	4
–		2.384	3			3	4	-2	3	4	-1
				2.392	8	3	4	-1	3	4	1
2.388	9										
		–		–		1	3	2	1	3	5
						2	5	0	2	5	2
2.340	7	2.328	2	2.333	6						
						2	5	-2	2	5	-2
–		2.306	2	–		4	1	-3	4	1	-2
2.256	4	2.245	3	2.246	6	2	4	1	2	4	4
2.239	3	2.225	6	2.220	5	0	5	2	0	5	4
2.169	2			–		3	4	0	3	4	3
		2.159	2								
–				–		3	0	-4	3	0	-5
2.158	5	–		2.162c	3	3	4	-3	3	4	-3
2.133	2	2.134	2	–		3	5	-2	3	5	-1
–		–				1	6	1	1	6	3
–				2.074	7	3	2	1	3	2	5
		2.069	2								
–		–				3	2	-4	3	2	-5
						2	2	2	2	2	6
				2.056	8						
						2	2	-4	2	2	-6
2.057	4	–									
		2.049	2	2.050	5	2	6	-2	2	6	-2
2.006	3	1.999	2	1.999	9	1	0	-4	1	0	-7
1.984	4	–		–		1	7	0	1	7	1

Samples from Casal Brunori, Rome, Italy; Montegalda, Vicenza, Italy; Indian Ocean (23°16′S., 74°59′E.). New data by the authors; Philips diffractometer, Ni filtered CuKα radiation ($\lambda = 1.54051$ Å), internal standard Pb(NO$_3$)$_2$ (cubic: $a = 7.8568$ Å), slits 1°, 0.1 mm, 1°, scanning speed 0.5°/min, indexing after Alberti's (1976) method

taking into account the diffraction intensities from single crystal measurements; cell parameters: Casal Brunori $a = 9.876(5)$, $b = 14.318(4)$, $c = 8.689(3)$ Å, $\beta = 124°19'(2)$; Montegalda $a = 9.904(6)$, $b = 14.228(4)$, $c = 8.692(4)$ Å, $\beta = 124°43'(2)$; Indian Ocean $a = 9.997(5)$, $b = 14.190(4)$, $c = 8.722(4)$ Å, $\beta = 125°5'(2)$ or for the pseudo-orthorhombic setting. Casal Brunori $a_0 = 9.876(5)$, $b_0 = 14.318(4)$, $c_0 = 14.387(3)$ Å, $\beta_0 = 89°31'(1)$; Montegalda $a_0 = 9.904(6)$, $b_0 = 14.228(4)$, $c_0 = 14.289(4)$ Å, $\beta_0 = 90°00'(1)$; Indian Ocean $a_0 = 9.997(5)$, $b_0 = 14.190(4)$, $c_0 = 14.275(4)$ Å, $\beta_0 = 90°07'(1)$.
* Primitive monoclinic setting; † pseudo-orthorhombic setting.
a: equally referable also to the monoclinic (1 1 2) or pseudo-orthorhombic (1 1 5) indices; b: equally referable also to the monoclinic (1 5 0) or pseudo-orthorhombic (1 5 1) indices; c: equally referable also to the monoclinic (3 4 0) or pseudo-orthorhombic (3 4 3) indices.

Table 3.2 III. X-ray powder pattern of wellsite

d(Å)	I/I	H	K	L*	H	K	L†	d(Å)	I/I	H	K	L*	H	K	L†
8.15	19	1	0	0	1	0	1	2.690	47	1	5	0	1	5	1
		1	0	-1	1	0	-1			2	2	-3	2	2	-4
7.13	100	0	2	0	0	2	0			3	1	0	3	1	3
6.41	19	0	1	1	0	1	2	2.669	20	3	1	-3	3	1	-3
5.36	24	1	2	0	1	2	1			1	2	2	1	2	5
		1	2	-1	1	2	-1	2.569	7	1	2	-3	1	2	-5
5.05	23	0	2	1	0	2	2			3	2	0	3	2	3
4.96	19	2	0	-1	2	0	0	2.537	6	3	2	-3	3	2	-3
4.30	16	1	0	1	1	0	3	2.528	9	0	4	2	0	4	4
		1	0	-2	1	0	-3	2.479	3	4	0	-2	4	0	0
4.11	65	1	1	1	1	1	3	2.388	5	3	4	-1	3	4	1
		1	1	-2	1	1	-3			3	4	-2	3	4	-1
		1	3	0	1	3	1	2.337	7	2	5	0	2	5	2
4.07	26	2	2	-1	2	2	0			2	5	-2	2	5	-2
3.962	5	0	3	1	0	3	2	2.311	4	4	1	-3	4	1	-2
3.922	5	2	1	0	2	1	2	2.250	7	2	4	1	2	4	4
3.684	3	1	2	-2	1	2	-3	2.228	4	0	5	2	0	5	4
		1	2	1	1	2	3			3	4	0	3	4	3
3.474	5	0	1	2	0	1	4	2.159	5	3	4	-3	3	4	-3
3.263	38	1	4	-1	1	4	-1			3	2	1	3	2	5
3.220	31	3	0	-1	3	0	1	2.072	5	3	2	-4	3	2	-5
3.189	100	0	4	1	0	4	2			2	2	2	2	2	6
3.138	58	3	1	-1	3	1	1	2.058	5	2	2	-4	2	2	-6
		3	1	-2	3	1	-1			1	0	-4	1	0	-7
3.093	7	2	3	-2	2	3	-2	2.003	3	3	5	-3	3	5	-3
2.934	23	3	2	-1	3	2	1			5	0	-2	5	0	1
		3	2	-2	3	2	-1	1.963	7	5	0	-3	5	0	-1
2.752	30	1	0	2	1	0	5								
		1	0	-3	1	0	-5								
2.702	27	1	1	2	1	1	5								
		1	1	-3	1	1	-5								

Sample from Mt. Calvarina, Verona, Italy. The data from Galli (1972) have been re-indexed by the authors after Alberti's (1976) method taking into account the diffraction intensities from single crystal measurements; cell parameters: $a = 9.914(3)$, $b = 14.247(3)$, $c = 8.706(2)$ Å, $\beta = 124°39'(1)$ or for the pseudo-orthorhombic setting $a_0 = 9.914(3)$, $b_0 = 14.247(3)$, $c_0 = 14.324(3)$ Å, $\beta_0 = 89°56'(1)$.
* Primitive monoclinic setting; † pseudo-orthorhombic setting.

Table 3.2IV. X-ray powder pattern of harmotome

d(Å)	I/I₀	H	K	L*	H	K	L†
8.12	60	1	0	-1	1	0	-1
7.16	65	0	0	1	0	0	2
7.05	19	1	1	0	1	1	1
		1	1	-1	1	1	-1
6.39	95	0	1	1	0	1	2
5.03	32	0	2	1	0	2	2
4.31	33	1	0	1	1	0	3
4.30	35	1	0	-2	1	0	-3
4.11	56	1	1	1	1	1	3
		1	1	-2	1	1	-3
4.07	67	1	3	-1	1	3	-1
4.05	59	2	2	-1	2	2	0
3.895	32	2	1	-2	2	1	-2
3.667	10	1	2	-2	1	2	-3
3.573	5	0	0	2	0	0	4
3.528	10	0	4	0	0	4	0
3.466	18	0	1	2	0	1	4
3.410	5	2	3	-1	2	3	0
3.241	62	1	4	-1	1	4	-1
3.195	28	0	2	2	0	2	4
3.169	70	0	4	1	0	4	2
3.126	100	3	1	-2	3	1	-1
3.075	31	2	3	0	2	3	2
		2	3	-2	2	3	-2
2.918	29	3	2	-2	3	2	-1
2.897	13	2	0	-3	2	0	-4
2.847	10	0	3	2	0	3	4
2.747	27	1	0	-3	1	0	-5
2.730	54	1	4	1	1	4	3
		1	4	-2	1	4	-3
2.697	61	1	1	2	1	1	5
		1	1	-3	1	1	-5
2.678	54	2	2	-3	2	2	-4
2.671	64	1	5	0	1	5	1
		1	5	-1	1	5	-1
2.628	15	0	5	1	0	5	2
2.561	9	1	2	-3	1	2	-5
2.529	33	3	2	-3	3	2	-3
2.515	20	0	4	2	0	4	4
2.470	11	4	0	-2	4	0	0
2.464	11	2	3	-3	2	3	-4
2.369	18	1	3	-3	1	3	-5
2.343	15	3	3	-3	3	3	-3
		4	0	-1	4	0	2
		2	5	0	2	5	2
2.320	23	2	5	-2	2	5	-2
2.299	17	4	1	-3	4	1	-2
2.261	10	1	6	0	1	6	1
		0	2	3	0	2	6
		2	4	1	2	4	4
2.241	15	2	4	-3	2	4	-4
		3	4	0	3	4	-3
2.151	18	3	4	-3	3	4	-3
2.147	19	2	0	-4	2	0	-6
2.127	7	2	6	-1	2	6	0
		1	6	1	1	6	3
2.066	17	1	6	-2	1	6	-3
2.059	18	2	2	2	2	2	6
2.053	16	2	2	-4	2	2	-6
2.025	10	2	5	1	2	5	4
		4	4	-2	4	4	0
		2	5	-3	2	5	-4
		4	1	-4	4	1	-4
2.005	8	1	0	3	1	0	7
		0	6	2	0	6	4
1.964	8	3	3	-4	3	3	-5

Sample from Andreasberg, Germany. New data by the authors; Philips diffractometer, Ni filtered CuKα radiation ($\lambda = 1.54051$ Å), internal standard $Pb(NO_3)_2$ (cubic: $a = 7.8568$ Å), slits 1°, 0.1 mm, 1°, scanning speed 0.5°/min, indexing after Alberti's (1976) method taking into account the diffraction intensities from single crystal measurements; cell parameters: $a = 9.876(5)$, $b = 14.130(4)$, $c = 8.680(4)$ Å, $\beta = 124°26'(2)$ or for the pseudo-orthorhombic setting $a_0 = 9.876(5)$, $b_0 = 14.130(4)$, $c_0 = 14.317(4)$ Å, $\beta_0 = 89°54'(2)$.

* Primitive monoclinic setting; † pseudo-orthorhombic setting.

Table 3.3II. X-ray powder pattern of merlinoite

d(A)	I/Io	H	K	L	d(A)	I/Io	H	K	L
10.02	12	1	1	0			{ 3	3	2
		0	1	1	2.770	16			
8.15	12	{					5	1	0
		1	0	1	2.730	27	3	4	1
7.12	90	0	2	0			4	3	1
7.08	88	2	0	0	2.720	30	{ 0	3	3
		1	2	1			5	0	1
5.36	40	{			2.670	11	4	2	2
		2	1	1	2.552	16	2	5	1
5.03	35	2	2	0	2.540	12	5	2	1
4.98	20	0	0	2			2	3	3
4.48	37	3	1	0	2.535	12	{		
4.29	28	0	3	1			3	2	3
4.07	8	2	0	2	2.507	11	4	4	0
3.655	18	3	2	1	2.435	8	1	5	2
3.556	5	0	4	0	2.428	8	5	3	0
3.526	4	4	0	0	2.391	4	1	4	3
3.336	2	3	3	0	2.386	3	4	1	3
3.258	44	1	4	1	2.354	2	6	0	0
3.241	41	4	1	1	2.181	6	5	3	2
		0	1	3			2	5	3
3.226	38	{			2.065	4	{		
		1	0	3			6	3	1
3.176	100	2	4	0	2.005	4	5	5	0
		1	2	3	1.977	4	7	0	1
2.935	34	{							
		2	1	3					

Sample from Cupaello, Rieti, Italy; data from Passaglia et al. (1977).

Table 3.4I. X-ray powder pattern of mazzite

d(A)	I/Io	H	K	L	d(A)	I/Io	H	K	L
15.93	35	1	0	0	2.865	10	5	1	0
9.20	60	1	1	0	2.681	12	5	1	1
7.96	35	2	0	0	2.643	16	3	2	2
6.89	25	1	0	1	2.552	20	5	2	0
6.02	53	2	1	0	2.511	5	6	0	1
5.53	12	2	0	1	2.446	1	5	0	2
5.31	17	3	0	0			2	0	3
4.73	50	2	1	1	2.422	9	{		
4.42	12	3	1	0			5	2	1
3.986	20	4	0	0	2.393	1	3	3	2
		3	1	1	2.302	22	4	4	0
3.824	95	{			2.298	22	3	0	3
		0	0	2			6	2	0
3.717	25	1	0	2	2.210	9	{		
3.655	47	3	2	0			3	1	3
3.531	90	1	1	2	2.147	10	4	0	3
3.474	12	4	1	0			6	2	1
3.452	10	2	0	2	2.123	17	{		
3.185	100	5	0	0			5	2	2
3.102	30	3	0	2			5	4	0
3.065	38	3	3	0	2.037	10	{		
3.010	40	4	2	0			7	1	1
		5	0	1	2.006	7	6	3	0
2.941	100	{					8	0	0
		2	2	2	1.991	10	{		
							5	0	3

Sample from Mont Semiol, Montbrison, Loire, France; data from Galli et al. (1974).

Table 3.5 III. X-ray powder pattern of paulingite

d(A)	I/Io	H	K	L	d(A)	I/Io	H	K	L
12.42	5	2	2	0	2.969	31	10	6	2
10.14	4	2	2	2	2.944	7	9	6	5
9.39	10	3	2	1	2.907	6	11	5	0
8.79	6	4	0	0			9	7	4
8.28	60	3	3	0	2.869	11	11	5	2
7.85	6	4	2	0	2.848	11	10	6	4
7.49	5	3	3	2	2.828	7	12	3	1
7.16	12	4	2	2			9	8	3
6.89	67	4	3	1	2.794	13	10	7	3
6.21	35	4	4	0	2.726	39	11	6	3
5.85	21	6	0	0	2.631	17	9	9	4
5.70	32	5	3	2	2.618	45	10	8	4
5.42	17	5	4	1	2.574	11	11	7	4
5.30	5	6	2	2	2.522	12	13	4	3
5.18	6	6	3	1			11	8	3
4.96	33	7	1	0	2.485	7	14	2	0
4.78	76	6	3	3			10	10	0
4.69	23	6	4	2	2.449	10	10	9	5
4.39	28	8	0	0	2.306	5	14	6	0
4.32	8	5	5	4	2.286	7	14	6	2
4.26	21	8	2	0			10	10	6
		6	4	4	2.259	7	11	11	0
4.08	25	7	4	3	2.178	4	14	8	0
3.974	4	7	5	2	2.137	5	14	7	5
3.877	32	8	3	3			11	10	7
3.700	11	9	3	0	2.106	6	11	11	6
3.620	20	9	3	2	2.063	8	17	1	0
3.583	60	8	4	4			15	7	4
3.444	7	10	2	0	2.049	12	12	11	5
3.377	12	10	2	2			17	2	1
3.349	69	10	3	1			14	7	7
3.261	89	10	4	0	2.034	5	17	3	0
3.231	8	9	6	1	2.008	10	13	11	4
3.204	6	10	4	2	1.945	4	14	11	3
3.179	18	9	5	4					
3.129	58	11	2	1					
		10	5	1					
3.081	100	11	3	0					
2.991	39	11	4	1					

Sample from Ritter, Grant Co., Oregon. New data by the authors; Philips diffractometer, Ni filtered CuKα radiation ($\lambda = 1.54051$ Å), internal standard $Pb(NO_3)_2$ (cubic: $a = 7.8568$ Å), slits 1°, 0.1 mm, 1°, scanning speed 0.5°/min, indexing after Alberti's (1976) method taking into account the diffraction intensities from single crystal measurements; cell parameters: $a = 35.114(2)$ Å.

Table 4.1 II. X-ray powder pattern of gmelinite

d(A)	I/I$_o$	H	K	L		d(A)	I/I$_o$	H	K	L
11.90	63	1	0	0				4	1	1
7.68	29	1	0	1		2.513	1			
6.88	16	1	1	0				0	0	4
5.95	9	2	0	0		2.400	1	3	2	2
5.12	23	2	0	1				1	1	4
5.03	28	0	0	2		2.355	1			
4.63	5	1	0	2				3	1	3
4.50	25	2	1	0		2.317	4	5	0	1
4.11	100	2	1	1		2.310	4	4	1	2
3.970	4	3	0	0		2.293	3	3	3	0
3.440	21	2	2	0				4	2	1
3.348	2	2	1	2		2.195	2			
3.302	6	3	1	0				2	1	4
3.227	41	1	0	3		2.124	2	3	0	4
3.138	1	3	1	1		2.118	2	3	2	3
3.116	1	3	0	2				5	1	1
2.978	55	4	0	0		2.086	12			
2.922	18	2	0	3				3	3	2
2.855	42	4	0	1				4	2	2
2.734	<1	3	2	0		2.054	4			
2.690	44	2	1	3				4	1	3
2.636	3	3	2	1				6	0	0
2.597	14	4	1	0		1.984	<1			
		4	0	2				1	0	5
2.561	2									
		3	0	3						

Sample from Montecchio Maggiore, Vicenza, Italy; data from Passaglia et al. (1978a).

Table 4.2 III. X-ray powder patterns of three chabazites

Fossil Canyon California d(Å)	I/Io	Nidda Germany d(Å)	I/Io	Casal Brunori Italy d(Å)	I/Io	H	K	L
9.26	70	9.36	73	9.44	61	1	0	1
6.85	22	6.90	12	–	–	1	1	0
6.30	8	6.37	4	–	–	0	1	2
5.51	32	5.56	33	5.56	15	0	2	1
4.96	38	5.02	39	5.13	24	0	0	3
4.64	4	4.68	15	4.71	8	2	0	2
4.29	100	4.33	95	4.33	50	2	1	1
4.02	5	–	–	–	–	1	1	3
3.957	5	3.980	13	3.974	17	3	0	0
3.844	20	3.875	38	3.887	29	1	2	2
3.549	47	3.594	47	3.660	22	1	0	4
3.427	21	3.446	18	3.442	21	2	2	0
3.217	10	3.237	9	3.233 }	14	1	3	1
3.152	11	3.183	13			0	2	4
3.056	2	–	–	3.142	6	3	0	3
3.011	1	–	–	–	–	3	1	2
2.885*	22	2.298 }	100	2.976	9	0	1	5
2.911	62			2.930 }	100	4	0	1
2.864	34	2.891	55			2	1	4
2.820	6	2.840	6	–	–	2	2	3
2.757	3	2.776	5	2.779	6	0	4	2
2.681	2	2.694	19	–	–	3	2	1
2.662	6	2.687	19	2.737	6	2	0	5
2.591	11	2.606	27	2.604	19	4	1	0
2.556	3	2.575	6	2.579	7	2	3	2
2.481	12	2.507	32	2.540	21	1	2	5
–	–	2.357 }	4	–	–	0	5	1
–	–			–	–	1	1	6
2.322	1	2.340	4	–	–	4	0	4
2.298	2	2.315	4	2.322	5	4	1	3
2.285	2	2.298	10	2.296	11	3	3	0
2.263	4	2.278	5	–	–	5	0	2
–	–	2.232	3	–	–	2	4	1
2.149	1	2.162	3	2.164 }	5	4	2	2
–	–	–	–			1	0	7
2.103	3			2.145	4	0	4	5
2.076	6	2.089	14	2.095	13	3	3	3
–	–	2.063	4	2.064	3	1	5	2
2.002	2	2.015	2	–	–	0	5	4
1.979	1	–	–	–	–	6	0	0
–	–	1.948	3	1.944	3	4	3	1

Samples from Fossil Canyon, San Bernardino County, California; Nidda, Hessen, Germany; Casal Brunori, Rome, Italy. Data for Fossil Canyon from Passaglia (1970); new data by the authors for the other two samples; Philips diffractometer, Ni filtered CuKα radiation (λ = 1.54051 Å), internal standard Pb(NO$_3$)$_2$ (cubic: a = 7.8568 Å), slits 1°, 0.1 mm, 1°, scanning speed 0.5°/min, indexing after Alberti's (1976) method taking into account the diffraction intensities from single crystal measurements; hexagonal cell parameters: Fossil Canyon a = 13.705(11), c = 14.870(8) Å; Nidda a = 13.791(2), c = 15.056(3) Å; Casal Brunori a = 13.773(3), c = 15.389(6) Å.

* Line out of sequence due to a shorter c value.

Table 4.2IV. X-ray powder pattern of willhendersonite

d(A)	I/Io	H	K	L*	d(A)	I/Io	H	K	L*
9.16	100	1	0	0			1	1	-3
5.18	30	1	1	1	2.907	60			
4.71	5	0	0	2			3	-1	0
4.57	5	2	0	0			1	3	-1
4.27	2	1	0	-2	2.804	50			
		2	1	0			1	-3	1
4.09	40				2.746	2	3	-1	1
		1	-2	0					
3.93	20	1	1	-2	2.674	1	2	2	-2
		1	2	-1			2	-2	2
3.82	20						2	-3	0
		1	-1	2	2.538	20			
		2	-1	1			2	0	3
3.71	30				2.508	20	0	3	2
		1	1	2	2.429	10			
		2	-1	2	2.264	15			
3.06	10				2.209	1			
		3	0	0	2.163	15			
		0	1	-3	2.078	10			
3.01	10	1	2	2	2.042	10			
		2	2	1	2.004	5			
					1.979	5			

Sample from San Venanzo, Terni, Italy; data from Peacor et al. (1984).
* Due to the pseudo-monoclinic dimensions of the sample, each d-spacing corresponds also to the set of indices with H and K interchanged.

Table 4.3III. X-ray powder pattern of levyne

d(A)	I/Io	H	K	L	d(A)	I/Io	H	K	L
10.32	28	1	0	1	2.445	3	0	4	5
8.15	78	0	1	2			4	1	3
7.66	17	0	0	3	2.395	14			
6.67	20	1	1	0			1	2	8
5.16	42	2	0	2			2	3	5
5.03	3	1	1	3	2.293	8			
4.27	46	0	1	5			1	3	7
4.08	100	1	2	2	2.250	3	1	0	10
3.850	26	3	0	0	2.223	11	3	3	0
3.591	7	2	0	5	2.173	5	2	4	1
3.475	23	2	1	4	2.144	7	0	5	4
3.438	9	3	0	3			3	3	3
3.332	19	2	2	0	2.133	12			
3.156	52	1	0	7			0	2	10
3.084	25	3	1	2	2.129	14	0	3	9
2.865	16	4	0	1	2.103	5	4	1	6
2.855	15	0	2	7	2.064	10	5	0	5
2.800	85	0	4	2			1	5	2
2.714	8	3	0	6	2.039	4			
2.623	40	2	1	7			2	4	4
2.581	6	2	3	2	1.972	3	4	2	5
2.521	14	4	1	0					

Sample from Pargate quarry, Co. Antrim, Northern Ireland. New data by the authors; Philips diffractometer, Ni filtered CuKα radiation ($\lambda = 1.54051$ Å), internal standard $Pb(NO_3)_2$ (cubic: $a = 7.8568$ Å), slits 1°, 0.1 mm, 1°, scanning speed 0.5°/min, indexing after Alberti's (1976) method taking into account the diffraction intensities from single crystal measurements; cell parameters: $a = 13.338(2)$, $c = 22.958(4)$ Å.

Table 4.4 III. X-ray powder pattern of erionite

(Å)	I/Io	H	K	L	(Å)	I/Io	H	K	L
11.56	76	1	0	0	3.153	26	2	0	4
9.18	5	1	0	1	3.122	11	3	1	1
7.55	9	0	0	2	2.936	13	3	1	2
6.65	48	1	1	0	2.880	76	4	0	0
6.31	6	1	0	2	2.851	100	2	1	4
5.76	28	2	0	0	2.826	65	4	0	1
5.39	20	2	0	1	2.688	28	3	0	4
4.60	26	1	0	3			4	0	2
4.58	30	2	0	2	2.513	33	0	0	6
4.35	78	2	1	0			4	1	0
4.18	28	2	1	1	2.494	28	2	2	4
3.839	60	3	0	0			3	2	2
3.771	94	0	0	4	2.215	19	3	3	0
		2	1	2	2.125	13	3	3	2
3.586	57	1	0	4	2.091	7	4	1	4
3.423	5	3	0	2			4	2	2
3.323	45	2	2	0	1.994	7	5	1	2
3.291	17	2	1	3					
3.277	13	1	1	4					
3.193	13	3	1	0					

Sample from Agate Beach, Oregon. New data by the authors; Philips diffractometer, Ni filtered CuKα radiation ($\lambda = 1.54051$ Å), internal standard $Pb(NO_3)_2$ (cubic: $a = 7.8568$ Å), slits 1°, 0.1 mm, 1°, scanning speed 0.5°/min, indexing after Alberti's (1976) method taking into account the diffraction intensities from single crystal measurements; cell parameters: $a = 13.295(2)$, $15.081(5)$ Å.

Table 4.5 II. X-ray powder pattern of offretite

d(Å)	I/Io	H	K	L	d(Å)	I/Io	H	K	L
11.61	70	1	0	0	3.046	2	2	2	1
7.60	13	0	0	1	2.946	17	3	1	1
6.67	17	1	1	0	2.881	72	4	0	0
6.36	7	1	0	1	2.862	94	2	1	2
5.77	40	2	0	0	2.694	18	4	0	1
4.59	30	2	0	1	2.644	3	3	2	0
4.36	78	2	1	0	2.516	32	4	1	0
3.842	50	3	0	0	2.499	16	3	2	1
3.784	100	0	0	2	2.307	3	5	0	0
		2	1	1	2.218	22	3	3	0
3.607	52	1	0	2	2.204	6	5	0	1
3.431	8	3	0	1	2.129	7	3	3	1
3.329	19	2	2	0	2.114	7	3	0	3
3.195	10	3	1	0	2.090	8	4	2	1
3.172	17	2	0	2	1.996	8	5	1	1

Sample from Mont Semiol, Montbrison, Loire, France. New data by the authors; Philips diffractometer, Ni filtered CuKα radiation ($\lambda = 1.54051$ Å), internal standard $Pb(NO_3)_2$ (cubic: $a = 7.8568$ Å), slits 1°, 0.1 mm, 1°, scanning speed 0.5°/min, indexing after Alberti's (1976) method taking into account the diffraction intensities from single crystal measurements; cell parameters: $a = 13.307(2)$, $c = 7.592(2)$ Å.

Table 4.6IV. X-ray powder pattern of faujasite

d(A)	I/Io	H	K	L		d(A)	I/Io	H	K	L
14.28	100	1	1	1				{ 9	1	1
8.74	19	2	2	0		2.710	11	{		
7.45	12	3	1	1				7	5	3
7.14	5	2	2	2		2.632	32	6	6	4
6.17	2	4	0	0		2.587	14	9	3	1
5.67	78	3	3	1		2.518	4	8	4	4
5.04	4	4	2	2				10	2	0
4.75	44	5	1	1		2.418	4	{ 8	6	2
4.37	52	4	4	0				10	2	2
4.18	3	5	3	1		2.374	23			
4.13	3	4	4	2				6	6	6
3.903	18	6	2	0		2.222	5	7	7	5
3.760	95	5	3	3		2.182	13	8	8	0
3.562	8	4	4	4				11	3	1
3.457	22	5	5	1		2.157	7			
3.301	67	6	4	2				9	7	1
3.210	19	7	3	1				11	3	3
3.083	3	8	0	0		2.095	8			
3.015	25	7	3	3				9	7	3
		{ 6	6	0		2.058	11	8	8	4
2.910	36	{				2.002	2	12	2	2
		8	2	2				11	5	3
2.850	78	5	5	5		1.983	4			
2.760	28	8	4	0				9	7	5

Sample from Limburg, Kaiserstuhl, Baden, Germany. New data by the authors; Philips diffractometer, Ni filtered CuKα radiation ($\lambda = 1.54051$ Å), internal standard $Pb(NO_3)_2$ (cubic: $a = 7.8568$ Å), slits 1°, 0.1 mm, 1°, scanning speed 0.5°/min, indexing after Alberti's (1976) method taking into account the diffraction intensities from single crystal measurements; cell parameter: $a = 24.684(3)$ Å.

Table 5.1III. X-ray powder pattern of mordenite

d(A)	I/Io	H	K	L		d(A)	I/Io	H	K	L
13.58	18	1	1	0		3.028	1	0	4	2
10.26	5	0	2	0		3.017	2	6	0	0
9.06	100	2	0	0		2.942	5	2	6	1
6.59	14	1	1	1		2.895	13	4	0	2
6.40	17	1	3	0		2.741	2	1	5	2
6.07	4	0	2	1		2.715	2	5	5	0
5.80	18	3	1	0		2.701	5	1	7	1
4.88	3	1	3	1		2.633	3	3	7	0
4.60	2	3	1	1		2.588	1	5	1	2
4.53	31	3	3	0		2.565	10	0	8	0
4.46	2	2	4	0		2.521	7	3	5	2
4.15	8	4	2	0		2.459	4	6	4	1
4.00	70	1	5	0				{ 5	3	2
3.842	7	2	4	1		2.436	2B	{		
3.765	4	0	0	2				2	6	2
3.629	3	4	2	1		2.294	1	6	2	2
3.568	4	5	1	0		2.279	1	5	7	0
3.532	2	0	2	2		2.263	1	8	0	0
3.476	43	2	0	2		2.232	2	4	8	0
3.417	11	0	6	0		2.166	2B	6	6	1
3.394	33	3	5	0		2.117	1	0	8	2
3.291	3	2	2	2		2.052	7	0	10	0
3.221	40	5	1	1		2.035	2	7	3	2
3.201	34	5	3	0		2.019	2	4	4	3
3.155	2	3	1	2		1.997	2	8	4	1
		{ 0	6	1						
3.101	4B	{								
		4	4	1						

Sample from Filone della Speranza, San Piero in Campo, Isola d'Elba, Italy; data from Passaglia (1975). B = broad.

Table 5.2 III. X-ray powder patterns of two dachiardites

Isola d'Elba Italy					Alpe di Siusi Italy	
d(Å)	I/Io	H	K	L	I/Io	d(Å)
9.79	10	0	0	1	16	9.72
8.90	50	2	0	0	100	8.84
6.91	50	1	1	0	4	6.88
6.00	35	1	1	-1	-	-
5.35	20	1	1	1	-	-
4.97	50	2	0	-2	15	5.00
4.88	50	0	0	2	100	4.88
4.61	10	4	0	-1	12	4.61
4.44	10	4	0	0	5	4.43
4.23	10	1	1	-2	-	-
3.932	50	4	0	-2	30	3.957
3.848	10	3	1	1	-	-
3.801	50	2	0	2	40	3.801
3.773	20	1	1	2	-	-
3.750	20	0	2	0	10	3.755
3.634	20	4	0	1	11	3.627
3.498	20	0	2	1	-	-
3.452	100	2	2	0	85	3.454
3.396	35	2	0	-3	-	-
3.375	20	2	2	-1	8	3.388
3.328	35	5	1	-1	4	3.332
		0	0	3	9	3.251
3.204	100	5	1	0	7	3.199
3.114	10	4	0	-3	4	3.131
3.077	10	1	1	-3	-	-
3.018	20	{3	1	-3	-	-
		3	1	2	-	-

Isola d'Elba Italy					Alpe di Siusi Italy	
d(Å)	I/Io	H	K	L	I/Io	d(Å)
2.964	50	6	0	-2	37	2.971
-	-	4	0	2 }	23	2.865
2.862	50	4	2	0)		
2.712	50	4	2	-2	7	2.726
2.666	50	2	2	2	8	2.673
2.607	10	4	2	1	-	-
2.550	50	2	0	-4	8	2.576
2.517	20	2	2	-3	-	-
2.472	20	1	3	0	-	-
2.449	20	0	2	3	4	2.462
2.416	10	1	3	-1	-	-
		{5	1	2	-	-
2.387	20	{1	1	-4	-	-
2.306	20	4	0	3	3	2.305
2.273	10	7	1	-3	-	-
2.234	10	2	2	3	-	-
2.216	10	1	1	4	-	-
2.185	10	2	0	4	2	2.186
2.170	10	1	3	2	-	-
2.067	10	4	2	-4	-	-
		{0	2	4	-	-
2.040	20	{2	0	-5	3	2.053
2.017	20	5	3	-2	-	-
1.992	35	3	3	2	-	-
		9	1	-1	2	1.985

Samples from Filone della Speranza, San Piero in Campo, Isola d'Elba, Italy and from Alpe di Siusi, Bolzano, Italy. The data from Galli (1965) and from Alberti (1975b) have been re-indexed by the authors after Alberti's (1976) method taking into account the diffraction intensities from single crystal measurements; cell parameter: Isola d'Elba $a = 18.647(11)$, $b = 7.489(3)$, $c = 10.227(5)$ Å, $\beta = 107°51'(2)$; Alpe di Siusi $a = 18.646(9)$, $b = 7.508(4)$, $c = 10.300(5)$ Å, $\beta = 108°19'(30)$.

Table 5.3II. X-ray powder pattern of epistilbite

d(A)	I/Io	H	K	L		d(A)	I/Io	H	K	L
8.90	100	0	2	0		2.984	5	3	1	-2
6.90	33	1	1	0		2.958	9	0	6	0
6.12	3	0	2	1		2.917	50	1	5	-2
4.92	55	1	1	-2		2.859	20	2	4	0
4.64	7	1	3	0		2.790	7	3	1	-3
4.49	9	2	0	-1		2.694	11	3	3	-2
4.44	10	0	4	0		2.680	15	1	3	2
4.33	20	1	1	1		2.556	13	2	0	-4
4.24	4	2	0	-2		2.457	4	2	2	-4
4.01	8	2	2	-1		2.425	13	2	6	-2
3.924	23	0	4	1				1	5	-3
3.870	73	1	3	-2		2.322	3	2	6	0
3.824	14	2	2	-2		2.215	6	2	4	-4
3.802	16	0	2	2		2.188	2	4	2	-3
3.737	8	2	0	0		2.094	3	1	3	3
3.446	92	2	2	0				4	0	-1
3.332	14	2	0	-3				4	2	-4
3.270	21	1	5	-1		2.058	3	2	6	1
3.262	21	1	1	-3						
3.208	83	1	5	0						
3.154	3	2	4	-1		2.004	5	4	4	-2
3.114	2	2	2	-3		1.996	5	2	4	2
3.061	7	2	4	-2						

Sample from Fonte del Prete, San Piero in Campo, Isola d'Elba, Italy. New data by the authors; Philips diffractometer, Ni filtered CuKα radiation ($\lambda = 1.54051$ Å), internal standard $Pb(NO_3)_2$ (cubic: $a = 7.8568$ Å), slits 1°, 0.1 mm, 1°, scanning speed 0.5°/min, indexing after Alberti's (1976) method taking into account the diffraction intensities from single crystal measurements; cell parameters: $a = 9.089(3)$, $b = 17.747(4)$, $c = 10.229(3)$ Å, $\beta = 124°39'(1)$.

Table 5.4 III. X-ray powder pattern of ferrierite

d(A)	I/I₀	H	K	L	d(A)	I/I₀	H	K	L
11.38	3	1	1	0	2.955	4	4	0	2
9.60	100	2	0	0	2.901	4	1	3	2
6.98	5	1	0	1	2.846	1	4	4	0
6.63	3	0	1	1	2.726	5	4	2	2
5.84	18	3	1	0	2.696	6	7	1	0
4.97	2	1	2	1	2.643	3	0	5	1
4.80	5	4	0	0			3	5	0
4.58	2	1	3	0	2.581	4			
4.01	21	3	2	1			7	0	1
3.974	38	4	2	0	2.572	3	0	4	2
3.888	14	4	1	1	2.432	4	6	0	2
3.797	20	3	3	0	2.372	9	7	3	0
3.708	31	5	1	0	2.319	1	4	5	1
3.562	14	1	1	2	2.255	2	8	1	1
3.535	26	0	4	0	2.212	2	3	2	3
3.493	22	2	0	2	2.190	2	7	1	2
3.416	8	5	0	1	2.115	4	4	6	0
3.318	7	2	4	0	2.109	3	9	1	0
3.310	6	0	2	2	2.051	3	9	0	1
3.199	9	6	0	0	2.021	3	8	0	2
3.151	10	1	4	1			5	2	3
3.130	4	2	2	2	2.006	3			
3.076	12	5	2	1			6	4	2
2.977	13	5	3	0	1.969	4	9	2	1

Sample from Albero Bassi, Vicenza, Italy. New data by the authors; Philips diffractometer, Ni filtered CuKα radiation ($\lambda = 1.54051$ Å), internal standard Pb(NO₃)₂ (cubic: $a = 7.8568$ Å), slits 1°, 0.1 mm, 1°, scanning speed 0.5°/min, indexing after Alberti's (1976) method taking into account the diffraction intensities from single crystal measurements; cell parameters: $a = 19.202(3)$, $b = 14.138(3)$, $c = 7.498(2)$ Å.

Table 5.5 III. X-ray powder pattern of bikitaite

d(A)	I/I₀	H	K	L	d(A)	I/I₀	H	K	L
7.87	80	1	0	0	2.523	20	2	1	1
6.93	50	0	0	1	2.479	90	0	2	0
6.73	30	1	0	-1	2.423	10	1	1	2
4.37	40	1	0	1			1	2	0
4.27	10	2	0	-1	2.364	20			
4.20	90	1	1	0			3	1	-2
4.02	20	0	1	1	2.337	10	0	2	1
3.991	10	1	1	-1	2.323	10	1	2	-1
3.926	10	2	0	0	2.316	10	3	1	0
3.806	30	1	0	-2	2.240	10	1	1	-3
3.462	100	0	0	2	2.167	10	3	0	1
3.371	100	2	0	-2	2.141	10	2	2	-1
3.284	40	1	1	1	2.097	10	2	2	0
3.215	40	2	1	-1	2.094	10	0	1	3
3.076	40	2	1	0	2.077	20	1	2	-2
3.023	10	1	1	-2	2.012	10	0	2	2
2.930	10	2	0	1	2.005	10	1	0	3
2.870	20	3	0	-1			2	1	2
2.794	10	2	1	-2	1.996	10			
2.739	10	1	0	2			2	2	-2
2.629	10	3	0	0					

Sample from Bikita, Zimbabwe (= Southern Rhodesia); data from Phinney and Stewart (1961).

Table 6.1IV. X-ray powder pattern of heulandite

d(A)	I/Io	H	K	L		d(A)	I/Io	H	K	L
8.96	100	0	2	0		3.075	12	1	3	-2
7.94	12	2	0	0		2.992	29	3	3	1
6.81	6	2	0	-1		2.972	91	1	5	1
6.65	5	0	0	1		2.890	4	4	0	1
5.94	4	2	2	0				5	3	0
5.59	3	1	3	0		2.806	23			
5.33	7	0	2	1				6	2	-1
5.26	10	3	1	-1		2.725	20	0	6	1
5.12	16	1	1	1		2.667	7	0	4	2
5.08	14	3	1	0		2.527	5	1	7	0
4.65	32	1	3	-1		2.519	10	5	5	-1
4.47	5	0	4	0		2.486	6	3	5	1
4.37	9	4	0	-1				6	4	-1
3.980	65	1	3	1		2.458	7			
3.897	43	2	4	0				4	0	-3
3.843	11	2	2	1		2.434	10	2	0	-3
3.735	9	2	4	-1		2.374	5	4	2	-3
3.717	9	2	0	-2		2.296	5	2	4	2
3.564	9	3	1	-2		2.229	9	0	8	0
3.479	8	5	1	-1		2.180	5	1	3	-3
3.429	21	2	2	-2		2.088	4	6	2	1
3.405	15	4	0	-2		2.075	3	5	7	-1
3.325	7	0	0	2				6	4	-3
3.181	19	4	2	-2		2.019	4			
3.125	22	4	4	-1				7	5	-2
						1.960	9	1	5	-3

Sample from Mossyrock Dam, Levis Co., Washington. New data by the authors; Philips diffractometer, Ni filtered CuKα radiation ($\lambda = 1.54051$ Å), internal standard $Pb(NO_3)_2$ (cubic: $a = 7.8568$ Å), slits 1°, 0.1 mm, 1°, scanning speed 0.5°/min, indexing after Alberti's (1976) method taking into account the diffraction intensities from single crystal measurements; cell parameters: $a = 17.739(7)$, $b = 17.885(5)$, $c = 7.430(3)$ Å, $\beta = 116°27'(2)$.

Table 6.1V. X-ray powder pattern of clinoptilolite

$d(Å)$	I/I_0	H	K	L
8.95	100	0	2	0
7.93	13	2	0	0
6.78	9	2	0	-1
5.94	3	2	2	0
5.59	5	1	3	0
5.24	10	3	1	-1
5.12	12	1	1	1
4.65	19	1	3	-1
4.35	5	4	0	-1
3.976	61	1	3	1
3.955	63	4	0	0
		3	3	0
3.905	48	2	4	0
3.835	7	2	2	1
3.738	6	2	4	-1
3.707	5	0	4	1
3.554	9	3	1	-2
3.513	4	1	1	-2
3.424	18	2	2	-2
3.392	12	4	0	-2
3.316	6	0	0	2
3.170	16	4	2	-2
3.120	15	4	4	-1
3.074	9	1	3	-2
2.998	18	3	5	-1
2.971	47	1	5	1

$d(Å)$	I/I_0	H	K	L
		5	3	0
2.795	16			
		6	2	-1
2.730	16	2	6	-1
		2	0	2
2.667	4	0	4	2
		6	2	0
2.527	6	1	7	0
		3	5	1
2.485	3	7	1	-1
2.458	3	6	4	-1
		2	6	1
2.437	8	5	1	1
2.422	5	4	4	1
2.319	2	3	7	-1
		3	7	-2
2.089	3	6	2	1
		3	7	1
2.056	2	1	1	3
2.016	2	6	4	-3
		7	5	-2
1.974	4	1	9	0

Sample from Agoura, California. New data by the authors; Philips diffractometer, Ni filtered CuKα radiation ($\lambda = 1.54051$ Å), internal standard $Pb(NO_3)_2$ (cubic: $a = 7.8568$ Å), slits 1°, 0.1 mm, 1°, scanning speed 0.5°/min, indexing after Alberti's (1976) method taking into account the diffraction intensities from single crystal measurements; cell parameters: $a = 17.671(11)$, $b = 17.912(6)$, $c = 7.410(3)$ Å, $\beta = 116°22'(3)$.

Table 6.2 II. X-ray powder pattern of stilbite

d(A)	I/Io	H	K	L
9.12	100	0	2	0
8.88	9	0	0	2
6.83	2	2	0	0
6.37	2	0	2	2
		2	2	0
5.46	1	2	0	-2
		1	3	-1
5.30	8			
		1	3	1
5.23	1	1	1	-3
4.68	16	2	2	-2
4.63	15	2	2	2
4.56	3	0	4	0
4.44	2	0	0	4
4.30	9	3	1	-1
4.27	7	3	1	1
		1	3	-3
4.06	58	0	4	2
		1	3	3
4.00	10	0	2	4
3.788	2	2	4	0
3.756	6	2	0	-4
3.699	6	2	0	4
3.504	3	2	4	-2
3.474	5	2	4	2
3.456	2	1	5	1
3.415	10	4	0	0
3.396	8	1	1	-5
3.368	5	1	1	5
		4	0	-2
3.199	13	4	2	0
		0	4	4
3.172	7	4	0	2
3.122	5	3	3	-3
3.088	2	3	3	3
		0	6	0
		1	5	-3
3.028	36			
		1	5	3
		4	2	-2
		1	3	-5
2.990	10	4	2	2
		1	3	5
2.965	4	0	0	6
2.877	3	0	6	2
2.811	4	3	5	1

d(A)	I/Io	H	K	L
2.780	21	2	6	0
2.751	1	3	1	5
2.730	8	4	0	-4
2.687	1	4	0	4
2.591	4	2	2	6
		4	2	4
2.577	6			
		3	5	-3
2.557	5	3	5	3
2.537	3	1	7	1
2.530	2	3	3	5
2.510	4	0	6	4
2.484	3	0	4	6
		5	1	-3
2.474	3			
		5	3	-1
		1	1	7
2.468	4			
		5	3	1
2.444	3	5	1	3
		1	7	-3
2.352	4	2	6	4
		1	7	3
		4	4	4
2.311	2B	5	3	-3
		1	3	7
2.276	1	6	0	0
2.255	1	4	0	-6
		3	7	1
2.239	2B			
		3	5	-5
2.216	2	3	1	-7
2.160	2	0	2	8
2.122	3	0	6	6
2.106	5	2	8	-2
2.101	4	2	8	2
		2	2	-8
2.066	3			
		1	5	-7
		1	5	7
2.056	4			
		5	3	-5
		0	8	4
2.028	4			
		5	3	5
1.995	1	4	4	6

Sample from Montresta, Nuoro, Sardinia, Italy; data from Passaglia et al. (1978b).

B = broad.

Table 6.2 III. X-ray powder pattern of stellerite

d(A)	I/I₀	H	K	L		d(A)	I/I₀	H	K	L
9.03	100	0	2	0		2.562	4	3	5	3
		0	0	2		2.546	1	3	3	5
6.37	<1	0	2	2		2.532	<1	1	7	1
5.44	2	2	2	0		2.512	1	0	6	4
5.41	3	2	0	2		2.508	1	1	5	5
5.29	4	1	3	1		2.494	2	0	4	6
4.65	15	2	2	2		2.485	2	1	1	7
4.56	4	0	4	0		2.452	1	5	1	3
4.47	2	0	0	4				2	6	4
4.28	6	3	1	1		2.351	2			
4.06	45	1	3	3				1	7	3
		0	4	2		2.318	1	1	3	7
4.01	6	0	2	4				6	0	0
3.784	1	2	4	0		2.267	1			
3.734	5	2	0	4				4	6	0
3.482	3	2	4	2				4	0	6
3.397	7	4	0	0		2.239	1			
		1	1	5				3	7	1
3.181	7	4	2	0		2.233	1	0	0	8
		4	0	2		2.223	1	3	5	5
3.100	3	3	3	3		2.206	1	3	1	7
3.028	23	1	5	3		2.201	1	6	2	0
3.003	10	1	3	5		2.168	<1	0	2	8
		4	2	2				5	5	1
2.976	1	0	0	6		2.161	<1			
2.875	2	0	6	2				2	8	0
2.827	1	0	2	6		2.124	1	0	6	6
2.804	1	3	5	1		2.120	1	2	0	8
2.771	8	2	6	0		2.097	1	2	8	2
2.703	2	4	0	4		2.078	<1	1	7	5
2.608	2	2	2	6				2	2	8
		4	4	2		2.066	2			
								1	5	7
						2.037	2	5	3	5
								6	4	0
						2.031	2	2	6	6
								0	8	4
						1.978	<1	6	4	2

Sample from Villanova Monteleone, Sardinia, Italy. The data from Galli and Passaglia (1973) have been re-indexed by the authors after Alberti's (1976) method taking into account the diffraction intensities from single crystal measurements; cell parameters: $a = 13.603(4)$, $b = 18.217(4)$, $c = 17.857(3)$ Å.

Table 6.2IV. X-ray powder pattern of barrerite

d(Å)	I/I₀	H	K	L
9.10	>100	0	2	0
6.83	4	2	0	0
5.30	6	1	3	1
5.23	5	1	1	3
4.66	21	2	2	2
4.55	6	0	4	0
4.46	3	0	0	4
4.29	12	3	1	1
4.05	100	1	3	3
4.01	13	0	2	4
3.732	12	2	0	4
3.566	1	3	3	1
3.483	12	2	4	2
3.408	11	4	0	0
3.393	11	1	1	5
3.189	16	4	0	2
3.103	7	3	3	3
3.028	78	1	5	3
3.004	25	4	2	2
		1	3	5
2.974	4	0	0	6
2.885	2	2	4	4
2.871	3	0	6	2
2.824	3	0	2	6
2.805	3	3	5	1
2.773	22	2	6	0
2.727	3	4	4	0
		2	0	6
2.709	3	4	0	4
2.609	4	2	2	6
2.563	6	3	5	3

d(Å)	I/I₀	H	K	L
2.524	1	1	7	1
		0	1	7
2.507	5	0	6	4
2.487	3	0	4	6
2.354	3	2	6	4
2.347	3	1	7	3
2.274	1	6	0	0
2.267	1	0	0	8
2.225	3	3	5	5
2.204	2	3	1	7
		6	2	0
		0	8	2
		0	2	8
2.164	1B	5	5	1
2.122	4	0	6	6
2.097	3	2	8	2
2.084	<1	3	3	7
2.075	1B	1	7	5
2.062	3	2	2	8
2.040	2	1	5	7
		5	3	5
2.034	3	6	4	0
2.025	3	2	6	6
		0	8	4
2.018	2	4	6	4
2.012	<1	4	4	6
1.893	3	1	9	3
		4	8	0

Sample from Capo Pula, Cagliari, Sardinia. The data from Passaglia and Pongiluppi (1974) have been re-indexed by the authors after Alberti's (1976) method taking into account the diffraction intensities from single crystal measurements; cell parameters: $a = 13.642(3)$, $b = 18.195(4)$, $c = 17.833(4)$ Å.
B = broad.

Table 6.3 III. X-ray powder pattern of brewsterite

d(A)	I/Io	H	K	L		d(A)	I/Io	H	K	L
8.78	31	0	2	0				0	6	0
7.73	3	0	0	1		2.920	100	1	5	-1
7.08	9	0	1	1				2	3	0
6.75	8	1	0	0		2.848	9	2	2	1
6.62	21	1	1	0		2.796	5	2	3	-1
5.80	3	0	2	1		2.732	24	0	6	1
5.30	4	1	0	-1		2.724	24	1	4	-2
5.07	14	1	1	-1				2	4	0
4.90	6	1	0	1		2.675	7			
4.72	12	1	1	1				2	3	1
4.66	71	0	3	1		2.610	3	1	4	2
4.53	20	1	2	-1		2.575	9	0	0	3
4.38	11	0	4	0		2.507	8	1	6	1
4.28	2	1	2	1		2.479	4	2	4	1
3.925	15	1	3	-1		2.408	3	2	3	-2
3.864	11	0	0	2		2.378	5	1	2	-3
3.812	13	0	4	1		2.347	18	1	7	0
3.772	15	0	1	2		2.260	7	2	3	2
3.677	10	1	4	0		2.221	3	0	4	3
3.534	18	0	2	2		2.211	6	2	6	0
3.404	7	1	1	-2		2.189	45	0	8	0
3.269	37	1	4	1		2.137	4	2	4	2
3.191	16	0	5	1		2.102	3	3	3	0
3.111	22	1	5	0		2.068	4	1	4	3
3.044	12	1	2	2		2.019	9	1	5	-3
2.967	6	2	1	1		2.015	8	3	0	-2
						1.972	3	3	4	-1

Sample from Strontian, Scotland. New data by the authors; Philips diffractometer, Ni filtered CuKα radiation ($\lambda = 1.54051$ Å), internal standard $Pb(NO_3)_2$ (cubic: $a = 7.8568$ Å), slits 1°, 0.1 mm, 1°, scanning speed 0.5°/min, indexing after Alberti's (1976) method taking into account the diffraction intensities from single crystal measurements; cell parameters: $a = 6.779(2)$, $b = 17.520(3)$, $c = 7.749(2)$ Å, $\beta = 94°28'(1)$.

Table 7.1 II. X-ray powder patterns of two cowlesites

Goble Oregon					Kuniga Japan		Goble Oregon					Kuniga Japan	
d(A)	I/Io	H	K	L	I/Io	d(A)	d(A)	I/Io	H	K	L	I/Io	d(A)
15.2	100	0	1	0	100	15.3	3.16	10	0	0	4	8	3.177
12.6	5	0	0	1	4	12.7	3.052	20	0	5	0	15	3.048
11.3	5	1	0	0	-	-	2.964	35	0	5	1		
8.40	10	1	0	1	2	8.47						25	2.953
7.62	15	0	2	0	10	7.64			1	5	0		
5.67	7	1	2	1	2	5.70	2.934	25	3	3	1	-	-
5.52	5	1	0	2	-	-			0	2	4	20	2.927
5.08	17	0	3	0	8	5.08	2.819	5	0	4	3		
4.70	7	0	3	1	3	4.71			4	0	0	8	2.825
4.50	3	1	2	2	6	4.50			4	1	0	2	2.773
4.16	5	2	0	2	-	-			1	4	3	3	2.738
3.81	35	0	4	0	30	3.811			2	2	4	3	2.600
3.75	15	3	0	0	15	3.763			2	4	3	2	2.529
3.65	10	3	1	0	10	3.651			3	5	2	2	2.220
-	-	1	4	1	4	3.461			3	3	4	2	2.189
-	-	0	4	2					3	6	2	5	1.996
					5	3.262							
3.25	5	3	2	1									

Samples from Goble, Oregon and from Kuniga, Oki Islands, Japan; data from Wise and Tschernich (1975) and from Matsubara et al. (1978b).

Table 7.2I. X-ray powder pattern of goosecreekite

d(Å)	I/I₀	H	K	L
8.75	5	0	2	0
7.19	50	1	0	0
5.59	50	1	2	0
4.91	50	1	2	-1
4.53	100	1	3	0
		0	3	1
4.36	5	1	1	1
4.17	15	1	3	-1
4.00	10	1	2	1
3.750	5	1	4	0
3.617	5	2	0	0
		1	0	-2
3.526	25	0	0	2
		1	4	-1
		1	1	-2
3.350	40	2	2	-1
		2	2	0
3.277	25	0	2	2
3.147	20	1	4	1
		0	5	1
		1	5	0
		2	3	0
3.073	25	1	3	-2
		0	3	2
3.022	25	1	5	-1
2.925	10	2	1	-2
		2	1	1
2.866	5	1	1	2
		1	4	-2
2.775	20	2	1	1
2.698	20			
2.610	15	2	3	1
2.501	5			
2.438	10			
2.395	5			
2.351	20			

Sample from Goose Creek quarry, Loudoun Co., Virginia; data from Dunn et al. (1980).

Table 7.3I. X-ray powder pattern of partheite

d(Å)	I/I₀	H	K	L
10.79	100	2	0	0
8.12	80	1	1	0
6.10	70	1	1	1
5.39	5	4	0	0
4.65	5	0	0	2
4.38	10	0	2	0
4.31	5	2	0	-2
4.05	20	2	2	0
		1	1	-2
3.870	10	5	1	0
3.740	50	2	2	-1
3.600	40	3	1	-2
		5	1	-1
3.532	15	3	1	2
3.400	15	4	2	0
3.190	40	0	2	2
3.046	30	2	2	2
3.000	10	5	1	-2
2.950	20	5	1	2
2.900	30	1	3	0
		1	1	3
2.790	10	7	1	-1
2.752	15	7	1	1
2.710	10	3	3	0
2.682	5	3	1	3
		6	2	-1
2.600	5	3	3	-1
		3	3	1
2.531	10	0	2	3
2.454	10	1	3	2
		2	2	3
		5	1	-3
2.427	5	5	3	0
2.332	10	5	3	1
		3	3	2
2.265	5	2	0	4
2.231	20	1	1	4
		9	1	1
2.200	5	0	4	0
2.161	10	3	1	-4
		5	3	-2
		5	3	2
2.134	5	0	4	1
		3	1	4
2.094	5	7	1	3
		2	4	-1
		6	2	-3
		9	1	-2
		2	4	1
2.051	10	0	2	4
		6	2	3
		9	1	2
		3	3	-3
		5	1	-4
2.010	5	2	2	4
		6	0	-4
1.974	15	10	0	-2
		5	1	4

Sample from Taurus Mts., Southwest Turkey; data from Sarp et al. (1979).

Appendix 2

Infra Red Spectra of Fibrous Zeolites
(here given mainly for identification purpose)

The first two digits of the figure codes correspond to the chapter in which the zeolite is described. The IR-spectra were obtained with a Perkin Elmer 180 spectrometer and KBr pellets method.

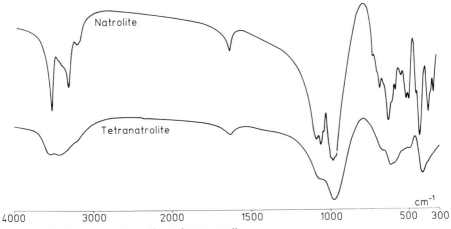

Fig. 1.1R. IR-spectra of natrolite and tetranatrolite

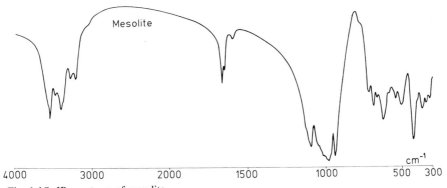

Fig. 1.1S. IR-spectrum of mesolite

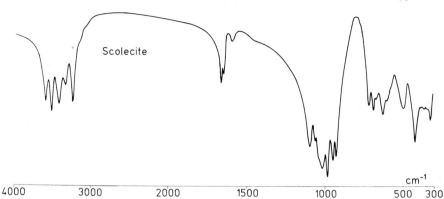

Fig. 1.1T. IR-spectrum of scolecite

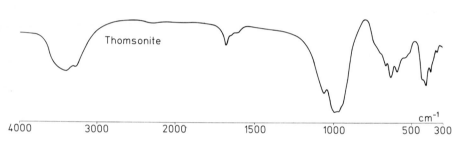

Fig. 1.2G. IR-spectrum of thomsonite

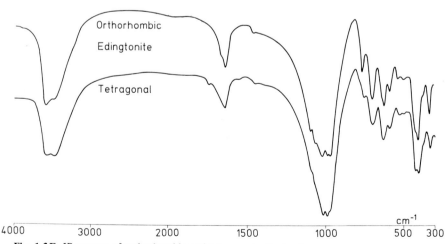

Fig. 1.3E. IR-spectra of orthorhombic and tetragonal edingtonite

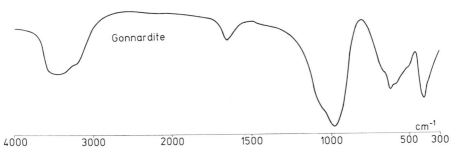

Fig. 1.4C. IR-spectrum of gonnardite

References

Abbona F, Franchini Angela M (1970) Sulla thomsonite di Saint Vincent (Valle d'Aosta). Atti Accad Sci Torino I Cl Sci Frs Mat Nat 104:371–380

Abe H, Aoki M (1976) Experiments on the interaction between $Na_2CO_3 – NaHCO_3$ solution and clinoptilolite tuff, with reference to analcimization around Kuroko-type mineral deposits. Chem Geol 17:89–100

Abe, Aoki M, Konno H (1973) Synthesis of analcime from volcanic sediments in sodium silicate solutions. Experimental studies in the water-rock interaction. Contrib Miner Petrol 42:81–92

Adiwidjaja G (1972) Strukturbeziehungen in der Natrolithgruppe und das Entwässerungsverhalten des Skolezits. Diss Univ Hamburg

Aiello R, Barrer RM (1970) Hydrothermal chemistry of silicates. Part XIV. Zeolite crystallization in presence of mixed bases. J Chem Soc (Lond) Sect A 1970:1470–1475

Aiello R, Franco E (1968) Formazione di zeoliti per trasformazione di halloysite e montmorillonite a bassa temperatura ed in ambiente alcalino. Rend Acad Sci Fis Mat Soc Naz Sci Lett Arti Napoli Ser 4 35:1–29

Aiello R, Barrer RM, Davies JA, Kerr IS (1970) Molecule sieving in relation to cation type and position in unfaulted offretite. Trans Faraday Soc 66:1610–1617

Akizuki M (1980) Origin of optical variation in analcime and chabazite. In: Rees LVC (ed) Proc 5th Int Conf Zeolites. Heyden, London, pp 171–178

Akizuki M (1981a) Origin of optical variation in analcime. Am Miner 66:403–409

Akizuki M (1981b) Origin of optical variation in chabazite. Lithos 14:17–21

Alberti A (1971) Le zeoliti di Val Nambrone (Gruppo dell'Adamello). Period Miner 40:289–294

Alberti A (1972) On the crystal structure of the zeolite heulandite. Tscherm Miner Petr Mitt 18:129–146

Alberti A (1973) The structure type of heulandite B (Heat-collapsed phase). Tscherm Miner Petr Mitt 19:173–184

Alberti A (1975a) The crystal structure of two clinoptilolites. Tscherm Miner Petr Mitt 22:25–37

Alberti A (1975b) Sodium-rich dachiardite from Alpe di Siusi, Italy. Contrib Miner Petrol 49:63–66

Alberti A (1976) The use of structure factors in the refinement of unit-cell parameters from powder diffraction data. J Appl Crystallogr 9:373–374

Alberti A (1979a) Chemistry of zeolites. Q-mode multivariate factor analysis. Chem Erde 38:64–82

Alberti A (1979b) Possible 4-connected frameworks with 4-4-1 unit found in heulandite, stilbite, brewsterite and scapolite. Am Miner 64:1188–1198

Alberti A, Gottardi G (1975) Possible structures in fibrous zeolites. Neues Jahrb Miner Monatsh 1975:396–411

Alberti A, Vezzalini G (1978) Crystal structures of the heat collapsed phases of Barrerite. In: Sand LB, Mumpton FA (eds) Natural zeolites. Pergamon, Oxford, pp 173–184

Alberti A, Vezzalini G (1979) The crystal structure of amicite, a zeolite. Acta Cryst B35:2866–2869

Alberti A, Vezzalini G (1981a) A partially disordered natrolite: relationships between cell parameters and Si-Al distribution. Acta Cryst B37:781–788

Alberti A, Vezzalini G (1981b) Crystal energies and coordination of ions in partially occupied sites: dehydrated mazzite. Bull Miner 104:5–9

Alberti A, Vezzalini G (1983a) The thermal behaviour of heulandites: structural study of the dehydration of Nadap heulandite. Tscherm Miner Petr Mitt 31:259 – 270

Alberti A, Vezzalini G (1983b) How the structure of natrolite is modified through the heating-induced dehydration. Neues Jahrb Miner Monatsh 1983:135 – 144

Alberti A, Vezzalini G (1984) Topological changes in dehydrated zeolites: breaking of T-O-T bridges. In: Bisio A, Olson DH (eds) Proc 6th Int Conf Zeolite. Butterworths, Guildford, pp 834 – 841

Alberti A, Rinaldi R, Vezzalini G (1978) Dynamics of dehydration in stilbite-type structures; stellerite phase B. Phys Chem Minerals 2:365 – 375

Alberti A, Hentschel G, Vezzalini G (1979) Amicite, a new natural zeolite. Neues Jahrb Miner Monatsh 1979:481 – 488

Alberti A, Vezzalini G, Tazzoli V (1981) Thomsonite: a detailed refinement with cross checking by crystal energy calculations. Zeolites 1:91 – 97

Alberti A, Pongiluppi D, Vezzalini G (1982a) The crystal chemistry of natrolite, mesolite and scolecite. Neues Jahrb Miner Abh 143:231 – 248

Alberti A, Galli E, Vezzalini G, Passaglia E, Zanazzi PF (1982b) Positions of cations and water molecules in hydrated chabazite. Natural and Na-, Ca-, Sr- and K-exchanged chabazite. Zeolites 2:303 – 309

Alberti A, Vezzalini G, Pecsi-Donath E (1983a) Some zeolites from Hungary. Acta Geol Acad Sci Hung 25:237 – 246

Alberti A, Cariati F, Erre L, Piu P, Vezzalini G (1983b) Spectroscopic investigation on the presence of OH in natural barrerite and its collapsed phases. Phys Chem Minerals 9:189 – 191

Aleksiev B, Djourova EG (1975) On the origin of zeolite rocks. C R Acad Bulg Sci 28:517 – 520

Alietti A (1959) Diffusione e significato dei minerali a strati misti delle serpentine mineralizzate a talco dell'Appennino Parmense. Period Miner 28:67 – 107

Alietti A (1967) Heulanditi e clinoptiloliti. Miner Petrogr Acta Bologna 13:119 – 138

Alietti A (1972) Polymorphism and crystal-chemistry of heulandites and clinoptilolites. Am Miner 57:1448 – 1462

Alietti A, Passaglia E, Scaini G (1967) A new occurrence of ferrierite. Am Miner 52:1562 – 1563

Alietti A, Gottardi G, Poppi L (1974) The heat behaviour of the cation exchanged zeolites with heulandite structure. Tscherm Miner Petr Mitt 21:291 – 298

Alietti A, Brigatti MF, Poppi L (1977) Natural Ca-rich clinoptilolites (heulandites of group 3): new data and review. Neues Jahrb Miner Monatsh 1977:493 – 501

Ames LL Jr (1960) The cation sieve properties of clinoptilolite. Am Miner 45:689 – 700

Ames LL Jr (1961) Cation sieve properties of the open zeolites chabazite, mordenite, erionite and clinoptilolite. Am Miner 46:1120 – 1131

Ames LL Jr (1962) Effect of base cation on the cesium kinetics of clinoptilolite. Am Miner 47:1310 – 1316

Ames LL Jr (1963) Synthesis of a clinoptilolite-like zeolite. Am Miner 48:1372 – 1381

Ames LL Jr (1964a) Some zeolites equilibria with alkali metal cations. Am Miner 49:127 – 145

Ames LL Jr (1964b) Some zeolites equilibria with alkaline earth metal cations. Am Miner 49:1099 – 1110

Ames LL Jr (1965) Self-diffusion of some cations in open zeolites. Am Miner 50:465 – 475

Ames LL Jr (1966) Exchange of alkali metal cations on a natural stilbite. Can Miner 8:582 – 592

Ames LL Jr, Mercer BW (1961) The use of clinoptilolite to remove potassium selectively from aqueous solutions of mixed salts. Econ Geol 56:1133 – 1136

Ames LL Jr, Sand LB (1958) Hydrothermal synthesis of wairakite and calcium-mordenite. Am Miner 43:476 – 480

Amirov ST (1979) Discovery of epistilbite in Azerbaidzhan (in Russian). In: Rundkvist DV (ed) Miner parageneзisy miner gornykh porod rud. Izd Nauka, Leningrad, pp 94 – 96

Amirov ST, Ilyukhin VV, Belov NV (1967) Crystal structure of Ca-zeolite-laumontite (leonhardite) $CaAl_2Si_4O_{12} \cdot nH_2O$ (2 < n < 4) (in Russian). Dokl Akad Nauk SSSR 174:667 – 670

Amirov ST, Ilyukhin VV, Belov NV (1971) Deciphering of laumontite structure from interaction peaks (in Russian). Zap Vses Miner Obshchest 100:20 – 30

Amirov ST, Asratkulu MA, Mamedov KhS, Belov NV (1972) Crystal structure of the zeolite gonnardite (in Russian). Dokl Akad Nauk SSSR 203:1299–1301

Amirov ST, Elchiev YM, Mamedov KS (1976) X-ray studies on the zeolite-bearing rocks of Azerbaidzhan (in Russian). Azerb Khim Zh 6:114–118

Amirov ST, Amiraslanov IR, Usubaliev BT, Mamedov KS (1978) Crystal structure and symmetry of the zeolite thomsonite (in Russian). Azerb Khim Zh 3:120–127

Antonín R (1942) Research on the minerals and rocks of the Vinařická Hill. Věstn Kral České Spol Nauk Trida Mat Prirodoved 1942:1–19

Aoki M (1978) Hydrothermal alteration of chabazite. J Japn Assoc Miner Petr Econ Geol 73:155–166

Aoki M, Minato H (1980) Lattice constants of wairakite as a function of chemical composition. Am Miner 65:1212–1216

Appleman D (1960) The crystal structure of bikitaite, $LiAlSi_2O_6 \cdot H_2O$ (Abstr.) Acta Crystallogr 13:1002

Arima M, Edgar AD (1980) Importance of time and H_2O contents on the analcime-H_2O system at 465 °C and 1 Kbar $P(H_2O)$. Neues Jahrb Miner Monatsh 1980:543–554

Arrhenius G, Bonatti E (1965) Neptunism and volcanism in the ocean. Prog Oceanogr 3:7–22

Ataman G, Beseme P (1972) Découverte de l'analcime sédimentaire en Anatolie du nord-ouest (Turquie): minéralogie, genèse, paragènese. Chem Geol 9:203–225

Aumento F (1964) X-ray studies of Nova Scotia zeolites. Can Min 8:132

Aumento F, Friedlander C (1966) Zeolites from North Mts., Nova Scotia. Miner Soc India IMA Vol:149–154

Aurisicchio C, De Angelis G, Dolfi D, Farinato R, Loreto L, Sgarlata F, Trigila R (1975) Sull'analcime di colore rosso nelle vulcaniti del complesso alcalino di Crowsnest (Alberta-Canada): notizie preliminari. Rend Soc Ital Miner Petrol 31:653–671

Baerlocher C, Barrer RM (1974) The crystal structure of synthetic zeolite F. Z Kristallogr 140:10–26

Baerlocher C, Meier WM (1970) Synthese und Kristallstruktur von Tetramethylammonium-Gismondin. Helv Chim Acta 53:1285–1293

Baerlocher C, Meier WM (1972) The crystal structure of synthetic zeolite Na-Pl, an isotope of gismondine. Z Kristallogr 135:339–354

Baldar NA, Whittig LD (1968) Occurrence and synthesis of soil zeolites. Proc Soil Sci Am 32:235–238

Balgord WD, Roy R (1971) Crystal chemical relationships in the analcite family. In: Molecular sieve zeolites I. Am Chem Soc Adv Chem Ser 101:140–147

Barabanov VF (1955) Stellerite from the occurrence of Bukuka (in Russian). Dokl Akad Nauk SSSR 100:151–154

Barbieri M, Dolfi D, Trigila R (1977) Sulla genesi dell'analcime delle vulcaniti di Crowsnest alla luce del comportamento geochimico del rubidio nelle blairmoriti e durante il processo di analcimizzazione della leucite fino a P = 3 Kbar. Rend Soc Ital Miner Petrol 33:55–62

Bargar KE, Beeson MH, Keith TEC (1981) Zeolites in Yellowstone National Park. Miner Record 12:29–38

Barić Lj (1965) Ferrierit aus dem Steinbruch Gotalovec in Hrvatsko Zagorje (Nordkroatien). Bull Sci Cons Acad RSF Yougosl Sect A 10:177–178

Barker DS, Long LE (1969) Felspathoidal syenite in a quartz diabase sill, Brookville, New Jersey. J Petrol 10:202–221

Barnes BA, Boles JR, Hickey J (1984) Zeolite occurrences in Triassic-Jurassic sedimentary rocks, Baja California Sur, Mexico. In: Bisio A, Olson DH (eds) Proc 6th Int Conf Zeolite, Butterworths, Guildford pp 584–594

Barrer RM (1938) The sorption of polar and non-polar gases by zeolites. Proc R Soc Lond Ser A 167:392–420

Barrer RM (1944) Sorption by gmelinite and mordenite. Trans Faraday Soc 40:555–564

Barrer RM (1948) Synthesis and reactions of mordenite. J Chem Soc (Lond) 1948:2158–2163

Barrer RM (1950) Ion-exchange and ion-sieve processes in crystalline zeolites. J Chem Soc (Lond) 1950:2342–2350

Barrer RM (1978) Zeolites and clay minerals as sorbents and molecular sieves. Academic, London

Barrer RM (1982) Hydrothermal chemistry of zeolites. Academic, London

Barrer RM, Baynham JW (1956) The hydrothermal chemistry of the silicates. Part VII. Synthetic potassium aluminosilicates. J Chem Soc (Lond) 1956:2882–2891

Barrer RM, Coughlan B (1968) Molecular sieves derived from clinoptilolite by progressive removal of framework charge: characterization by sorption of CO_2 and krypton. In: Molecular sieves. Society of Chemical Industry, London, pp 141–148

Barrer RM, Denny PJ (1961) Hydrothermal chemistry of the silicates. Part X. A partial study of the field $CaO-Al_2O_3-SiO_2-H_2O$. J Chem Soc (Lond) 1961:983–1000

Barrer RM, Hinds L (1953) Ion-exchange in crystals of analcite and leucite. J Chem Soc (Lond) 1953:1879–1883

Barrer RM, Kerr IS (1959) Intracrystalline channels in levynite and some related zeolites. Trans Faraday Soc 55:1915–1923

Barrer RM, Klinowski J (1975) Hydrogen mordenite and hydronium mordenite. Ion exchange and thermal stability. J Chem Soc Faraday Trans I 1975:690–698

Barrer RM, Mainwaring DE (1972) Chemistry of soil minerals. Part XIII. Reactions of metakaolinite with single and mixed bases. J Chem Soc Dalton Trans 1972:2534–2546

Barrer RM, Marcilly C (1970) Hydrothermal chemistry of silicates. Part XV. Synthesis and nature of some salt-bearing aluminosilicates. J Chem Soc (Lond) Sect A 1970:2735–2745

Barrer RM, Marshall DJ (1964a) Hydrothermal chemistry of silicates. Part XII. Synthetic strontium aluminosilicates. J Chem Soc (Lond) 1964:485–497

Barrer RM, Marshall DJ (1964b) Hydrothermal chemistry of silicates. Part XIII. Synthetic barium aluminosilicates. J Chem Soc (Lond) 1964:2296–2305

Barrer RM, Marshall DJ (1965) Synthetic zeolites related to ferrierite and yugawaralite. Am Miner 50:484–489

Barrer RM, McCallum N (1953) Hydrothermal chemistry of silicates. Part IV. Rubidium and caesium aluminosilicates. J Chem Soc (Lond) 1953:4029–4035

Barrer RM, Munday BM (1971) Cation exchange reactions of a sedimentary phillipsite. J Chem Soc (Lond) Sect A 1971:2904–2909

Barrer RM, Sieber W (1977) Hydrothermal chemistry of silicates. Part XXI. Zeolites from reaction of lithium and caesium ions with tetramethylammonium aluminosilicate solutions. J Chem Soc Dalton Trans 1977:1020–1026

Barrer RM, Townsend RP (1976) Transition metal ion exchange in zeolites. Part II. Ammines of Co^{3+}, Cu^{2+} and Zn^{2+} in clinoptilolite, mordenite and phillipsite. J Chem Soc Faraday Trans I 1976:2650–2660

Barrer RM, Vaughan DEW (1969) Sorption and diffusion of rare gases in heulandite and stilbite. Surface Sci 14:77–92

Barrer RM, Vaughan DEW (1971) Trapping and diffusion of rare gases in phillipsite, zeolite K-M and other silicates. Trans Faraday Soc 67:2129–2136

Barrer RM, White EAD (1952) The hydrothermal chemistry of silicates. Part II. Synthetic crystalline sodium aluminosilicates. J Chem Soc (Lond) 1952:1561–1571

Barrer RM, Baynham JW, Bultitude FW, Meier WM (1959) Hydrothermal chemistry of the silicates. Part VIII. Low-temperature crystal growth of aluminosilicates, and some gallium and germanium analogues. J Chem Soc (Lond) 1959:195–208

Barrer RM, Papadopoulos R, Rees LVC (1967) Exchange of sodium in clinoptilolite by organic cations. J Inorg Nucl Chem 29:2047–2063

Barrer RM, Cole JF, Sticher H (1968) Chemistry of soil minerals. Part V. Low temperature hydrothermal transformation of kaolinite. J Chem Soc (Lond) Sect A 1968:2475–2485

Barrer RM, Davies JA, Rees LVC (1969) Thermodynamics and thermochemistry of cation exchange in chabazite. J Inorg Nucl Chem 31:219–232

Barrer RM, Beaumont R, Colella C (1974) Chemistry of soil minerals. Part XIV. Action of some basic solutions on metakaolinite and kaolinite. J Chem Soc Dalton Trans 1974:934–941

Barrows KJ (1980) Zeolitization of Miocene volcanoclastic rocks, southern Desatoya Mts., Nevada. Geol Soc Am Bull 91:199–210

References

Bartl H (1970) Strukturverfeinerung von Leonhardit Ca[Al$_2$Si$_4$O$_{12}$] · 3 H$_2$O, mittels Neutronen-Beugung. Neues Jahrb Miner Monatsh 1970:298 – 310

Bartl H (1973) Neutronenbeugungsuntersuchung des Zeolithes Heulandit. Z Kristallogr 137:440 – 441

Bartl H, Fischer KF (1967) Untersuchung der Kristallstruktur des Zeolithes Laumontit. Neues Jahrb Miner Monatsh 1967:33 – 42

Bass NN, Moberley R, Rhodes JM, Shih C, Church SE (1973) Volcanic rocks cored in the Central Pacific, Leg 17, Deep Sea Drilling Project. Initial Rep DSDP 17:429 – 503

Batiashvili TV, Gvakhariya GV (1968) Erionite found for the first time in Georgia (in Russian). Dokl Akad Nauk SSSR 179:426 – 427

Bauer J, Hřichová R (1962) Ptilolite (mordenite) from Teigarhorn, Iceland (in Czech). Sb Vys Šk Chem Technol Miner 6:27 – 36

Baur WH (1964) On the cation and water positions in faujasite. Am Miner 49:697 – 704

Belitskiy IA (1972) Thermography of cation-exchanged forms (Li, Ag, NH$_4$, K, Tl, Rb, Cs) of natural natrolite (in Russian). Zap Vses Miner Obshchest 101:52 – 61

Belitskiy IA, Bukin GV (1968) First find of erionite in the USSR (in Russian). Dokl Akad Nauk SSSR 178:169 – 172

Belitskiy IA, Fedorov A (1976) Special trends of the ion exchange Na↔K in natrolite (in Russian). In: Godovikov AA, Ptitsyn AB (eds) Eksperiment Issled Miner. Akad Nauk SSSR Inst Geol Geofiz, Novosibirsk, pp 46 – 54

Belitskiy IA, Gabuda SP (1968) Water diffusion in ion-exchanged forms of some natural zeolites. Chem Erde 27:79 – 90

Belitskiy IA, Pavlyuchenko VS (1967) A new synthetic barium aluminosilicate (in Russian). Dokl Akad Nauk SSSR 173:654 – 657

Bellinzona G (1923) Thomsonite associata a cabasite dei Monzoni. Rend R Accad Naz Lincei Cl Sci Fis Mat Nat Ser 5 32:346 – 348

Bennet JAE (1945) Some occurrences of leucite in East Lothian. Trans Edinb Geol Soc 14:34 – 52

Bennett JM, Gard JA (1967) Non-identity of the zeolites erionite and offretite. Nature (Lond) 214:1005 – 1006

Bennett JM, Grose RW (1978) Characterization of the offretite-levynite intergrowth from Beech Creek, Oregon, by adsorption and electron diffraction. In: Sand LB, Mumpton FA (eds) Natural zeolites. Pergamon, Oxford, pp 77 – 84

Bergerhoff G, Koyama H, Nowacki W (1956) Zur Kristallstruktur der Mineralien der Chabazit-und der Faujasit-Gruppe. Experientia (Basel) 12:418 – 419

Bergerhoff G, Baur WH, Nowacki W (1958) Über die Kristallstrukturen des Faujasits. Neues Jahrb Miner Monatsh 1958:193 – 200

Berman H (1925) Notes on dachiardite. Am Miner 10:421 – 428

Bernat M, Church TM (1978) Deep-sea phillipsites: trace geochemistry and modes of formation. In: Sand LB, Mumpton FA (eds) Natural zeolites. Pergamon, Oxford, pp 259 – 268

Betz V (1981) Zeolites from Iceland and the Faeroes. Miner Record 12:5 – 26

Betz V (1982) Mont Semiol, seltene Zeolithe aus Frankreich. Lapis 7, 12:9 – 11

Betz V, Hentschel G (1978) Offretit und Erionit von Gedern (Vogelsberg). Geol Jahrb Hessen 106:419 – 421

Beus AA (1960) Geochemistry of beryllium and the genetic types of beryllium deposits (in Russian). Izdatel Akad Nauk SSSR, Moscow

Beutell A, Blaschke K (1915) Der Basenaustausch beim Desmin. Zentralbl Miner 1915:142 – 144

Birch WD (1976) Chabazite or gmelinite (or both) in basalts from Flinders, Victoria. Aust Miner 1:2 – 3

Birch WD (1979) Levyne from Clunes, Victoria: a new occurrence. Aust Miner 25:119 – 120

Birch WD (1980) Offretite and levyne from Flinders, Victoria. Aust Miner 31:154

Birch WD, Morvell G (1978) Ferrierite from Phillip Island, Victoria. Aust Miner 15:75 – 76

Bitter PH von, Plint-Geberl HA (1980) Phillipsite altered to analcime in basal Windsor (shallow-water marine) carbonates of Mississipian age in Nova Scotia. Can J Earth Sci 17:1096 – 1100

Black PM (1969) Harmotome from the Tokatoka District, New Zealand. Miner Mag 37:453 – 458

Blum JR, Delffs W (1843) Leonhardit, ein neues Mineral; Analyse des Leonhardits. Poggendorf Ann Phys Chem 59:336 – 342

Bobonich FM (1979) Rock-forming clinoptilolite in meteoritic crater sediments (in Russian). Dokl Akad Nauk SSSR 248:710 – 714

Bøggild OB (1953) The mineralogy of Greenland. Medd om Grønland 149, no 3:1 – 442

Boles JR (1971) Synthesis of analcime from natural heulandite and clinoptilolite. Am Miner 56:1724 – 1734

Boles JR (1972) Composition, optical properties, cell dimensions and thermal stability of some heulandite group zeolites. Am Miner 57:1463 – 1493

Boles JR, Wise WS (1978) Nature and origin of deep-sea clinoptilolite. In: Sand LB, Mumpton FA (eds) Natural Zeolites. Pergamon, Oxford, pp 235 – 243

Bonardi M (1979) Composition of type dachiardite from Elba: a re-examination. Miner Mag 43:548 – 549

Bonardi M, Roberts AC, Sabina AP, Chao GY (1981) Sodium-rich dachiardite from Francon quarry, Montreal Island, Quebec. Can Miner 19:285 – 289

Bonatti E (1963) Zeolites in Pacific pelagic sediments. Trans NY Acad Sci Ser II 25:938 – 948

Bonatti E (1965) Palagonite, hyaloclastites and alteration of volcanic glass in the ocean. Bull Volcanol 28:3 – 15

Bonatti E, Nayudu YR (1965) The origin of manganese nodules on the ocean floor. Am J Sci 263:17 – 39

Bonatti S (1942) Ricerche sulla dachiardite. Mem Soc Tosc Sci Nat 50:14 – 25

Bonatti S, Gottardi G (1960) Dati ottici e strutturali sulla dachiardite. Period Miner 29: 103 – 107

Bond WL (1943) A mineral survey for piezo-electric materials. Bell Syst Tech J 22:145 – 152

Born I von (1772) Lithophylacium Bornianum 1, Prague, p 46

Bosc D'Antic L (1792) Mémoire sur la chabazie. J Hist Nat 2:181

Bosmans HJ, Tambuyzer E, Paenhuis J, Ylen L, Vancluisen J (1973) Zeolite formation in the system $K_2O-Na_2O-Al_2O_3-SiO_2-H_2O$. In: Meier WM, Ujetterhoeven JB (eds) Molecular sieves. Am Chem Soc Adv Chem Ser 121:179 – 188

Boussaroque JL, Maury R (1972) Etude de l'association dawsonite-analcite. C R Acad Sci Paris Ser D 275:1839 – 1841

Bower CE, Turner DT (1982) Ammonia removal by clinoptilolite in the transport of ornamental freshwater fishes. Prog Fish-Cult 44:19 – 23

Bramlette MN, Posniak E (1933) Zeolitic alteration of pyroclastics. Am Miner 18:167 – 171

Breck DW [to Union Carbide Corp] (1962, 1964) Crystalline zeolite Y. Belg Pat 617,598; US Pat 3,130,007

Breck DW (1974) Zeolite molecular sieves. Wiley, New York

Breck DW, Flanigen EM (1968) Synthesis and properties of Union Carbide zeolites X and Y. In: Molecular sieves. Society of Chemical Industry, London, pp 47 – 61

Breck DW, Eversole WJ, Milton RM, Reed TB, Thomas TL (1956) Crystalline zeolites. I. The properties of a new synthetic zeolite, type A. J Am Chem Soc (Lond) 78:5963 – 5971

Breithaupt A (1818) Ergänzungen und Berichtigungen zu dem applikativen Theil. In: Hoffmann CAS (ed) Handbuck der Mineralogie, vol 4, Abt 2. Freiberg, p 41

Bresciani-Pahor N, Calligaris M, Nardin G, Randaccio L, Russo E, Comin-Chiaramonti P (1980) Crystal structure of a natural and a partially silver-exchanged heulandite. J Chem Soc Dalton Trans 1980:1511 – 1514

Bresciani-Pahor N, Calligaris M, Nardin G, Randaccio L (1981) Location of cations in metal ion-exchanged zeolites. Part II. Crystal structures of a fully silver-exchanged heulandite. J Chem Soc Dalton Trans 1981:2288 – 2291

Brewster D (1821) Account of comptonite, a new mineral from Vesuvius. Edinb Phil J 4:131 – 133

Brewster D (1825a) Description of gmelinite, a new mineral species. Edinb J Sci 2:262 – 267

Brewster D (1825b) Description of levyne, a new mineral species. Edinb J Sci 2:232 – 334

Brooke HJ (1820) On mesotype, needlestone, and thomsonite. Ann Phil 16:193 – 194

Brooke HJ (1822) On the comptonite of Vesuvius, the brewsterite of Scotland, the stilbite and the heulandite. Edinb Phil J 6:112 – 115

Brown G, Catt JA, Weir AN (1969) Zeolites of the clinoptilolite-heulandite type in sediments of South-East England. Miner Mag 37:480 – 488

Brownell JM (1938) Zeolites at the Sherritt Gordon mine. Univ Toronto Stud Geol Ser 41: 19 – 22

Burns VM, Burns RG (1978) Authigenic todorokite and phillipsite inside deep-sea manganese nodules. Am Miner 63:827 – 831

Bur'yanova EZ (1954) Analcime-bearing sediments from Tuva (in Russian). Dokl Akad Nauk SSSR 98:261 – 264

Bur'yanova EZ, Bogdanov VV (1967) Distribution of the authigenic zeolites laumontite and heulandite in the sedimentary rocks of the Tarbagatai coal deposits (in Russian). Litol Polez Iskop 1967, no 2:59 – 68

Byström AM (1956) Harmotome penetration of a scapolite partly altered to argillic material in Ultevis, North Sweden. Geol Foren Stockh Forh 78:645 – 653

Caglioti V (1927) Ricerche su alcune zeoliti delle leucititi dei dintorni di Roma; la gismondite di Capo di Bove e la pseudophillipsite di Acquacetosa. Rend Accad Sci Fis Mat Cl Soc r Ser 3 Napoli 33:163 – 177

Čaikova M, Haramlova S (1959) Einige Zeolith-Vorkommen in Andesithohlräumen in der Slovakei. Geol Pr (Bratisl) 54:191 – 224

Callegari E (1964) Rapporti fra distribuzione di analcime e fenomeni di albitizzazione nella "pietra verde" degli strati di Livinallongo della Regione dolomitica. Stud Trent Sci Nat 61:25 – 43

Calleri M, Ferraris G (1964) Struttura dell'analcime. Atti Accad Sci Torino I Cl Sci Fis Met Nat 98:821 – 846

Calligaris M, Nardin G (1982) Cation site location in hydrated chabazites. Crystal structure of barium- and cadmium-exchanged chabazites. Zeolites 2:200 – 204

Calligaris M, Nardin G, Randaccio L (1982) Cation site location in a natural chabazite. Acta Cryst B38:602 – 605

Calligaris M, Nardin G, Randaccio L (1983) Cation site location in hydrated chabazites. Crystal structure of potassium- and silver-exchanged chabazites. Zeolites 3:205 – 208

Candel Vila R (1949) Notas sobre minerales des mediodia de Francia de la coleccion regional del Laboratorio de Mineralogia de la Facultad de Ciencias de Toulouse. Pirineos, Rev Inst Estud Pirenáicos, Zaragoza 5:657 – 696

Capdecomme L (1952) Laumontite du Pla des Aveillans (Pyrénées-Orientales). Bull Soc Hist Nat Toulouse 87:299 – 304

Capola A (1948) L'herschelite di Palagonia (Catania). Notizie Miner Sicil Calabr 2:63 – 69

Cerio V (1980) Le zeoliti nelle metamorfiti carbonatiche di San Giorgio (Novate Mezzola, Sondrio). Natura (Milan) 71:118 – 120

Černý P (1960) Milarite and wellsite of Vežná (in Czech). Acta Acad Sci Czeck Basis Brunensis 32, op 399:1 – 16

Černý P, Povondra P (1965a) Harmotome from desilicated pegmatites at Hrubšice, Western Moravia. Acta Univ Carolinae Geol, pp 31 – 43

Černý P, Povondra P (1965b) New occurrences of strontian chabazite. Acta Univ Carolinae Geol, pp 163 – 167

Černý P, Povondra P (1966) Re-examination of two Moravian natrolites. Acta Univ Carolinae Geol, pp 113 – 122

Černý P, Povondra P (1969) A polycationic strontian heulandite; comments on crystal chemistry and classification of heulandite and clinoptilolite. Neues Jahrb Miner Monatsh 1969:349 – 361

Černý P, Rinaldi R, Surdam RC (1977) Wellsite and its status in the phillipsite-harmotome group. Neues Jahrb Miner Abh 128:312 – 330

Chang WC, Lee H, Vaughan DEW (1978) Evaluation of natural ferrierite as a Claus tail-gas catalyst. In: Sand LB, Mumpton FA (eds) Natural zeolites. Pergamon, Oxford, pp 495 – 502

Chao GY (1980) Paranatrolite, a new zeolite from Mont St. Hilaire, Quebec. Can Miner 18:85 – 88

Chatterjee AC (1971) Levyne in the Deccan Traps. Miner Mag 38:527 – 528

Chen NY, Reagan WJ, Kokotailo GT, Childs LP (1978) A survey of catalitic properties of North American clinoptilolites. In: Sand LB, Mumpton FA (eds) Natural zeolites. Pergamon, Oxford, pp 411 – 420

Chen TT, Chao GY (1980) Tetranatrolite from Mont St. Hilaire, Quebec. Can Miner 18: 85 – 88

Chumakov IS, Shumenko SI (1977) Heulandite authigène des dépôts d'embouchure pliocènes du Nil. C R Acad Sci Paris Ser D 284:333 – 336

Ciric J [to Mobil Oil Corp] (1966, 1968) Synthetic crystalline aluminosilicate catalysts. Fr Pat 1, 502, 289; Br Pat 1, 117, 568

Clarke FW, Steiger G (1900) Die Einwirkung von Ammoniumchloride auf Natrolith, Skolecit, Prehnit und Pectolith. Z Anorg Chem 24:139 – 147

Cocco G, Garavelli C (1958) Riesame di alcune zeoliti elbane. Mem Soc Tosc Sci Nat Ser A 65:262 – 283

Colella C, Aiello R (1971) Utilizzazione di prodotti naturali per la sintesi di zeoliti. Nota III. Zeoliti potassiche da vetro riolitico. Rend Accad Sci Fis Mat Soc Naz Sci Lett Arti Ser 4 Napoli 38:243 – 258

Colella C, Aiello R (1975) Sintesi idrotermale di zeoliti da vetro riolitico in presenza di basi miste sodico-potassiche. Rend Soc Ital Miner Petrol 31:641 – 652

Colella C, Aiello R, Sersale R (1973) Genesis, occurrence and properties of zeolitic tuff. Rend Soc Ital Miner Petrol 29:439 – 451

Colella C, Aiello R, Di Ludovico V (1977) Sulla merlinoite sintetica. Rend Soc Ital Miner Petrol 33:511 – 518

Comin-Chiaramonti P (1977) The Theic Dam analcimite (eastern Azerbaijan, north-western Iran). Bull Soc Fr Miner Crist 100:113

Comin-Chiaramonti P, Pongiluppi D, Vezzalini G (1979) Zeolites in shoshonitic volcanics of the north-eastern Azerbaijan (Iran). Bull Miner 102:386 – 390

Coombs DS (1952) Cell size, optical properties and chemical composition of laumontite and leonhardite. Am Miner 37:812 – 830

Coombs DS (1955) X-ray observations of wairakite and non-cubic analcime. Miner Mag 30:699 – 708

Coombs DS (1965) Sedimentary analcime rocks and sodium-rich gneisses. Miner Mag 34:144 – 158

Coombs DS, Whetten JT (1967) Composition of analcime from sedimentary and burial metamorphic rocks. Geol Soc Am Bull 78:269 – 282

Coombs DS, Ellis AJ, Fyfe WS, Taylor AM (1959) The zeolite facies, with comments on the interpretation of hydrothermal syntheses. Geochim Cosmochim Acta 17:53 – 107

Coombs DS, Nakamura NY, Vuagnat M (1976) Pumpellyite-actinolite facies schists of the Taveyanne formation near Loeche, Valais, Switzerland. J Petrol 17:440 – 471

Cormier WE, Sand LB (1976) Synthesis and meta-stable phase transformations of Na-, (Na,K)-, and K-ferrierites. Am Miner 61:1259 – 1266

Cortelezzi CR (1973) Estudio de wairakita y levynita en Rocas del Cerro China Muerta, Provincia de Nequen, Republica Argentina. Rev Mus La Plata (Nueva Serie) Sect Geol 9:1 – 7

Cortesogno L, Lucchetti G, Penco AM (1975) Associazioni a zeoliti nel Gruppo di Voltri: caratteristiche mineralogiche e significato genetico. Rend Soc Ital Miner Petrol 31:673 – 710

Coulsell R (1980) Notes on the zeolite-collecting area of Flinders, Victoria, Australia. Aust Miner 33:163 – 167

Cronstedt AF (1756) Observation and description of an unknown kind of rock to be named zeolites (in Swedish). Kongl Vetenskaps Acad Handl Stockh 17:120 – 123

Cross W, Eakins LG (1886) On ptilolite, a new mineral. Am J Sci 132:117 – 121

Cross W, Shannon EV (1927) The geology, petrography and mineralogy of the vicinity of Italian Mountain, Gunnison County, Colorado. Proc US Nat Mus 71, art 18:1 – 42

Cundari A, Graziani G (1964) Prodotti di alterazione della leucite nelle vulcaniti vicane. Period Miner 33:35 – 52

Currie J (1907) The minerals of the Faroes, arranged topographically. Trans Edin Geol Soc 9:1 – 68

D'Achiardi G (1906) Zeoliti del filone della Speranza presso S. Piero in Campo (Elba). Mem Soc Tosc Sci Nat 22:150 – 165

Damour AA (1842) Description de la faujasite, nouvelle espèce minérale. Ann Mines Ser 4 1:395 – 399

Dana ES (1914) The system of mineralogy of JD Dana: descriptive mineralogy, with appendices I and II, 6th edn. Wiley, New York

Davis RJ (1958) Mordenite, ptilolite, flökite and arduinite. Miner Mag 31:887 – 888

Deffeyes KS (1959) Erionite from Cenozoic tuffaceous sediments, Central Nevada. Am Miner 44:501 – 509

De Gennaro M, Franco E (1976) La K-cabasite di alcuni "Tufi del Vesuvio". Rend Accad Naz Lincei Cl Sci Fis Mat Nat Ser 8 60:490 – 497

Deisinger H (1973) Einige Zeolithe aus dem Rossbergbasalt von Roßdorf bei Darmstadt. Aufschluß 24:443 – 446

Demartin F, Stolcis T (1979) Nuovo giacimento di dachiardite in Valle di Fassa. Riv Miner Ital 4:93 – 95

Dent LS, Smith JV (1958) Crystal structure of chabazite, a molecular sieve. Nature (Lond) 181:1794 – 1796

Deriu M (1954) Mesolite di Rio Cambone (Montiferro-Sardegna Centro-Occidentale). Period Miner 23:37 – 47

Des Cloizeaux ALO (1847) Mémoire sur la christianite, nouvelle espece minerale. Ann Mines Ser 4 12:373 – 381

Di Franco S (1926) L'analcite e il basalto analcitico dell'isola dei Ciclopi. Boll Soc Geol Ital 45:1 – 7

Di Franco S (1932) La thomsonite dell'isola dei Ciclopi. Period Miner 3:197 – 201

Di Franco S (1933) Nuovi minerali dell'isola dei Ciclopi. Atti Accad Gioenia Sci Nat Catania Ser 5 19:1 – 16

Djourova EG (1976) Analcime zeolites from the North-Eastern Rhodopes. C R Acad Bulg Sci 29:1023 – 1025

Doelter C (1921) Handbuch der Mineralchemie, vol II, part III. Steinkopff, Dresden

Domine D, Quobex J (1968) Synthesis of mordenite. In: Molecular Sieves. Society of Chemical Industry, London, pp 78 – 84

Donnelly TW (1962) Wairakite in West Indian spilitic rocks. Am Miner 47:794 – 802

Drysdale DJ (1971) A synthesis of bikitaite. Am Miner 56:1718 – 1723

Duffy SC, Rees LVC (1974, 1975) Ion exchange in chabazite. Part I. Sodium and potassium tracer diffusion studies in pure and mixed ionic forms of chabazite. J Chem Soc Faraday Trans I 70:777 – 786; Part II. Comparison of experimental and theoretical rates of Na^+/K^+ exchange. J Chem Soc Faraday Trans I 71:602 – 609

Dufrénoy A (1859) Traité de minéralogie, 2nd edn, Vol 4. Dalmont et Dunod, Paris

Dunham KC (1933) Crystal cavities in lavas from the Hawaiian islands. Am Miner 18:369 – 385

Dunn PJ, Peacor DR, Newberry N, Ramik RA (1980) Goosecreekite, a new calcium aluminium silicate hydrate possibly related to brewsterite and epistilbite. Can Miner 18:323 – 327

Dzotzenidse GS (1948) The evolution of chemistry of Paleozoic to Miocene volcanic rocks in relation to geotectonic history of Georgia (in Russian). Bull Soc Nat Moscou Sect Géol (New Ser 53) 23:73 – 87

Eakle AS (1898) Erionite a new zeolite. Am J Sci 6:66 – 68

Eakle AS (1899) Erionit, ein neuer Zeolith. Z Kristallogr 30:176 – 178

Eberlein GD, Erd RC, Weber F, Beatty LB (1971) New occurrence of yugawaralite from the Chena Hot Springs area, Alaska. Am Miner 56:1699 – 1717

Eberly PE Jr (1964) Adsorption properties of naturally occurring erionite and its cationic-exchanged forms. Am Miner 49:30 – 41

Echle W (1975) Zusammensetzung und Entstehung sedimentärer Analcime in jungtertiären pyroklastischen Gesteinen nördlich Mihaliçcik, Westanatolien, Türkei. Neues Jahrb Miner Abh 124:128 – 146

Elter P, Gratziu C, Labesse B (1964) Sul significato dell'esistenza di una unità alloctona costituita da formazioni terziarie nell'Appennino Settentrionale. Boll Soc Geol Ital 83:373 – 394

England BM, Ostwald J (1978) Ferrierite: an Australian occurrence. Miner Mag 42:385 – 389

England BM, Ostwald J (1979) Levyne-offretite intergrowths from Tertiary basalts in the Merriwa district, Hunter Valley, New South Wales, Australia. Aust Miner 25:117 – 119

Erdélyi J (1940) New contributions to the mineralogical knowledge of the quarry of Nadap (in Hungarian). Mat Term Tud Ertesito 59:1039 – 1061 (abstr in Chem Zentralbl 1941, II: 1494)

Erdélyi J (1942) Die hydrothermalen Mineralien des Andesitbruches bei Satoros, Foldt Kozl 72:271 – 294

Erdélyi J (1943) Epidesmin aus dem Steinbruch des Malomvolgy bei Szob. Foldt Kozl 73:493 – 497

Erdélyi J (1955) Beiträge zur mineralogischen Kenntnis des Gebirges von Velence. Acta Miner Petrogr Szeged 8:3 – 11

Eskola P (1960) Laumontite in Finland. Indian Miner 1:29 – 41

Estéoule J, Estéoule-Choux J (1972) Minéralogie de la phase argileuse et carbonatée de quelques échantillons secondaires et tertiaires prélevés en Manche centrale et orientale. Mem BRGM 79:166 – 169

Estéoule J, Estéoule-Choux J, Louail J (1971) Sur la présence de clinoptilolite dans les dépôts marno-calcaires du Crétacé supérieur de l'Anjou. C R Acad Sci Paris Ser D 272:1569 – 1572

Faessler C, Badollet MS (1947) The epigenesis of the minerals and rocks of the serpentine belt, Eastern Township, Quebec. Can Miner J 3:157 – 167

Fagnani G (1948) Prehnite e laumontite del Lago Bianco in Val Bavona (Canton Ticino). Atti Soc Ital Sci Nat Mus Civ Stor Nat Milano 87:189 – 195

Fälth L, Hansen S (1979) Structure of scolecite from Poona, India. Acta Crystallogr 35:1877 – 1880

Feoktistov GD, Ushchapovskaya ZF, Lakhno TA (1969) A find of garronite in the USSR (in Russian). Dokl Akad Nauk SSSR 188:670 – 672

Feoktistov GD, Ushchapovskaya ZF, Kashaev AA, Lakhno TA (1971) On the finding of levyne in the traps of Siberian Platform (in Russian). Zap Vses Miner Obshchest 100:745 – 748

Ferraris G, Jones DW, Yerkess J (1972) A neutron diffraction study of the crystal structure of analcime. Z Kristallogr 135:240 – 252

Fersman AE (1909) Études sur les zeolites de la Russie. I. Leonhardite et laumontite dans les environs de Simferopolis (Crimee). Trav Musée Géol Pierre le Grand Acad Imp Sci St Pétersbourg 2:103 – 150 (abstr in Z Kristallogr 50:75 – 76

Fersman AE (1922) Materials for the investigation of the zeolite of Russia. IV. General review of the zeolites of Russia (in Russian). Trav Musée Géol Miner Pierre le Grand Acad Sci St Pétersbourg Ser 2 2:263 – 374

Fersman AE, Labuntzov A (1925) Comptes Rendus de l'expedition aux gisements d'uranium en Carelie russe (in Russian). Dokl Akad Nauk SSSR Sect A 1925:147 – 150

Fischer K (1963) The crystal structure determination of the zeolite gismondite. Am Miner 48:664 – 672

Fischer K (1966) Untersuchung der Kristallstruktur von Gmelinit. Neues Jahrb Miner Monatsh 1966:1 – 13

Fischer K, O'Daniel H (1956) Bemerkungen zur Struktur der Würfelzeolithe. Naturwissenschaften 43:348

Fischer K, Schramm V (1971) Crystal structure of gismondite, a detailed refinement. In: Molecular sieve zeolites I. Am Chem Soc Adv Chem Ser 101:250 – 258

Flanigen EM, Kellberg ER [to Union Carbide Corp] (1967) Method for the preparation of new synthetic crystalline zeolites. Dutch Pat Appl 6, 710, 729

Flanigen EM, Kellberg ER [to Union Carbide Corp] (1968, 1970) Crystalline zeolite molecular sieves. Fr Pat 1, 548, 382; Br Pat 1, 178, 186

Fornaseri M, Penta A (1960) Elementi alcalini minori negli analcimi e loro comportamento nel processo di analcimizzazione della leucite. Period Miner 29:85 – 102

Foster MD (1965) Studies of the zeolites: composition of zeolites of the natrolite group and compositional relations among thomsonites gonnardites, and natrolites. US Geol Surv Prof Pap 504-D, E:D1 – E10

Franco E, Aiello R (1968) Trasformazione dell'halloysite per trattamento idrotermale in ambiente alcalino. Rend Soc Ital Miner Petrol 24:251 – 269

Franco E, Aiello R (1969) Zeolitizzazione di vetri naturali. Rend Accad Sci Fis Mat Soc Naz Sci Lett Arti Ser 4 Napoli 36:3 – 25

Franco RR (1952) Zéolitas dos basaltos do Brazil Meridional. Boll Fac Fil Ciê Let Univ São Paulo Miner 10 150:3 – 70

Franco RR (1953) Scolecita, Rio Pelotas, RS. Boll Fac Fil Ciê Let Univ São Paulo Miner 12 163:24 – 30

Frankart R, Herbillon AJ (1970) Présence et genèse d'analcime dans les sols sodiques de la Basse Ruzizi (Burundi). Bull Groupe Fr Argiles 22:79 – 89

Fricke G (1971) Die Zeolithe Chabasit und Phillipsit in der essexitischen Gesteinfamilie des Kaiserstuhls. Aufschluß 22:201 – 204

Fuchs JN, Gehlen AF (1816) Über die Zeolithe. Schweigg J Chem Phys 18:1 – 29

Füchtbauer H (1967) Der Einfluß des Ablagerungsmilieus auf die Sandsteindiagenese im Mittleren Buntsandstein. Sediment Geol 1:159 – 179

Füchtbauer H (1977) Tertiary lake sediments of the Ries, research bore hole 1973. Geol Bavarica 75:18 – 22

Fuller AO, Moir GJ (1967) Analcite in the Upper Karroo sediments, Herschel district, South Africa. J Sediment petrol 37:1236 – 1237

Furbish WJ (1965) Laumontite-leonhardite from Durham County, N.C. Southeastern Geol 6:189 – 200

Furman MJ, Karner FR (1971) Description of authigenic analcite in the Eocene Golden Valley formation, southwestern South Dakota. Proc North Dakota Acad Sci 24:52 – 61

Galabova IM, Haralampiev GA, Aleksiev B (1978) Oxygen enrichment of air using Bulgarian clinoptilolite. In: Sand LB, Mumpton FA (eds) Natural Zeolites. Pergamon, Oxford, pp 431 – 437

Gallezot P (1984) Physical and catalytic properties of noble metals in zeolites. In: Bisio A, Olson DH (eds) Proc 6th Int Zeolite Conf, Butterworths, Guildford, pp 352 – 367

Galli E (1965) Lo spettro di polvere della dachiardite. Period Miner 34:129 – 135

Galli E (1971) Refinement of the crystal structure of stilbite. Acta Cryst B27:833 – 841

Galli E (1972) La phillipsite barifera ("wellsite") di M. Calvarina (Verona). Period Miner 41:23 – 33

Galli E (1974) Mazzite, a zeolite. Cryst Struct Comm 3:339 – 344

Galli E (1975a) Le zeoliti. Rend Soc Ital Miner Petrol 31:549 – 564

Galli E (1975b) Crystal structure refinement of mazzite. Rend Soc Ital Miner Petrol 31:599 – 612

Galli E (1976) Crystal structure refinement of edingtonite. Acta Cryst B32:1623 – 1627

Galli E (1980) The crystal structure of roggianite, a zeolite-like silicate. In: Rees LVC (ed) Proc 5th Int Conf Zeolites. Heyden, London, pp 205 – 213

Galli E, Alberti A (1975a) The crystal structure of stellerite. Bull Soc Fr Miner Crist 98:11 – 18

Galli E, Alberti A (1975b) The crystal structure of barrerite. Bull Soc Fr Miner Crist 98:331 – 340

Galli E, Gottardi G (1966) The crystal structure of stilbite. Miner Petrogr Acta Bologna 12:1 – 10

Galli E, Loschi Ghittoni AG (1972) The crystal structure of phillipsites. Am Miner 57:1125 – 1145

Galli E, Passaglia E (1973) Stellerite from Villanova Monteleone, Sardinia. Lithos 6:83 – 90

Galli E, Rinaldi R (1974) The crystal chemistry of epistilbites. Am Miner 59:1055 – 1061

Galli E, Passaglia E, Pongiluppi D, Rinaldi R (1974) Mazzite, a new mineral, the natural counterpart of the synthetic zeolite Omega. Contrib Miner Petrol 45:99 – 105

Galli E, Gottardi G, Mazzi F (1978) The natural and synthetic phases with the leucite framework. Miner Petrogr Acta Bologna 22:185 – 193

Galli E, Gottardi G, Pongiluppi D (1979) The crystal structure of the zeolite merlinoite. Neues Jahrb Miner Montsh 1979:1 – 9

Galli E, Rinaldi R, Modena C (1981) Crystal chemistry of levynes. Zeolites 1:157 – 160

Galli E, Passaglia E, Zanazzi PF (1982) Gmelinite: structural refinement of sodium-rich and calcium-rich natural crystals. Neues Jahrb Miner Monatsh 1982:145 – 155

Galli E, Gottardi G, Mayer H, Preisinger A, Passaglia E (1983) The structure of potassium exchanged heulandite at 293, 373 and 593 K. Acta Cryst B39:189 – 197

Gallitelli P (1928) La laumontite di Toggiano. Rend R Accad Naz Lincei Cl Sci Mat Fis Nat Ser 6 8:82 – 87

Gard JA, Tait JM (1972) The crystal structure of the zeolite offretite. Acta Cryst B28:825 – 834

Gard JA, Tait JM (1973) Refinement of the crystal structure of erionite. In: Uytterhoeven JB (ed) Proc 3rd Int Conf Mol Sieves. Leuven Univ Press, Leuven, pp 94 – 99

Gardiner G (1966) Mineral hunting in the Parrsboro, Nova Scotia area. Rocks Minerals 41:85 – 91

Gärtner HR von, Machatschki F (1927) Der Thomsonit aus dem Basalt von Disko, Grönland. Zentralbl Miner Abt A 1927:365 – 366

Gehlen AF, Fuchs JN (1813) Über Werners Zeolith, Hauys Mesotype und Stilbite. Schweigg J Chem Phys 8:353 – 366

Gellens LR, Price GD, Smith JV (1982) The structural relation between svetlozarite and dachiardite. Miner Mag 45:157 – 162

Gennaro V (1929) Thomsonite e scolecite dell'alta Valle di Ayas e delle Valli di Lanzo. Atti R Accad Sci Torino 64:133 – 143

Ghent ED, Miller BE (1974) Zeolite and clay-carbonate assemblages in the Blaimore group (Cretaceous) Southern Alberta Foothills, Canada. Contrib Miner Petrol 44:313 – 329

Gianello A, Gottardi G (1969) Sulla zeolitizzazione del livello cineritico detto "Tripoli di Contignaco". Miner Petrogr Acta Bologna 15:5 – 8

Giot D, Jacob C (1972) Présence d'analcime et de clinoptilolite (zéolites) dans les formations sédimentaires oligocènes de la Limagne de Clermont-Ferrand (Puy-de-Dôme, France). C R Acad Sci Ser D Paris 274:166 – 169

Girod M (1965) Données pétrographiques sur les basanites à analcime. Bull Soc Fr Miner Crist 88:58 – 65

Gismondi G (1817) Osservazioni sopra alcuni fossili particolari de' contorni di Roma. G Enciclop Napoli Anno XI 2:1 – 15

Glaccum R, Boström K (1976) (Na,K)-phillipsite its stability conditions and geochemical role in the deep sea. Mar Geol 21:47 – 58

Gonnard F (1890) Sur l'offrétite, espèce minérale nouvelle. C R Acad Sci Paris 111:1002 – 1003

Goodwin JH (1973) Analcime and K-feldspar in tuffs of the Green River formation, Wyoming. Am Miner 58:93 – 105

Gordon EK, Samson S, Kamb WB (1966) Crystal structure of the zeolite paulingite. Science (Wash DC) 154:1004 – 1007

Gordon SG (1924) Minerals obtained in Greenland on the second Academy-Vaux expedition, 1923. Proc Acad Nat Sci Phila 76:249 – 268

Görgey R (1910) Ein Beitrag zur topographischen Mineralogie der Faröer. Neues Jahrb Miner Beil 29:269 – 315

Goto Y (1977) Synthesis of clinoptilolite. Am Miner 62:330 – 332

Gottardi G (1960) Sul dimorfismo mordenite-dachiardite. Period Miner 29:183 – 191

Gottardi G (1978) Mineralogy and crystal chemistry of zeolites. In: Sand LB, Mumpton FA (eds) Natural zeolites. Pergamon, Oxford, pp 31 – 44

Gottardi G (1979) Topologic symmetry and real symmetry in framework silicates. Tscherm Miner Petr Mitt 26:39 – 50

Gottardi G, Alberti A (1974) Domain structure in garronite: a hypothesis. Miner Mag 39:898 – 899

Gottardi G, Meier WM (1963) The crystal structure of dachiardite. Z Kristallogr 119:53 – 64

Gottardi G, Obradović J (1978) Sedimentary zeolites in Europe. Fortschr Miner 56:316 – 366

Graham RPD (1918) On ferrierite, a new zeolitic mineral from British Columbia; with notes on some other Canadian minerals. Trans R Soc Can Ser 3 12:185 – 201

Grämlich V (1971) Untersuchung und Verfeinerung pseudosymmetrischer Strukturen. Diss no 4633 ETH, Zürich

Grice JD, Gault RA (1981) Edingtonite and natrolite from Ice River, British Columbia. Miner Record 12:221 – 226

Grivakov AG (1967) Occurrences of non-cubic hydrothermal analcime in the Crimea (in Russian). Zap Vses Miner Obshchest 96:724 – 728

Groben M (1970) Natrolite and analcime from Eckman Creek quarry, Waldport, Oregon. Miner Record 1:154, 173, 176, 178

Grütter O (1931) Ein Skolezitfund in der Valle Maggia (Tessin). Schweiz Miner Petr Mitt 11:266

Gschwind M, Brandenberger E (1932) Über zwei neue Zeolithvorkommen in Tessin. Schweiz Miner Petr Mitt 12:445 – 449

Gude AJ 3rd, Sheppard RA (1966) Silica-rich chabazite from the Barstow formation, San Bernardino County, Southern California. Am Miner 51:909 – 915

Gude AJ 3rd, Sheppard RA (1978) Chabazite in siliceous tuffs of a Pliocene lacustrine deposit near Durkee, Baker County, Oregon. J Res US Geol Surv 6:467 – 472

Gude AJ 3rd, Sheppard RA (1981) Woolly erionite from the Reese river zeolite deposit, Lauder County, Nevada, and its relationship to other erionites. Clays Clay Minerals 29:378 – 384

Gupta AK, Fyfe WS (1975) Leucite survival: the alteration to analcime. Can Miner 13:361 – 363

Guseva LD, Menshikov YP, Romanova TS, Bussen IV (1975) Tetragonal natrolite from the Lovozero alkaline massif (in Russian). Zap Vses Miner Obshchest 104:66 – 69

Guyer A, Ineichen M, Guyer P (1957) Über die Herstellung von künstlichen Zeolithen und ihre Eigenschaften als Molekelsiebe. Helv Chim Acta 40:1603 – 1611

Haidinger W (1825) Description of edingtonite, a new mineral species. Edinb J Sci (Brewster) 3:316 – 320

Hamilton DL, Ryabchikov I (1976) Analcite and jadeite stabilities. Prog Exp Petrol NERC 3:28 – 36

Hanley FB (1939) New accessibility of Thomsonite Beach, Minnesota. Am Miner 24:726 – 727

Harada K, Katsutoshi T (1967) A sodian stilbite from Onigajo, Mié Pref., Japan. Am Miner 52:1438 – 1450

Harada K, Nakao K (1969) Two natrolites from Japan. J Geol Soc Jpn 75:343 – 344

Harada K, Sudo T (1976) A consideration on the wairakite-analcime series. – Is valid a new mineral name for sodium analogue of monoclinic wairakite? Miner J 8:246 – 251

Harada K, Tomita K, Sudo T (1966) Note on the conversion of stilbite and heulandite into wairakite under very low water saturated vapor pressure. Proc Jpn Acad 42:925 – 928

Harada K, Iwamoto S, Kihara K (1967) Erionite, phillipsite and gonnardite in the amygdales of altered basalt from Maze, Niigata Pref., Japan. Am Miner 52:1785 – 1794

Harada K, Hara M, Nakao K (1968) Mineralogical notes on mesolite and scolecite from Japan. Miner J 5:309 – 320

Harada K, Nakao K, Nagashima K (1969a) Ionic substitutions in natural mesolites. J Jpn Assoc Miner Petr Econ Geol 61:112 – 115

Harada K, Umeda M, Nakao K, Nagashima K (1969b) Mineralogical notes on thomsonite, laumontite and heulandite in altered basaltic and andesitic rocks in Japan. J Geol Soc Jpn 75:551 – 552

Harada K, Tanaka K, Nagashima K (1972) New data on the analcime-wairakite series. Am Miner 57:924 – 931

Harris PG, Brindley GW (1954) Mordenite as an alteration product of a pitchstone glass. Am Miner 39:819 – 824

Haüy RJ (1801) Traité de minéralogie, Vol 3. Ches Louis, Paris, p 180

Haüy RJ (1809) Tableau comparatif des résultats de cristallographie et de l'analyse chimique relativement à la classification des minéraux. Courcier, Paris (see also J Physique 49:56 – 77)

Hawkins DB (1967a) Zeolite studies. I. Synthesis of some alkaline earth zeolites. Mat Res Bull 2:951 – 958

Hawkins DB (1967b) Zeolite studies. II. Ion-exchange properties of some synthetic zeolites. Mat Res Bull 2:1021 – 1028

Hawkins DB (1981) Kinetics of glass dissolution and zeolite formation under hydrothermal conditions. Clays Clay Minerals 29:331 – 340

Hawkins DB, Sheppard RA, Gude AJ 3rd (1978) Hydrothermal synthesis of clinoptilolite and comments on the assemblage phillipsite-clinoptilolite-mordenite. In: Sand LB, Mumpton FA (eds) Natural zeolites. Pergamon, Oxford, pp 337 – 344

Hay RL (1963a) Stratigraphy and zeolitic diagenesis of the John Day formation of Oregon. Calif Univ Pubs Geol Sci 42:199 – 262

Hay RL (1963b) Zeolitic weathering in Olduvai Gorge, Tanganyika. Geol Soc Am Bull
    74:1281 – 1286
Hay RL (1964) Phillipsite of saline lakes and soils. Am Miner 49:1366 – 1387
Hay RL (1966) Zeolites and zeolitic reactions in sedimentary rocks. Geol Soc Am Spec Pap (Reg
    Stud) 85:1 – 130
Hay RL (1976) Geology of Olduvai Gorge. Univ Calif Press, Berkeley
Hay RL (1978) Geologic occurrences of zeolites. In: Sand LB, Mumpton FA (eds) Natural Zeo-
    lites. Pergamon, Oxford, pp 135 – 144
Hay RL, Iijima A (1968a) Petrology of palagonite tuffs of Koko craters, Oahu, Hawaii. Contrib
    Miner Petrol 17:141 – 154
Hay RL, Iijima A (1968b) Nature and origin of palagonite tuffs of the Honolulu Group on Oahu,
    Hawaii. Geol Soc Am Mem 116:331 – 376
Hayakawa N, Suzuki S (1969) Zeolitized tuffs and occurrence of ferrierite in Tadami-Machi, Fu-
    kushima Pref., Japan (in Japanese). Mining Geol 20:295 – 305
Hayhurst DT (1978) The potential use of natural zeolites for ammonia removal during coal-gasi-
    fication. In: Sand LB, Mumpton FA (eds) Natural Zeolites. Pergamon, Oxford, pp 503 – 508
Hayhurst DT, Willard JM (1980) Effects of feeding clinoptilolite to roosters. In: Rees LVC (ed)
    Proc 5th Int Conf Zeolites. Heyden, London, pp 805 – 812
Hazen RM, Finger LW (1979) Polyhedral tilting: a common type of pure displacive phase transi-
    tion and its relationship to analcite at high pressure. Phase Transitions 1:1 – 22
Headden WP (1893) Kehoeite, a new phosphate from Galena, Lawrence County, South Dakota.
    Am J Sci Ser 3 46:22 – 25
Heddle MF (1855) Analysis of the mineral "edingtonite". The London, Edinburgh and Dublin
    Phil Mag and J Sci 4th Ser 9:179 – 181
Heling D (1978) Contact metamorphic zeolites and smectite in Oligocene bituminous shale from
    the Southern Rhinegraben. Neues Jahrb Miner Abh 133:33 – 40
Henderson EP, Glass JJ (1933) Additional notes on laumontite and thomsonite from Table Mts.,
    Colorado. Am Miner 18:402 – 406
Hentschel G (1978) Einige Funde ungewöhnlicher Minerale aus quartären Vulkanvorkommen der
    Eifel. Mainzer Geowiss Mitt 7:151 – 154
Hentschel G (1980a) Weitere Offretit-Vorkommen im Vogelsberg (Hessen). Geol Jahrb Hessen
    108:171 – 176
Hentschel G (1980b) Weitere bemerkenswerte Mineralfunde aus quartären Vulkanvorkommen
    der Eifel. Mainzer Geowiss Mitt 8:169 – 172
Hentschel G (1982) Neue Mineralfunde aus Vulkanvorkommen der Eifel. Mainzer Geowiss Mitt
    11:87 – 90
Hentschel G (1983) Der Schellkopf bei Brenk. Lapis 8, no 9:18 – 25
Hentschel G, Vollrath R (1977) Die Zeolithe im Basalt von Ober-Widdersheim, Vogelsberg. Auf-
    schluß 28:409 – 412
Hesse KF (1983) Refinement of a partially disordered natrolite. Z Kristallogr 163:67 – 74
Hewett DF, Shannon EV, Gonyer FA (1928) Zeolites from Ritter, Hot Spring, Giant County,
    Oregon. Proc US Nat Mus 73:1 – 18
Hey MH (1932a) Studies on the zeolites. Part II. Thomsonite (including faroelite) and gonnard-
    ite. Miner Mag 23:51 – 125
Hey MH (1932b) Studies on the zeolites. Part III. Natrolite and metanatrolite. Miner Mag
    23:243 – 249
Hey MH (1933) Studies on the zeolites. Part V. Mesolite. Miner Mag 23:421 – 447
Hey MH (1934) Studies on the zeolites. Part VI. Edingtonite. Miner Mag 23:483 – 494
Hey MH (1936) Studies on the zeolites. Part IX. Scolecite and metascolecite. Miner Mag
    24:227 – 253
Hey MH (1959) A new occurrence of erionite. Miner Mag 32:343
Hey MH (1962) An index of mineral species and varieties arranged chemically, 2nd edn. Trustess
    of the British Museum (Natural History), London
Hey MH, Bannister FA (1933) Studies on the zeolites. Part IV. Ashcroftine (kalithomsonite of
    SG Gordon). Miner Mag 23:305 – 308

Hey MH, Bannister FA (1934) Studies on the zeolites. Part VII. "Clinoptilolite", a silica-rich variety of heulandite. Miner Mag 23:556 – 559

Hey MH, Fejer EE (1962) The identity of erionite and offretite. Miner Mag 33:66 – 67

Hibsch JE (1915) Der Marienberg bei Aussig und seine Minerale. Tscherm Miner Petr Mitt 33:340 – 348

Hibsch JE (1917) V. Geologische Karte des Böhmischen Mittelgebirges. Tscherm Miner Petr Mitt 34:73 – 201

Hibsch JE (1927) Erläuterungen zur Geologischen Karte der Umgebung von Böhm. Kamnitz, Prag

Hibsch JE (1929) Erläuterungen zur Geologischen Karte der Umgebung von Böhm. Kamnitz, Prag

Hibsch JE (1939) Einige neue Mineralfunde im Böhmischen Mittelgebirge. Tscherm Miner Petr Mitt 50:487 – 488

Hintze C (1897) Handbuch der Mineralogie, vol 2 (Silikate und Titanate). Von Veit, Leipzig

Hodge-Smith T (1929) The occurrence of zeolites at Kyogle, New South Wales. Rec Aust Mus 17:279 – 290

Hoffmann GC (1901) On some new mineral occurrences in Canada. Amer J Sci Ser IV 12:447 – 448

Hoffmann GC (1903) Mineralogical news. Annu Rep Geol Surv Can 11

Hogart DD, Griffin WL (1980) Contact-metamorphic lapis lazuli: the Italian Mountain deposits, Colorado. Can Miner 18:59 – 70

Höller H (1970) Untersuchungen über die Bildung von Analcim aus natürlichen Silikaten. Contrib Miner Petrol 27:80 – 94

Höller H, Wirsching U (1978) Experiments on the formation of zeolites by hydrothermal alteration of volcanic glasses. In: Sand LB, Mumpton FA (eds) Natural zeolites. Pergamon, Oxford, pp 329 – 336

Höller H, Wirsching U (1980) Experiments on the hydrothermal formation of zeolites from nepheline and nephelinite. In: Rees LVC (ed) Proc 5th Int Conf Zeolites. Heyden, London, pp 164 – 170

Höller H, Wirsching U, Fakhuri M (1974) Experimente zur Zeolithbildung durch hydrothermale Umwandlung. Contrib Miner Petrol 46:49 – 60

Hollis JD (1979) Some little known zeolite localities in central Victoria. Aust Miner 21:101 – 102

Honda S, Muffler LJP (1970) Hydrothermal alteration in core from research drill hole Y-1, Upper Geyser Basin, Yellowstone National Park, Wyoming. Am Miner 55:1714 – 1737

Honnorez J (1978) Generation of phillipsites by palagonitisation of basaltic glass in sea water and the origin of K-rich deep-sea sediments. In: Sand LB, Mumpton FA (eds) Natural zeolites. Pergamon, Oxford, pp 245 – 248

Hoover DL (1968) Genesis of zeolites, Nevada Test Site. Geol Soc Am Mem 110:275 – 284

How DCL (1864) On mordenite, a new mineral from the traps of Nova Scotia. J Chem Soc (Lond) 2:100

Hudson MJ, Hudson S (1978) Occurrence of ferrierite from the south coast area, New South Wales. Aust Miner 16:80

Hurlbut CS Jr (1957) Bikitaite, a new mineral from Southern Rhodesia. Am Miner 42:792 – 797

Hurlbut CS Jr (1958) Additional data on bikitaite. Am Miner 43:768 – 770

Iijima A (1978) Geological occurrences of zeolite in marine environments. In: Sand LB, Mumpton FA (eds) Natural zeolites. Pergamon, Oxford, pp 175 – 198

Iijima A, Harada K (1969) Authigenic zeolites in zeolitic palagonite tuffs on Oahu, Hawaii. Am Miner 54:182 – 197

Iijima A, Hay RL (1968) Analcime composition in tuffs of the Green River formation of Wyoming. Am Miner 53:184 – 200

Iijima A, Utada M (1966) Zeolites in sedimentary rocks, with reference to the depositional environments and zonal distribution. Sedimentology 7:327 – 357

Iijima A, Utada M (1971) Present-day zeolitic diagenesis of the Neogene geosynclinal deposits of the Niigata oil field, Japan. In: Molecular sieve zeolites I. Am Chem Soc Adv Chem Ser 101:342 – 349

Imelik B, Naccache C, Ben Taarit Y, Vedrine JC, Coudurier G, Praliaud H (eds) (1980) Studies in surface science and catalysis, vol 5 (catalysis by zeolites). Elsevier Scientific Publishing Co, Amsterdam

Irrera G (1949) L'herschelite in un basalto palagonitico da un sondaggio a Ficarazzi (Catania). Notizie Miner Sicil Calabr 3:40 – 47

Ivanov OK, Mozzherin YV (1982) Partheite from gabbro-pegmatites of Denerhkin Kamen, Urals (first occurrence in the USSR) (in Russian). Zap Vses Miner Obshchest 11:209 – 214

Ivanov PP, Gurevich LP (1975) Experimental study of T-X (CO_2) boundaries of metamorphic zeolite facies. Contrib Miner Petrol 53:55 – 60

Jacobs PA, Uytterhoeven JB, Beyer HK, Kiss A (1979) Preparation and properties of hydrogen form of stilbite, heulandite and clinoptilolite zeolites. J Chem Soc Faraday Trans I 75:883 – 891

Jakob H (1982) Neue mineralogische Beobachtungen im Basaltbruch des Zeilberges bei Maroldsweisach/Unterfranken. Aufschluß 33:344 – 345

Jakobsson S (1977) Icelandic zeolites (in Icelandic). In: Arbok Ferdafelags Islands. Ferdafelag, Reykjavík, pp 190 – 201

Jarkow O, Klaska KH (1981) $Rb_2[Ga_2Ge_3O_{10}]$ ein wasserfreier Zeolith vom Natrolith-Typ. Z Kristallogr 156:64 – 65

Jérémine E (1934) Note sur les zéolites des Monts Lubur et du Mont Elgon. Bull Soc Fr Miner Crist 57:240 – 243

Johnson JH, Waldschmidt WA (1925) Famous Colorado mineral localities: Table Mountain and its zeolites. Am Miner 10:118 – 120

Joulia F, Bonifas M, Camez T, Millot G, Weil R (1958) Analcimolites sedimentaires dans le continental intercalaire du Sahara Central (Bassin du Niger, AOF). Serv Carte Geol Alsace-Lorraine Bull 11:67

Juan VC, Lo HJ (1969) Phase relation in the system $NaAlSi_3O_8$-$CaAl_2Si_2O_8$-H_2O at low temperatures and pressures. Proc Geol Soc China (Formosa) 12:21 – 29

Juan VC, Lo HJ (1971) The stability fields of natural laumontite and wairakite and their bearing on the zeolite facies. Proc Geol Soc China (Formosa) 14:34 – 44

Juan VC, Lo HJ (1973) Stability field of stilbite. Proc Geol Soc China (Formosa) 16: 37 – 49

Juan VC, Liu JG, Jahn BM (1964) A preliminary study of minerals in the zeolite group in Taiwanite from Taitung, Taiwan. Proc Geol Soc China (Formosa) 8:85 – 90

Judin MI (1963) Jadeite and natrolite-rocks in ultrabasics from Borus Mts. (West Sajans) and their genesis (in Russian). Izv Acad Nauk SSSR Ser Geol 4:78 – 98

Kalb G (1939) Über Mineralien und Gesteine der niederrheinischen Vulkangebiete. Decheniana 98A:173 – 183

Kal'berg EA, Levando EP (1962) On analcime- and zeolite-bearing rocks in the northern Onezhskiy bauxite (in Russian). Dokl Akad Nauk SSSR 142:919 – 921

Kaley ME, Hanson RF (1955) Laumontite and leonhardite cement in Miocene sandstone from a well in San Joaquin Valley, California. Am Miner 40:923 – 925

Kalló D, Papp J, Valyon J (1982) Adsorption and catalytic properties of sedimentary clinoptilolite and mordenite from the Tokaj Hills, Hungary. Zeolites 2:13 – 16

Kamb WB, Oke WC (1960) Paulingite, a new zeolite, in association with erionite and filiform pyrite. Am Miner 45:79 – 91

Kanazawa Y, Endo Y (1981) Drawing of crystal and twin figures. Miner J 10:279 – 295

Karsten DLG (1808) Mineralogische Tabellen mit Rücksicht auf die neuesten Entdeckungen, 2nd edn. Rottmann, Berlin

Karup-Møller S (1976) Gmelinite and herschelite from the Ilimaussaq intrusion in South Greenland. Miner Mag 40:867 – 873

Kastner M, Stonecipher SA (1978) Zeolites in pelagic sediments of the Atlantic, Pacific and Indian Oceans. In: Sand LB, Mumpton FA (eds) Natural zeolites. Pergamon, Oxford, pp 199 – 200

Kato A (1973) Sakurai mineral collections (in Japanese). Committee for the commemoration of the 60th birthday of Dr K Sakurai, Tokyo, p 112

Kawahara A, Curien H (1969) La structure cristalline de l'érionite. Bull Soc Fr Miner Crist 92:250–256

Kerr GT [to Mobil Oil Corp] (1969) Synthetic zeolite and method for preparing the same. US Pat 3, 459, 676

Kerr IS (1964) Structure of epistilbite. Nature Lond 202:589

Kerr IS, Williams DJ (1967) The crystal structure of yugawaralite. Z Kristallogr 125:220–225

Kerr IS, Williams DJ (1969) The crystal structure of yugawaralite. Acta Cryst B25:1183–1190

Kerr IS, Gard JA, Barrer RM, Galabova IM (1970) Crystallographic aspects of the co-crystallization of zeolite L, offretite and erionite. Am Miner 55:441–454

Khomyakov AP, Katayeva ZT, Kurova TA, Rudnitskaya Ye S, Smol'yaninova NN (1970) First find of brewsterite in the USSR (in Russian). Dokl Akad Nauk SSSR 190:1192–1195

Khomyakov AP, Sandomirskaya SM, Malinovskii YA (1980) Kalborsite, a new mineral (in Russian). Dokl Akad Nauk SSSR 252:1465–1468

Khomyakov AP, Kurova TA, Muravishkaya GI (1981) Merlinoite, first occurrence in the USSR (in Russian). Dokl Akad Nauk SSSR 256:172–174

Khomyakov AP, Cherepivskaya GE, Kurova TA, Kaptsov VV (1982) First occurrence of amicite in the USSR (in Russian). Dokl Akad Nauk SSSR 263:978–980

Kim KT, Burley BJ (1971) Phase equilibria in the system $NaAlSi_3O_8$-$NaAlSiO_4$-H_2O with special emphasis on the stability of analcite. Can J Earth Sci 8:311–338, 549–558, 558–572

Kim KT, Burley BJ (1980) A further study of analcime solid solutions in the system $NaAlSi_3O_8$-$NaAlSiO_4$-H_2O, with particular note of an analcime phase transformation. Miner Mag 43:1035–1045

Kinoshita K (1922) Harmotome from Udo, Shimane Pref (Abstr). Jpn J Geol Geogr 1:19

Kirov GN (1967) Calcium-rich clinoptilolite from the Eastern Rhodopes. Annu Univ Sofia Fac Géol Geogr Livre 1 Géol 60:193–201

Kirov GN (1974) Zeolites in sedimentary rocks from Bulgaria (in Bulgarian with English summary). Annu Univ Sofia Fac Géol Geogr Livre 1 Géol 66:171–185

Kirov GN, Filizova L (1966) Calcium ferrierite from the Eastern Rhodopes. Annu Univ Sofia Fac Géol Geogr Livre 1 Géol 59:237–246

Kirov GN, Pechigargov V (1977) Hydrothermal modelling of zeolitization in volcanic tuffs. I. Synthesis of clinoptilolite, mordenite and analcime (in Bulgarian). Geokhim Miner Petrol 7:75–84

Kirov GN, Pechigargov V, Georgiev VM (1975) Synthesis of analcime from natural zeolites, perlite and expanded perlite (in Bulgarian). Geokhim Miner Petrol 2:51–60

Kirov GN, Pechigargov V, Landzheva E (1979) Experimental crystallization of volcanic glasses in a thermal gradient field. Chem Geol 26:17–28

Kiss K, Page HT (1969) Electron microscopic identificatiion of single crystals of wairakite, a rare component in clays. Clays Clay Minerals 17:31–35

Klaproth MH (1803) XV. Chemische Untersuchung des Natroliths. Ges Naturforsch Freunde Berlin Neue Schriften 4:243–248

Klitchenko MA, Suprichev VA (1974) First find of ferrierite in the USSR (in Russian). Zap Vses Miner Obshchest 103:140–141

Klopp G, Suto J, Szebenji I (1980) Molecular sieve sorbents from Hungarian clinoptilolite. In: Rees LVC (ed) Proc 5th Int Conf Zeolites. Heyden, London, pp 841–849

Knowles CR, Rinaldi FF, Smith JV (1965) Refinement of the crystal structure of analcime. Indian Miner 6:127–140

Koch LH (1917) A new occurrence of ptilolite. Am Miner 2:143–144

Koch S (1978) Die Zeolithe von Ungarn. Lapis 3, no 1:28–31

Kocman V, Gait RI, Rucklidge J (1974) The crystal structure of bikitaite. Am Miner 59:71–78

Koizumi M (1953) The differential thermal analysis curves and the dehydration curves of zeolites. Miner J 1:36–47

Koizumi M, Kiriyama R (1957) Hydrothermal study of dehydrated natrolite (Abstr). Geol Soc Am Bull 68:1755

Koizumi M, Roy R (1960) Zeolite studies. I. Synthesis and stability of calcium zeolites. J Geol 68:41–53

Kokotailo GT, Sawruk S, Lawton SL (1972) Direct observation of stacking faults in the zeolite erionite. Am Miner 57:439 – 444

Koptewa WW (1955) A zeolite occurrence in the region of Simferopol, Crimea (in Russian). Tr Mosk Geol Rasv Inst 28:51 – 58

Koritnig S (1940) Zeolithe aus dem Moränegeroll des Jamgletschers in der Silvretta. Zentralbl Miner Abt A 1940:177 – 181

Kostov I (1962) The zeolites in Bulgaria analcime, chabazite, harmotome (in Bulgarian with English summary). Annu Univ Sofia Fac Biol Geol Geogr Livre 2 Geol 55:159 – 174

Kostov I (1969) Zoning in the development of the volcanogenic zeolites. Neues Jahrb Miner Abh 111:264 – 278

Kostov I (1970) Tectonomagmatic significance of zeolites in the Srednogorian zone and the Rhodopes. Bull Geol Inst Bulgar Acad Sci Ser Geochim Miner Petr 19:235 – 241

Kostov I, Mavrudchiev B, Kunov A (1967) Distribution of zeolitic minerals in the western Srednogorie (in Bulgarian). Bull Geol Inst Bulgar Acad Sci Ser Geochim Miner Petr 16:61 – 93

Koyama K, Takéuchi Y (1977) Clinoptilolite: the distribution of potassium atoms and its role in thermal stability. Z Kristallogr 145:216 – 239

Kralik J (1971) Montmorillonite and zeolites from the variagated beds of the Ostrava-Karvina Basin (in Czech with English summary). Sb Ved Pr Vys Sk Banske Rada Horn-Geol 17, no 3:1 – 28

Kräutner HG (1966) The zeolite facies of the banatitic hydrothermal metamorphism and its evolution in Ruschita area (in Romanian). Stud Cercet Geol Geofiz Geogr Ser Geol 11:207 – 216

Kristmannsdóttir H, Tómasson J (1978) Zeolite zones in geothermal areas Iceland. In: Sand LB, Mumpton FA (eds) Natural zeolites. Pergamon, Oxford, pp 277 – 284

Krogh Andersen E, Danø M, Petersen OV (1969) The mineralogy of Ilimaussaq. XIII. A tetragonal natrolite. Medd Grønland 181, no 10:1 – 19

Kühl GH (1969) Synthetic phillipsite. Am Miner 54:1607 – 1612

Kühl GH (1977) Acidity of mordenite. In: Katzer JR (ed) Molecular Sieves II. Am Chem Soc Symp Ser 40:96 – 107

Kühl GH, Miale JN (1978) Thermal stability of natural gmelinite and some of its ion-exchanged forms. In: Sand LB, Mumpton FA (eds) Natural zeolites. Pergamon, Oxford, pp 421 – 429

Kuzurenko MV (1950) Chalcedony-like natrolite in pegmatites of alkalic magmas (in Russian). Dokl Acad Nauk SSSR 72:767 – 770

Kvick A, Smith JV (1983) A neutron diffraction study of the zeolite edingtonite. J Chem Phys 79:2356 – 2362

Labuntzov AN (1927) The zeolites from Khibinsky and Lovozersky Mts., Russian Lapland (in Russian). Trav Musée Minér Acad Sci USSR 2:91 – 100

Lacroix A (1915) Les zéolites et les produits siliceux des basaltes de l'archipel de Kerguelen. Bull Soc Fr Miner Crist 38:134 – 137

Lacroix ML (1896) Sur la gonnardite. Bull Soc Miner 19:426 – 427

Larsen ES, Berman H (1934) The microscope determination of non-opaque minerals. US Geol Surv Bull 848:1 – 266

Laurent G (1941) Luminescence of minerals with special reference to quartz and nepheline from Swedish localities. Geol Foren Stockh Forh 63:59 – 83

Leavens PB, Hurlbut CS Jr, Nelen JA (1968) Eucryptite and bikitaite from King's Mountain, North Carolina. Am Miner 53:1202 – 1207

Leimer HW, Slaughter M (1969) The determination and refinement of the crystal structure of yugawaralite. Z Kristallogr 130:88 – 111

Lenzi G, Passaglia E (1974) Fenomeni di zeolitizzazione nelle formazioni vulcaniche della regione sabatina. Boll Soc Geol Ital 93:623 – 645

Leonhard KC von (1817) Taschenbuch für die gesamte Mineralogie mit Hinsicht auf die neuesten Entdeckungen. Hermann, Frankfurt, p 168

Leonhard KC von (1821) Handbuch der Oryktognosie, 1st edn. Mohr und Winter, Heidelberg, p 448

Lévy A (1825) Descriptions of two new minerals – herschelite – phillipsite. Ann Phil New Series 10:361 – 363

Leyerzapf H (1978) Aus Blasen und Mandeln. Lapis 3, no 1:20 – 23

Lin SB (1979) Amygdaloidal zeolites in the Penghu basalts, Penghu island, Taiwan. Proc Geol Soc China (Formosa) 22:68 – 83

Liou JG (1970) Synthesis and stability relations of wairakite $CaAl_2Si_4O_{12} \cdot 2H_2O$. Contrib Miner Petrol 27:259 – 282

Liou JG (1917a) Stilbite-laumontite equilibrium. Contrib Miner Petrol 31:171 – 177

Liou JG (1971b) P-T stabilities of laumontite, wairakite, lawsonite and related minerals in the system $CaAl_2Si_2O_8 - SiO_2 - H_2O$. J Petrol 12:379 – 411

Liou JG (1971c) Analcime equilibria. Lythos 4:389 – 402

Lippmann F, Rothfuss H (1980) Tonminerale in Taveyannaz-Sandsteinen. Schweiz Miner Petr Mitt 60:1 – 29

Lo HJ (1981) Hydrothermal synthesis of epistilbite from materials with the compositions of calcium oxide · aluminium oxide · 2-10-silicon dioxide ($CaO \cdot Al_2O_3 \cdot 2$-$10SiO_2$). Chung-Kuo Ti Chih Hsueh Hui Hui Kan 24:9 – 20

Lonsdale JT (1940) Igneous rocks from the Terlingua-Solitario region, Texas. Geol Soc Am Bull 51:1539

Lorent G (1933) Der Limburgit von Sasbach am Kaiserstuhl und seine hydrothermale Neuführung. Ber Naturforsch Ges Freib I Br 32:2

Lu G, Xu G, Liu G, Zhang F, Zhao F, Zhang H, Cai Z (1981) Study of natural mordenite catalyst for disproportionation of toluene (in Chinese). Ranliao Huaxue Xuebao 9:206 – 215

Lucchetti G (1976) Caratteristiche mineralogiche di una zeolite della serie phillipsite-harmotomo presente in vene nei diaspri di Cassagna (Liguria Orientale). Doriana 5:1 – 7

Lucchetti G, Massa B, Penco AM (1982) Strontian heulandite from Campegli (Eastern Ligurian Ophiolites, Italy). Neues Jahrb Miner Monatsh 1982:541 – 550

Machatschki F (1926) Ein Harmotomvorkommen in Steiermark. Zentralbl Miner Abt A 1926:115 – 119

Maggiore R, Solarino L, Crisafulli C, Schenbari G (1981) Italian natural zeolites as catalysts-isopropanol dehydration. Ann Chim 71:697 – 706

Maglione G, Tardy Y (1971) Néoformation pédogénétique d'une zéolite, Ca mordenite, associée aux carbonates de sodium dans une dépression interdunaire des bords du lac Tchad. C R Acad Sci Paris Ser D 272:772 – 774

Majer V (1953) Chabazite and stilbite from Bor (East Serbia, Yugoslavia) (in Serbo-Croatian with English summary). Jugosl Akad Znan Umjet 1953:175 – 192

Malashevich LN, Komarov VS, Pis'mennaya AV, Rat'ko AI (1981) Synthesis and physicochemical and absorption properties of high-silica zeolite ferrierite (in Russian). Zh Prikl Khim (Leningrad) 54:1032 – 1035

Maleev MN (1976) Svetlozarite, a new high-silica zeolite (in Russian). Zap Vses Miner Obshchest 105:449 – 453

Malinowskii YA, Belov NV (1980) Crystal structure of kalborsite (in Russian). Dokl Akad Nauk SSSR 252:611 – 615

Mallet JW (1859) On brewsterite. The London, Edinburgh and Dublin Phil Mag and J Sci 4th Ser 18:218 – 220

Manchester JG (1919) The minerals of the Bergen Archways. Am Miner 4:107 – 116

Martini J, Vuagnat M (1965) Présence de faciès à zéolites dans la formation des "grès" de Taveyanne (Alpes franco-suisses). Bull Suisse Miner Petrogr 45:281 – 293

Martini J, Vuagnat M (1970) Metamorphose niedrigst temperierten Grades in den Westalpen. Fortschr Miner 47:52 – 64

Mason B (1946a) Analcite, apophyllite and natrolite from the Pahau river, North Canterbury. New Zeal J Sci Technol Sect B 28:53 – 54

Mason B (1946b) The occurrence of laumontite at Bluff, Southland, New Zealand. N Z J Sci Technol Sect B 28:130

Mason B (1962) Herschelite – a valid species? Am Miner 47:985 – 987

Mason B, Greenberg SS (1954) Zeolites and associated minerals from Southern Brazil. Ark Mineral Geol 1:519 – 526

Mason B, Sand LB (1960) Clinoptilolite from Patagonia. The relationship between clinoptilolite and heulandite. Am Miner 45:341 – 350

Matsubara S, Tiba T, Kato A (1978a) Erionite in welded tuff from Ashio, Tochigi Prefecture, Japan. Bull Nat Sci Mus Ser C 4:1 – 6

Matsubara S, Tiba T, Kato A (1978b) The occurrence of cowlesite from Kuniga, Oki islands, Japan. Bull Nat Sci Mus Ser C 4:33 – 36

Matsubara S, Kato A, Tiba T, Saito Y, Nomura M (1979) Pectolite, analcime, natrolite and thomsonite in altered gabbro from Yani, Shinshiro, Aichi Pref., Japan. Mem Nat Sci Mus (Tokyo) 12:13 – 22

Mattinen V (1952) Stellerite from Stillböle, Finland. C R Soc Geol Finl 25:147 – 148

Mauritz B (1931) Die Zeolithmineralien der Basalte des Plattenseegebietes in Ungarn. Neues Jahrb Miner Abt A 1931:477 – 494

Mauritz B (1938) Die Mineralien in den Hohlräumen der Basalte von Halap und Gulacs im Plattenseegebiet (Ungarn). Tscherm Miner Petr Mitt 50:93 – 106

Mauritz B (1955) Recent observations dealing with the zeolite-minerals of the basaltic rocks in the highlands of Lake Balaton. Acta Miner Petrogr Szeged 8:37 – 40

Mazzi F, Galli E (1978) Is each analcime different? Am Miner 63:448 – 460

Mazzi F, Galli E (1983) The tetrahedral framework of chabazite. Neues Jahrb Miner Monatsh 1983:461 – 480

Mazzi F, Galli E, Gottardi G (1984) Structure refinements of two tetragonal edingtonites. Neues Jahrb Miner Monatsh 1984:373 – 382

McClellan HW (1926) Laumontite from Southern Oregon. Am Miner 11:287 – 288

McConnell D (1952) Viséite, a zeolite with the analcite structure and containing linked SiO_4, PO_4 and H_xO_4 groups. Am Miner 37:609 – 617

McConnell D, Foreman DW Jr (1974) The structure of kehoeite. Can Miner 12:352 – 353

McCulloh TH, Frizzell VA Jr, Stewart RJ, Barnes I (1981) Precipitation of laumontite with quartz, thenardite, and gypsum at Sepse Hot Springs, Western Transverse Ranges, California. Clays Clay Minerals 29:353 – 364

McLintock WFP (1915) On the zeolites and associated minerals from the Tertiary lavas around Ben More, Mull. Trans R Soc Edinb 51:1 – 33

McNamara MJ (1966) Zeolite occurrences in Swedish glacial clays. Geol Foren Stockh Forh 88:273 – 275

Meier AE (1939) Association of harmotome and barium feldspar at Glen Riddle, Pennsylvania. Am Miner 24:540 – 560

Meier WM (1960) The crystal structure of natrolite. Z Kristallogr 113:430 – 444

Meier WM (1961) The crystal structure of mordenite (ptilolite). Z Kristallogr 115:439 – 450

Meier WM (1968) Zeolite structures. In: Molecular Sieves. Society of Chemical Industry, London, pp 10 – 27

Meier WM, Olson DH (1978) Atlas of zeolite structure types. Polycrystal Book Service, Pittsburg

Meixner H, Hey MH, Moss AA (1956) Some new occurrences of gonnardite. Miner Mag 31:265 – 271

Mélon J (1943) La viséite, nouvelle espèce minerale. Ann Soc Geol Belg Bull 66:B53 – B56

Mercer BW, Ames LL Jr (1978) Zeolite ion exchange in radioactive and municipal wastewater treatment. In: Sand LB, Mumpton FA (eds) Natural Zeolites. Pergamon, Oxford, pp 451 – 462

Merkle AB, Slaughter M (1968) Determination and refinement of the structure of heulandite. Am Miner 53:1120 – 1138

Merlino S (1965) Struttura dell'epistilbite. Mem Soc Tosc Sci Nat Ser A 72:480 – 483

Merlino S (1972) Orizite discredited (= epistilbite). Am Miner 57:592 – 593

Merlino S (1975) Le strutture dei tettosilicati. Rend Soc Ital Miner Petrol 31:513 – 540

Merlino S (1976) Framework silicates. Jzvj Jugosl Centr Krist (Zagreb) 11:19 – 37

Merlino S (1984) Recent results in structural studies of zeolites and zeolite-like materials. In: Bisio A, Olson DH (eds) Proc 6th Int Conf Zeolite, Butterworths, Guildford, pp 747 – 759

Merlino S, Galli E, Alberti A (1975) The crystal structure of levyne. Tscherm Miner Petr Mitt 22:117 – 129

Merritt CA (1958) Igneous geology of the Lake Altas area, Oklahoma. Okla Geol Surv Bull
76:56 – 58

Mezösi J (1965) New occurrences of zeolite in the Matra Mountains. Acta Miner Petrog Szeged
17:29 – 38

Mikhailov AS, Vlasov VV, Varfolomeeva EK, Shlyapkina EN (1975) Distribution of zeolites in
volcanic-sedimentary formations of the URSS and some physical methods for their study (in
Russian). In: Kossovskaya AG (ed) Kristallokhim miner geol probl. Nauka, Moskow, pp
177 – 184

Miller BE, Ghent ED (1973) Laumontite and barian-strontian heulandite from the Blairmore
groups (Cretaceous), Alberta. Can Miner 12:188 – 192

Milton C, Davidson N (1950) An occurrence of natrolite, andradite, and allanite in the Franklin
Furnace quadrangle, New Jersey. Am Miner 35:500 – 507

Milton RM [to Union Carbide Corp] (1959) Molecular-sieve adsorbents. US Pat 2, 882, 244

Milton RM [to Union Carbide Corp] (1961) Zeolite W. Br Pat 864, 707; US Pat 3, 012, 853

Minato H, Takano Y (1964) An occurrence of potassium-clinoptilolite from Itaya, Yamagata
Pref., Japan (in Japanese). J Clay Sci Soc Jpn 4:12 – 22

Minato H, Tamura T (1978) Production of oxygen and nitrogen with natural zeolites. In: Sand
LB, Mumpton FA (eds) Natural zeolites. Pergamon, Oxford, pp 509 – 516

Minato H, Utada M (1971) Clinoptilolite from Japan. In: Molecular sieve zeolites I. Am Chem
Soc Adv Chem Ser 101:311 – 316

Mirodatos C, Beaumont R, Barthomeuf D (1975) Proprietés catalytiques d'une zéolithe de type
offrétite. C R Acad Sci Paris Ser C 281:959 – 961

Mitchell RH, Platt RG (1982) Mineralogy and petrology of nepheline syenites from the Coldwell
alkaline complex, Ontario, Canada. J Petrol 23:186 – 214

Moiola RJ (1970) Authigenic zeolites and K-feldspar in the Esmeralda Formation, Nevada. Am
Miner 55:1681 – 1691

Monese A, Sacerdoti M (1970) L'analcime contenuto nelle lave basiche Triassiche dell'alta Val
Duron (Val di Fassa – Trento). Ann Univ Ferrara Sci Miner Petr 1:41 – 46

Moncure GK, Surdam RC, McKagne HL (1981) Zeolite diagenesis below Pahute Mesa, Nevada
Test Site. Clays Clay Minerals 29:385 – 396

Monticelli T, Covelli N (1825) Prodromo della mineralogia Vesuviana, vol I. Tramater, Napoli, p
224

Morbidelli L (1964) Contributo alla conoscenza della venanzite: la facies differenziata di Podere
Pantano e la phillipsite che l'accompagna. Period Miner 33:199 – 221

Morelli GL, Novelli L (1967) Sulla presenza di analcime sedimentario nella Libia centro-meridio-
nale. Natura (Milan) 58:113 – 120

Morozewicz J (1909) Über Stellerit, ein neues Zeolithmineral. Bull Acad Sci Cracoviae Cl Sci Mat
Nat, pp 344 – 359

Morozewicz J (1925) The Commander islands. A study (geography and natural history). Instytut
Popierania Nauki, Warsawa

Mörtel H (1971) Foide und Zeolithe (Restkristallisate) basaltischer Gesteine des Vogelsberges
(Hessen). Neues Jahrb Miner Abh 115:54 – 97

Mortier WJ (1982) Compilation of extra framework sites in zeolites. Zeolites Spec Issue

Mortier WJ, Pearce JR (1981) Thermal stability of the heulandite-type framework: crystal structure of the calcium/ammonium form dehydrated at 483K. Am Miner 66:309 – 314

Mortier WJ, Pluth JJ, Smith JV (1975, 1976, 1978) Positions of cations and molecules in zeolites
with mordenite-type framework. I Dehydrated Ca-exchanged ptilolite. Mat Res Bull
10:1037 – 1046; II Dehydrated hydrogen-ptilolite. Mat Res Bull 10:1319 – 1326; III Rehydrated
Ca-exchanged ptilolite. Mat Res Bull 11:15 – 22; IV Dehydrated and rehydrated K-exchanged
ptilolite. In: Sand LB, Mumpton FA (eds) Natural Zeolites. Pergamon, Oxford, pp
53 – 75

Mortier WJ, Pluth JJ, Smith JV (1976a) The crystal structure of dehydrated offretite with stacking faults of erionite type. Z Kristallogr 143:319 – 332

Mortier WJ, Pluth JJ, Smith JV (1976b) Crystal structure of natural zeolite offretite after carbon
monoxide absorption. Z Kristallogr 144:32 – 41

Mortier WJ, Pluth JJ, Smith JV (1977) Positions of cations and molecules in zeolites with the chabazite framework. I. Dehydrated Ca-exchanged chabazite. Mat Res Bull 12:97–102; III Dehydrated Na-exchanged chabazite. Mat Res Bull 12:241–250

Muchi M (1977) Saponite and related thomsonite from Iwano, Saga Pref., Japan. Bull Fukuoka Univ Educ Part III Nat Sci 26:103–115

Müller G (1961) Die rezenten Sedimente im Golf von Neapel. II. Mineral-, Neu- und Umbildungen in den rezenten Sedimenten des Golfes von Neapel. Ein Beitrag zur Umwandlung vulkanischer Gläser zur Halmyrolyse. Beitr Miner Petr 8:1–20

Müller U von, Deisinger H (1971) Strontium-Thomsonit aus dem Roßbergbasalt von Roßdorf bei Darmstadt. Aufschluß 22:145–148

Mumpton FA (1960) Clinoptilolite redefined. Am Miner 45:351–369

Mumpton FA (1975) Commercial utilization of natural zeolites. In: Lefond SJ (ed) Industrial minerals and rocks, 4th edn. Am Inst Mining Metall Petrol Engrs Inc, New York, pp 1262–1274

Mumpton FA, Ormsby WC (1976) Morphology of zeolites in sedimentary rocks by SEM. Clays Clay Minerals 24:1–23

Mumpton FA, Ormsby WC (1978) Morphology of zeolites in sedimentary rocks by SEM. In: Sand LB, Mumpton FA (eds) Natural zeolites. Pergamon, Oxford, pp 113–134

Murata KJ, Whiteley KR (1973) Zeolites in the Miocene Briones Sandstone and related formations of the Central Coast Ranges, California. J Res US Geol Surv 1:255–265

Murphy CO, Hrycyk O, Gleason WT (1978) Natural zeolites: novel uses and regeneration in wastewater treatment. In: Sand LB, Mumpton FA (eds) Natural zeolites. Pergamon, Oxford, pp 471–478

Murray J, Renard AF (1891) Report on the scientific results of the voyage of "HMS Challenger" during the years 1873–1876. Neill, Edinburgh

Naccache C, Ben Taarit Y (1980) Recent developments in catalysis by zeolites. In: Rees LVC (ed) Proc 5th Int Conf Zeolites. Heyden, London, pp 592–606

Nacken R, Wolff W (1921) Über die Adsorption von Gasen durch Chabasit. Zentralbl Miner 1921:364–372, 388–394

Nakajima W (1973) Mordenite solid solution in the system $Na_2Al_2Si_{10}O_{24}$-$CaAl_2Si_{10}O_{24}$-H_2O (in Japanese). Bull Fac Educ Kobe Univ 48:91–98

Nakajima W (1983) Disordered wairakite from Hikihara, Haga Town, Hyogo Prefecture. Bull Fac Educ Kobe Univ 70:39–46

Nambu M (ed) (1970) Introduction to Japanese minerals. Geol Surv Jpn, Tokyo

Nanne JM, Post MFM, Stork WHJ [to Shell Internationale Research Maatschappij BV] (1980) Ferrierite, and its use as catalyst or catalyst carrier for converting hydrocarbons, and for separating hydrocarbons. Eur Pat Appl 12, 473

Nasedkina VKh (1980) Zeolites in the bauxite beds of the Northern Onega region (in Russian). In: Kossovskaya AG (ed) Prir Tseolity [Mater Vses Semin]. Nauka, Moscow, pp 166–170, 192, 211–214

Nawaz R (1980a) Crystallography and optical properties of heulandite. J Earth Sci R Dublin Soc 2:185–187

Nawaz R (1980b) Morphology, twinning and optical orientation of gismondine. Miner Mag 43:841–844

Nawaz R (1983) New data on gobbinsite and garronite. Miner Mag 47:567–568

Nawaz R, Malone JF (1981) Orientation and geometry of the thomsonite unit cell: a re-study. Miner Mag 44:231–235

Nawaz R, Malone JF (1982) Gobbinsite, a new zeolite mineral from Co. Antrim, North Ireland. Miner Mag 46:365–369

Needham CE (1938) Zeolites in New Mexico. Am Miner 23:285–287

Negishi T (1970) Laumontite-tuff from Kozawa, Iwanai district, Hokkaidô (in Japanese). Nendo Kagaku 10:110–116

Neumann H (1944) Silver deposits at Kongsberg. Norg Geol Unders 162:1–133

Nikolina VYa, Knysh LI, Dolenko VA, Kirkach LI, Lipkind BA, Kanakova OA, Slepneva AT (1981) Erionite (in Russian). USSR Pat 833,499 from Otkrytiya, Izobret, Prom Obraztsy, Tovarnye znaki 1981:43

Nishido H, Otsuka R (1981) Chemical composition and physical properties of dachiardite group zeolites. Miner J 10:371 – 384

Nishido H, Otsuka R, Nagashima K (1979) Sodium-rich dachiardite from Chichijima, the Ogasawara Islands, Japan. Bull Sci Eng Res Lab Waseda Univ 87:29 – 37

Nordenskiöld O (1895) Note sur l'edingtonite de Böhlet (Suède). Bull Soc Miner 18:396 – 398

Norin E (1955) The mineral composition of the Napolitan yellow tuff. Geol Rundschau 43:526 – 534

Novak F (1970) Some new data for edingtonite. Acta Univ Carolinae Geol, pp 237 – 251

Novotna B (1926) Contribution to the knowledge of Moravian zeolites (in Czech). Cas Moravske-ho Zem Musea 24:133 – 144

Nowacki H, Koyama H, Mladek MH (1958) Über die Kristallstruktur des Zeolithes Chabasit. Experientia (Basel) 14:396

Obradović J (1977) The review on the occurrences of zeolites in sedimentary rocks of Yugoslavia (in Serbo-Croatian with English summary). Ann Géol Péninsule Balk 41:293 – 302

Obradović J (1979) Authigenic minerals in Middle Triassic volcanoclastic rocks in Dinarides. Bull Mus Hist Nat Belgrade Ser A 34:13 – 36

Obradović J, Stojanovic D (1972) Occurrence of "pietra verde", similar to that occurring in Italian Dolomites, in volcanic sedimentary serie in the Budva-Bar area (Montenegro, Jugoslavia). Stud Trent Sci Nat Sez A 49:161 – 172

Ohtani S, Iwamura T, Sando K, Matsumura K [to Ray Industries Inc] (1972) Conversion catalyst for hydrocarbon. Jpn Pat 72 46, 667

Olson DH, Sherry HS (1968) X-ray study of strontium-sodium ion exchange in Linde X. Example of a two-phase zeolite system. J Phys Chem 72:4095 – 4104

Olsson RW, Rollmann LD (1977) Crystal chemistry of dealuminized mordenites. Inorg Chem 16:651 – 654

Olver MD (1983) The effect of feeding clinoptilolite (zeolite) to laying hens. S Afr J Anim Sci 13:107 – 110

Paar W (1971) Ein neues Zeolithvorkommen von Stubachtal/Salzburg. Aufschluß 22:74 – 78

Pabst A (1971) Natrolite from the Green River formation, Colorado, showing an intergrowth akin to twinning. Am Miner 56:560 – 569

Pagliani G (1948) Le zeoliti del granito di Baveno. Period Miner 17:175 – 188

Parker RL (1922) Über einige schweizerische Zeolithparagenesen. Schweiz Miner Petr Mitt 2:290 – 298

Passaglia E (1966) Zeoliti ed altri minerali dei basalti del Monte Baldo (Trentino). Period Miner 35:187 – 195

Passaglia E (1969) Roggianite, a new silicate mineral. Clay Minerals 8:107 – 111

Passaglia E (1970) The crystal chemistry of chabazites. Am Miner 55:1278 – 1301

Passaglia E (1972) Le zeoliti di Albero Bassi (Vicenza). Period Miner 28:237 – 243

Passaglia E (1975) The crystal chemistry of mordenite. Contrib Miner Petrol 50:65 – 77

Passaglia E (1978a) Lattice-constant variations in cation-exchanged chabazites. In: Sand LB, Mumpton FA (eds) Natural zeolites. Pergamon, Oxford, pp 45 – 52

Passaglia E (1978b) New data on ferrierite from Weitendorf near Wildon, Styria, Austria. Mitt Bl Abt Miner Landesmuseum Joanneum 46:565 – 566

Passaglia E (1980) The heat behaviour of cation exchanged zeolites with the stilbite framework. Tscherm Miner Petr Mitt 27:67 – 78

Passaglia E, Bertoldi G (1983) Harmotome from Selva di Trissino (Vicenza, Italy). Period Miner 52:75 – 82

Passaglia E, Moratelli E (1972) Zeoliti ed altri minerali di neoformazione di Bulla (Bolzano). Atti Soc Ital Sci Nat 113:274 – 280

Passaglia E, Pongiluppi D (1974) Sodian stellerite from Capo Pula, Sardegna. Lithos 7:69 – 73

Passaglia E, Pongiluppi D (1975) Barrerite, a new natural zeolite. Miner Mag 40:208

Passaglia E, Sacerdoti M (1982) Crystal structural refinement of Na-exchanged stellerite. Bull Miner 105:338 – 342

Passaglia E, Vezzalini G (to be published 1985) The crystal chemistry of zeolites in sedimentary deposits of Italy. Contrib Miner Petrol

Passaglia E, Galli E, Rinaldi R (1974) Levynes and erionites from Sardinia, Italy. Contrib Miner Petrol 43:253 – 259

Passaglia E, Pongiluppi D, Rinaldi R (1977) Merlinoite a new mineral of the zeolite group. Neues Jahrb Miner Monatsh 1977:355 – 364

Passaglia E, Pongiluppi D, Vezzalini G (1978a) The crystal chemistry of gmelinites. Neues Jahrb Miner Monatsh 1978:310 – 324

Passaglia E, Galli E, Leoni L, Rossi G (1978b) The crystal chemistry of stilbites and stellerites. Bull Miner 101:368 – 375

Pauling L (1930) The structure of some sodium and calcium aluminosilicates. Proc Nat Acad Sci USA 16:453 – 459

Pawloski JA (1965) Two interesting zeolites from Connecticut. Rocks Minerals 40: 494

Peacor DR (1973) High temperature, single crystal X-ray study of natrolite. Am Miner 58:676 – 680

Peacor DR, Dunn PJ, Simmons WB, Tillmanns E, Fischer RX (1984) Willhendersonite, a new zeolite isostructural with chabazite. Am Miner 69:186 – 189

Pearce TH (1970) The analcite-bearing volcanic rocks of the Crowsnest Formation, Alberta. Can J Earth Sci 7:46 – 66

Pechar F (1982) Precise definition of the cristalline structure of natural mesolite (in Czech). Acta Mont 59:143 – 152

Pecora WT, Fisher B (1946) Drusy vugs in a monzonite dyke, Bear paw-Montaine, Montana. Am Miner 31:370 – 385

Pelloux A (1931) Armotomo ed altri minerali del giacimento di grafite di Cerisieri (Valle del Chisone). Period Miner 2:281 – 285

Pelloux A (1949) Sopra alcune zeoliti della Valle del Varenna presso Pegli con descrizione di un minerale nuovo per l'Italia. Ann Mus Civ Stor Nat Genova 63:1 – 8

Pereyron A, Guth JL, Wey R (1971) Etude du diagramme Na_2O, Li_2O, Al_2O_3, SiO_2, H_2O dans le domaine de formation de zéolites. C R Acad Sci Paris Ser D 272:181 – 184

Perrotta AJ (1967) The crystal structure of epistilbite. Miner Mag 36:480 – 490

Perrotta AJ (1976) A low-temperature synthesis of a harmotome-type zeolite. Am Miner 61:495 – 496

Perrotta AJ, Smith JV (1964) The crystal structure of brewsterite, $(Sr,Ba,Ca)(Al_2Si_6O_{16}) \cdot 5 H_2O$. Acta Crystallogr 17:857 – 862

Petrova VV, Trukhin YuP (1975) On the zeolite formation in a recent geothermal field at Paratunskoe (Kamchatka) (in Russian). Nov Vidy Nemet Polezn Iskop Nauka, Moskow, pp 110 – 117

Petzing J, Chester R (1979) Authigenic zeolites and their relationship to global volcanism. Mar Geol 29:253 – 271

Phillips WJ (1968) The crystallization of the teschenite from the Lugar Sills, Ayrshire. Geol Mag 105:23 – 34

Phinney WC, Stewart DB (1961) Some physical properties of bikitaite and its dehydration and decomposition products. US Geol Surv Prof Pap 424:353 – 357

Pierkarska E, Gawel A (1954) Heulandite from Rudno (Cracow district) (in Polish with English summary). Ann Soc Geol Pologne 22:353 – 362

Pillard F, Maury RC, Tournemire R, Massal P (1980) Évolution hydrothermale de l'hawaiite d'Éspalion (Aveyron). Bull Miner 103:101 – 106

Pipping F (1966) The dehydration and chemical composition of laumontite. Miner Soc India IMA Vol:159 – 166

Pirsson LV (1890) On mordenite. Am J Sci 40:232 – 237

Poitevin E (1936) Thomsonite from the Eastern Townships, Quebec. Univ Toronto Stud Geol Ser 40:63 – 65

Poitevin E (1938) Natrolite from the Eastern Townships, Quebec. Univ Toronto Stud Geol Ser 41:57 – 58

Pongiluppi D (1974) Ulteriori notizie sulle zeoliti della Sardegna. Rend Soc Ital Miner Petrol 30:1201 – 1205

Pongiluppi D (1976) Offretite, garronite and other zeolites from "Central Massif", France. Bull Soc Fr Miner Crist 99:322–327

Pongiluppi D (1977) A new occurrence of yugawaralite at Osilo, Sardinia. Can Miner 15:113–114

Pongiluppi D, Passaglia E, Galli E (1974) Su alcune zeoliti della Sardegna. Rend Soc Ital Miner Petrol 30:77–100

Ponomareva TM, Tomilov NP, Berger AS (1974) Hydrothermal synthesis and properties of the gallium analog of natrolite (in Russian). Geokhimiya 6:925–931

Pough FH (1936) Bertrandite and epistilbite from Bedford, New York. Am Miner 21:264–265

Poutsma ML (1976) Mechanistic considerations of hydrocarbon transformations catalyzed by zeolites. In: Rabo JA (ed) Zeolite chemistry and catalysis. Am Chem Soc Monog 171:437–528

Povarennykh AS (1954) On the question of zeolitization of alkalic rocks (in Russian). Dokl Akad Nauk SSSR 94:761–764

Pratt JH, Foote HW (1897) On wellsite, a new mineral. Am J Sci 153:443–448

Pullé G, Capacci CW (1874) Un viaggio nell'arcipelago toscano. Le Monnier, Firenze

Raade G (1969) Cavity minerals from the Permian biotite granite at Nedre Eiker church. Norsk Geol Tidsskr 49:227–239

Raatz F, Dexpert H, Freund E, Marcilly C (1984) Comparison between small port and large port mordenites. 6th Int Conf Zeolite Poster Session. 11–15 July, 1983. Reno, Nevada

Rabinowitsch E (1932) Über Gasaufnahme durch Zeolithe. Z Phys Chem Abt B 16:43–71

Rabinowitsch E, Wood WC (1933a) Ion exchange and sorption of gases by chabazite. Nature (Lond) 132:540

Rabinowitsch E, Wood WC (1933b) Über die Elektrizitätleitung in Zeolithen. Z Elektrochem 39:562–566

Rabo JA (ed) (1976) Zeolite chemistry and catalysis. Am Chem Soc Monog 171

Rabo JA, Poutsma ML (1971) Structural aspects of catalysis with zeolites: cracking of cumenes and hexanes. In: Molecular sieve zeolites II. Am Chem Soc Adv Chem Ser 102:284–314

Ramasamy R (1981) Gonnardite from carbonatite complex of Tiruppattur, Tamil Nadu. Curr Sci (Bangalore) 50:271–272

Ramdohr P (1978) "Versteinerten Tropfen gleich ..." Der Natrolith von Hohentwiel. Lapis 3, no 1:18–19

Ramdohr P, Strunz H (1978) Klockmanns Lehrbuch der Mineralogie. Enke, Stuttgart

Rao AB, Cunha e Silva J (1963) Mineralogia de alguns zeolitos do Nordeste. Arq Geol Univ Recife 4:33–47

Read PB, Eisbacher GH (1974) Regional zeolite alteration of the Sustut Group, North-Central British Columbia. Can Miner 12:527–541

Reed JC (1937) Amygdales in Columbia River lavas near Freedom, Idaho. Nat Res Council Trans Am Geophys Union, 18th Annu Meet, pp 239–243

Reeuwijk LP van (1971) The dehydration of gismondite. Am Miner 56:1655–1659

Reeuwijk LP van (1972) High-temperature phases of zeolites of the natrolite group. Am Miner 57:499–510

Reeuwijk LP van (1974) The thermal dehydration of natural zeolites. Meded Landbouwhogesch Wageningen 74–9:1–88

Regis AJ (1970) Occurrences of ferrierite in altered pyroclastics in central Nevada (Abstr). Geol Soc Am Annu Meet Programs 2, no 7:661

Regis AJ, Sand LB (1967) Lateral gradation of chabazite to herschelite in the San Simon basin. Clays Clay Minerals 27:193

Reichert R (1924) Laumontit aus dem Graf Cziraky Steinbruche von Nadap (Komitat Fejer). Foldt Kozl 54:187–189

Reichert R, Erdélyi J (1935) Über die Minerale des Csodi-Berges bei Dunabogdany (Ungarn). Tscherm Miner Petr Mitt 46:237–255

Rengarten NV (1950) Laumontite and analcime from lower Devonian deposits in northern Caucasus (in Russian). Dokl Akad Nauk SSSR 70:485–488

Resegotti G (1929) Heulandite e stilbite di Prascorsano presso Cuorgnè (Piemonte). Atti R Accad Sci Torino 64:105–116

Rinaldi R (1976) Crystal chemistry and structural epitaxy of offretite-erionite from Sasbach, Kaiserstuhl. Neues Jahrb Miner Monatsh 1976:145 – 156

Rinaldi R, Pluth JJ, Smith JV (1974) Zeolites of the phillipsite family. Refinement of the crystal structure of phillipsite and harmotome. Acta Cryst B30:2426 – 2433

Rinaldi R, Smith JV, Jung G (1975a) Chemistry and paragenesis of faujasite, phillipsite and offretite from Sasbach, Kaiserstuhl, Germany. Neues Jahrb Miner Monatsh 1975:433 – 443

Rinaldi R, Pluth JJ, Smith JV (1975b) Crystal structure of mazzite dehydrated at 600°C. Acta Cryst B31:1603 – 1608

Rinne F (1890) Über die Umänderungen welche die Zeolithe durch Erwärmen bei und nach dem Trübwerden erfahren. Sitzungber Preuss Akad Wiss 46:1163 – 1207

Rinne F (1923) Bemerkungen und Röntgenographische Erfahrungen über die Umgestaltung und den Zerfall von Kristallstrukturen. Z Kristallogr 59:230 – 248

Romé de L'isle JBL (1783) Crystallographie, ou description des formes propres à tous les corps due règne minéral, 2nd edn. Paris, p 40

Roques M (1947) Édifices mimétiques et symétrie triclinique de l'analcime des îles de Los (Guniée française). C R Acad Sci Paris 225:946

Rose G (1826) Über den Epistilbit, eine neue zur Familie der Zeolithe gehörige Mineralgattung. Ann Phys Chem (Leipzig) 6:183 – 189

Rosický V, Thugutt SJ (1913) Epidesmin, ein neuer Zeolith. Zentralbl Miner 1913: 422 – 426

Ross CS (1928) Sedimentary analcite. Am Miner 13:195 – 197

Ross CS (1941) Sedimentary analcite. Am Miner 26:627 – 629

Ross CS, Shannon EV (1924) Mordenite and associated minerals from near Challis, Custer County, Idaho. Proc US Nat Mus 64, art 19

Roux J, Hamilton DL (1976) Primary igneous analcite – an experimental study. J Petrol 17:244 – 257

Rubin MK [to Mobil Oil Corp] (1969) Crystalline aluminosilicate zeolites. German Pat 1, 806, 154

Russell A (1946) An account of the Struy lead mines Inverness-shire, and of wulfenite, harmotome, and other minerals which occur there. Miner Mag 27:147 – 154

Rychly R, Ulrych J (1980) Mesolite from Horni Jilove, Bohemia. Tscherm Miner Petr Mitt 27:201 – 208

Rychly R, Danek M, Siegl J (1982) Structural epitaxy of offretite-erionite from Prackovice nad Labem, Bohemia. Chem Erde 41:263 – 268

Sacerdoti M, Gomedi I (1984) Crystal structure refinement of Ca-exchanged Barrerite. Bull Miner 107:799 – 804

Sadanaga R, Marumo F, Takéuchi Y (1961) The crystal structure of harmotome. Acta Crystallogr 4:1153 – 1163

Saebø PC, Reitan PH (1959) An occurrence of zeolites at Kragerø, Southern Norway. Norg Geol Unders 205:174 – 180

Saebø PC, Sverdrup TL (1959) Note on stilbite from a pegmatite at Elveneset, Innhavet in Nordland Co, Northern Norway. Norg Geol Unders 205:181 – 183

Saebø PC, Reitan PH, Gaul JJC (1959) Stilbite, stellerite and laumontite at Honningvag, Northern Norway. Norg Geol Unders 205:171 – 173

Saether E (1949) A find of laumontite in the gneiss area at Møre (in Norwegian). Norsk Geol Tidsskr 28:50 – 51

Saha P (1959) Geochemical and X-ray investigation of natural and synthetic analcime. Am Miner 44:300 – 313

Saha P (1961) The system nepheline-albite-H_2O. Am Miner 46:859 – 884

Sahama TG, Lehtinen M (1967) Harmotome from Korsnäs, Finland. Miner Mag 36:444 – 448

Sakurai K, Hayashi A (1952) Yugawaralite, a new zeolite. Sci Rep Yokohama Natl Univ Ser 2 1:69 – 77

Šamajová E (1977) Mordenite occurrence in andesites of central Vihorlát Mountains (in Slovak with English summary). Acta Geol Geogr Univ Comenianae Ser Geol 32:147 – 155

Šamajová E (1979) Zeolites in Neogene volcanoclastic of Slovakia. Geol Zb Geol Carpathice (Bratislava) 30:363 – 378

Sameshima J (1929) Sorption of gas by mineral. I. Heulandite and chabazite. Bull Chem Soc Jpn 4:96 – 103

Sameshima J, Hemmi H (1934) Sorption of gas by mineral. IV. Zeolites and bentonite. Bull Chem Soc Jpn 9:27 – 41

Sameshima T (1969) Yugawaralite from Shinoda, Shizuoka Pref., Central Japan. Geosci Mag 20:71 – 78

Sameshima T (1978) Zeolites in tuff beds of the Miocene Waitamata group, Auckland province, New Zealand. In: Sand LB, Mumpton FA (eds) Natural zeolites. Pergamon, Oxford, pp 309 – 318

Sand LB (1968) Synthesis of large-port and small-port mordenites. In: Molecular sieves. Society of Chemical Industry, London, pp 71 – 77

Sands CD, Drewer JI (1978) Authigenic laumontite in deep-sea sediments. In: Sand LB, Mumpton FA (eds) Natural zeolites. Pergamon, Oxford, pp 269 – 273

Sanero E (1938) Stilbite nella tonalite di Val Nambrone (Gruppo dell'Adamello). Period Miner 9:205 – 212

Sarp H, Deferne J, Bizouard J, Liebich BW (1979) La parthéite, $CaAl_2Si_2O_8 \cdot 2 H_2O$, un nouveau silicate naturel d'aluminium et de calcium. Schweiz Miner Petr Mitt 59:5 – 13

Sato M (1979) Derivation of possible framework structures formed from parallel four- and eight-membered rings. Acta Crystallogr Sect A Cryst Phys Diff Theor Gen Crystallogr 35:547 – 552

Sato M, Gottardi G (1982) The slipping scheme of the double crankshaft structures in tectosilicates and its mineralogical implication. Z Kristallogr 161:187 – 194

Sawatzki G, Vuagnat M (1971) Sur la présence du faciès à zéolites dans les grès de Taveyanne du synclinal de Thônes (Haute-Savoie, France). C R Séances Soc Phys Hist Nat Genève 6:69 – 79

Scaini G, Nardelli M (1952) La stilbite dell'alta Val Malenco. Soc Ital Sci Nat 91:25 – 30

Schaller WT (1932a) Ptilolite from Utah. Am Miner 17:125 – 127

Schaller WT (1932b) The mordenite-ptilolite group; clinoptilolite a new species. Am Miner 17:128 – 134

Scherillo A (1938) Su alcune zeoliti dell'Eritrea. Period Miner 9:61 – 69

Schlenker JL, Pluth JJ, Smith JV (1977a) Dehydrated natural erionite with stacking faults of the offretite type. Acta Cryst B33:3265 – 3268

Schlenker JL, Pluth JJ, Smith JV (1977b) Refinement of the crystal structure of brewsterite. Acta Cryst B33:2907 – 2910

Schlenker JL, Pluth JJ, Smith JV (1978, 1979) Positions of cations and molecules in zeolites with the mordenite-type framework. V Dehydrated Rb-mordenite. Mat Res Bull 13:77 – 82; VI Dehydrated Ba-mordenite. Mat Res Bull 13:169 – 174; VII Dehydrated Cs-mordenite. Mat Res Bull 13:901 – 905; VIII Dehydrated sodium-exchanged mordenite. Mat Res Bull 14:751 – 758; IX Dehydrated H-mordenite via acid exchange. Mat Res Bull 14:848 – 856; X Dehydrated calcium hydrogen mordenite. Mat Res Bull 14:961 – 966

Schorr K (1980) Erster Ferrierit-Fund in Deutschland. Aufschluß 31:96 – 98

Schramm V, Fischer KF (1970) Refinement of the crystal structure of laumontite. In: Molecular sieve zeolites. I. Am Chem Soc Adv Chem Ser 101:259 – 265

Seager AF (1969) Zeolites and other minerals from Dean quarry, the Lizard, Cornwall. Miner Mag 37:147 – 148

Seager AF (1978) Zonal dissolution in analcime and pseudomorhs of adularia after analcime from the Lizard, Cornwall. Miner Mag 42:245 – 249

Section of Crystal Structure Analysis, Academy of Geological Science and Academia Sinica (1973) The crystal structure of hsianghualite. Acta Geol Sin, pp 240 – 242

Sedlacek M (1949) Neue Mineralvorkommen in niederösterreichischen Waldviertel. Verh Geol Bundesanst 1949:133 – 136

Seebach K von (1862) Notiz über ein neues Vorkommen des Analzims. K Ges Wiss Göttingen Nachr, pp 334 – 335

Seki Y (1966) Wairakite in Japan. J Jpn Assoc Miner Petr Econ Geol 55:254 – 261; 56: 30 – 39

Seki Y (1968) Synthetized wairakites their difference from natural wairakites. J Geol Soc Jpn 74:457 – 458

Seki Y (1970) Alteration of bore-hole cores to mordenite-bearing assemblages in Atosanupuri active geothermal area, Hokkaidô, Japan. J Geol Soc Jpn 76:605 – 611

Seki Y (1973) Distribution and modes of occurrence of wairakites in the Japanees island arc. J Geol Soc Jpn 79:521 – 527

Seki Y, Haramura H (1966) On chemical composition of yugawaralite (in Japanese). J Jpn Assoc Miner Petr Econ Geol 59:107 – 111

Seki Y, Oki Y (1969) Wairakite-analcime solid solutions from low-grade metamorphic rocks of the Tanzawa Mountains, Central Japan. Miner J 6:36 – 45

Seki Y, Okumura K (1968) Yugawaralite from Onikobe active geothermal area, Northeast Japan. J Jpn Assoc Miner Petr Econ Geol 60:27 – 33

Seki Y, Onuki H (1971) Alteration of wallrocks into wairakite-bearing assemblages at the Kawaji damsite, Central Japan. J Geol Soc Jpn 77:483 – 488

Seki Y, Onuki H, Okumura K, Takashima I (1969a) Zeolite distribution in the Katayama geothermal area, Onikobe, Japan. Jpn J Geol Geogr 40:63 – 79

Seki Y, Oki Y, Matsuda T, Mikami K, Okumura K (1969b) Metamorphism in the Tanzawa Mountains, Central Japan. J Jpn Assoc Miner Petr Econ Geol 61:50 – 75

Seki Y, Oki Y, Onuki H, Odaka S (1971) Metamorphism and vein minerals of North Tanzawa Mountains, Central Japan. J Jpn Assoc Miner Petr Econ Geol 66:1 – 21

Seki Y, Oki Y, Okada S, Ozawa K (1972) Stability of mordenite in zeolite facies metamorphism of the Oyama-Isehara district, east Tanzawa Mts., Central Japan. J Geol Soc Jpn 78:145 – 160

Senderov EE (1963) Crystallization of mordenite (in Russian). Geokhimija 9:820 – 829

Senderov EE (1974) On the influence of pH and CO_2 on the transformation of zeolites into clay minerals (in Russian). Litol Polesn Iskopajom 1974:80 – 87

Senderov EE (1980) Estimation of Gibbs energy for laumontite and wairakite from conditions of their formation in geothermal areas. In: Rees LVC (ed) Proc 5th Int Conf Zeolites. Heyden, London, pp 56 – 63

Senderov EE, Khitarov NI (1966) Conditions of natrolite formation (in Russian). Geokhimija 1966:1398 – 1412

Senderov EE, Khitarov NI (1971) Synthesis of thermodynamically stable zeolites in the $Na_2O – Al_2O_3 – SiO_2 – H_2O$ system. In: Molecular Sieve Zeolites. I. Am Chem Soc Adv Chem Ser 101:149 – 154

Sersale R (1958) Genesi e costituzione del tufo giallo napoletano. Rend Accad Sci Fis Mat Soc Naz Sci Lett Arti Ser 4 Napoli 25:6 – 31

Sersale R (1959a) Sulla natura zeolitica del tufo cosidetto "lionato" della regione vulcanica dei Colli Albani. Rend Accad Sci Fis Mat Soc Naz Sci Lett Arti Ser 4 Napoli 26:110 – 116

Sersale R (1959b) Analogie costituzionali fra il "trass" renano ed il tufo giallo napoletano. Rend Accad Sci Fis Mat Soc Naz Sci Lett Arti Ser 4 Napoli 26:117 – 125

Sersale R (1960a) Sulla natura zeolitica dei tufi rossi a scorie nere della regione Sabazia e Cimina. Rend Accad Sci Fis Mat Soc Naz Sci Lett Arti Ser 4 Napoli 27:306 – 319

Sersale R (1960b) Sulla natura zeolitica del tufo "carpato" della regione vulcanica del Monte Vulture. Rend Accad Sci Fis Mat Soic Naz Sci Lett Arti Ser 4 Napoli 27:543 – 556

Sersale R (1962) Sulla "pseudomorfosi" della leucite in analcime. Period Miner 31:337 – 368

Sersale R, Aiello R, Frigione G (1963) Sulla presenza di orizzonti zeolitici nella serie oligo-miocenica di Garbagna (Alessandria). Atti Acad Sci Torino I Cl Sci Fis Mat Nat 97:1 – 15

Sersale R, Aiello R, Vero E (1965) Ricerche sulla preparazione e sul comportamento termico della phillipsite. Period Miner 34:419 – 434

Shannon EV (1925) The so-called halloysite of Jones Falls, Maryland. Am Miner 10:159 – 161

Shashkina WP (1958) Zeolites from the basalts of Volhynia (in Russian). Miner Sb Lvov 12:380 – 395

Shepard AO, Starkey HC (1966) The effect of exchanged cations and the thermal behaviour of heulandite and clinoptilolite. Miner Soc India IMA Vol:155 – 158

Sheppard RA (1971) Zeolites in sedimentary deposits of the United States – a review. In: Molecular sieve zeolites. I. Am Chem Soc Adv Chem Ser 101:279 – 310

Sheppard RA, Gude AJ 3rd (1964) Reconnaissance of zeolite deposits in tuffaceous rocks of the western Mojave Desert and vicinity, California. US Geol Surv Prof Pap 501C:114 – 116

Sheppard RA, Gude AJ 3rd (1968) Distribution and genesis of authigenic silicate minerals in tuffs of Pleistocene Lake Tekopa, Inyo County, California. US Geol Surv Prof Pap 597

Sheppard RA, Gude AJ 3rd (1969a) Diagenesis of tuffs in the Barstow formation, Mud Hills, San Bernardino County, California. US Geol Surv Prof Pap 634

Sheppard RA, Gude AJ 3rd (1969b) Chemical composition and physical properties of the related zeolites offretite and erionite. Am Miner 54:875 – 886

Sheppard RA, Gude AJ 3rd (1970) Calcic siliceous chabazite from the John Day formation, Grant County, Oregon. US Geol Surv Prof Pap 700-D:D176 – D180

Sheppard RA, Gude AJ 3rd (1971) Sodic harmotome in lacustrine Pliocene tuffs near Wikieup, Mohave County, Arizona. US Geol Surv Prof Pap 750D:D50 – D55

Sheppard RA, Gude AJ 3rd (1973) Zeolites and associated authigenic silicate minerals in tuffaceous rocks of the Big Sandy formation, Mohave County, Arizona. US Geol Surv Prof Pap 830

Sheppard RA, Gude AJ 3rd (1983) Harmotome in a basaltic volcaniclastic sandstone from a lacustrine deposit near Kirkland Junction, Yavapai County, Arizona. Clays Clay Minerals 31:57 – 59

Sheppard RA, Gude AJ 3rd, Griffin JJ (1970) Chemical composition and physical properties of phillipsite from the Pacific and Indian Oceans. Am Miner 55:2053 – 2062

Sheppard RA, Gude AJ 3rd, Desborough GA, White JS Jr (1974) Levyne-offretite intergrowths from basalt near Beech Creek, Grant County, Oregon. Am Miner 59:837 – 842

Sheppard RA, Gude AJ 3rd, Mumpton FA (1983) Zeo trip '83. Int Committee on Natural Zeolites, Brockport

Sheridan MF, Maisano MD (1976) Zeolite and sheet silicate zonation in a Late-Tertiary geothermal basin near Hassayampa, central Arizona. Proc 2nd UN Symp on Dev and Use of Geothermal Resources, San Francisco, 20 – 29 May 1975, pp 597 – 607

Sherman JD (1977) Identification and characterization of zeolites synthetized in the $K_2O - Al_2O_3 - SiO_2 - H_2O$ system. In: Katzer JR (ed) Molecular sieves II. Am Chem Soc Symp Ser 40:30 – 42

Sherman JD (1978a) Ion exchange separations with molecular sieve zeolites. In: Sherman JD (ed) Adsorption and ion exchange separations. Am Inst Chem Eng Symp Ser 74, no 179:98 – 116

Sherman JD (1978b) Ionsiv F-80 and ionsiv W-85: molecular sieve zeolite. NH_4 ion exchangers for removal of urea nitrogen. In: Chang TMS (ed) Artificial kidney, artificial liver and artificial cells. Plenum, New York

Sherman JD, Bennett JM (1973) Framework structures related to the zeolite mordenite. In: Meier WM, Ujetterhoeven JB (eds) Molecular sieves. Am Chem Soc Adv Chem Ser 121:52 – 65

Sherman JD, Ross RJ [to Union Carbide Corp] (1982) Phillipsite-type zeolites for ammonia adsorption. US Pat 4, 344, 851

Sherry HS (1966) The ion-exchange properties of zeolites. I. Univalent ion exchange in synthetic faujasite. J Phys Chem 70:1158 – 1168

Sherry HS (1967) Barium ion exchange of Linde 13-X. J Phys Chem 71:780 – 782

Sherry HS (1968a) The ion-exchange properties of zeolites. IV. Alkaline earth ion exchange in the synthetic zeolites Linde X and Y. J Phys Chem 72:4086 – 4094

Sherry HS (1968b) The ion-exchange properties of zeolites. III. Rare earth ion exchange of synthetic faujasites. J Colloid Interface Sci 28:288 – 292

Sherry HS (1979) Ion-exchange properties of the natural zeolite erionite. Clays Clay Minerals 27:231 – 237

Shibue Y (1981) Cation-exchange reactions of siliceous and aluminous phillipsites. Clays Clay Minerals 2:397 – 402

Shikunov BI, Lafer LI, Yakerson VI, Rubinstein AM (1971) Determination of hydroxonium ion in mordenite by an infrared spectroscopic method (in Russian). Izv Akad Nauk SSSR Ser Khim 1971:1595

Shimazu M, Kawakami T (1967) Distribution of zeolite and other minerals in the Mazé Basalts, Niigata Pref., Japan. Sci Rep Niigata Univ Ser E Geol Mineral 1:17 – 32

Shimazu M, Mizota T (1972) Levyne and erionite from Chojabaru, Iki island, Nagasaki Pref., Japan. J Jpn Assoc Miner Petr Econ Geol 67:418 – 424

Shkabara MN (1940) On the zeolites of the Crimea (in Russian). Dokl Akad Nauk SSSR 26:659 – 661

Shkabara MN (1941) Zeolites of the Mama-Vitim mica-bearing region (in Russian). Dokl Akad Nauk SSSR 32:420 – 423

Shkabara MN (1948) Thomsonite from the teschenite of Kursebi (Caucasus) (in Russian). Dokl Akad Nauk SSSR 59:1161 – 1163

Shkabara MN, Shturm EA (1940) Zeolites of the Lower Tunguska region (in Russian). Zap Vses Miner Obshchest 69:63 – 66

Simonot-Grange MH (1979) Thermodynamic and structural features of water sorption in zeolites. Clays Clay Minerals 27:423 – 428

Simonot-Grange MH, Cointot A (1969) Évolution des propriétés d'adsorption de l'eau par la heulandite en relation avec la structure cristalline. Bull Soc Chim Fr 1969:421 – 427

Simonot-Grange MH, Wattelle-Marion G, Cointot A (1968) Caractères physico-chimiques de l'eau dans l'heulandite. Étude diffratométrique des phases observées au cours de la déshydratation et de la réhydratation. Bull Soc Chim Fr 1968:2747 – 2754

Simonot-Grange MH, Cointot A, Thrierr-Sorel A (1970) Étude des systemes divariants zeolite-eau. Cas de la stilbite. Comparaison avec l'heulandite. Bull Soc Chim Fr 1970:4286 – 4297

Simpson ES (1931) Contributions to the mineralogy of Western Australia. J R Soc West Aust 17:137 – 149

Slaughter M (1970) Crystal structure of stilbite. Am Miner 55:387 – 397

Slaughter M, Kane WT (1969) The crystal structure of a disordered epistilbite. Z Kristallogr 130:68 – 87

Slawson CB (1925) The thermo-optical properties of heulandite. Am Miner 10:305 – 331

Smith JV (1962) Crystal structures with a chabazite framework. I. Dehydrated Ca-chabazite. Acta Crystallogr 15:835 – 845

Smith JV (1971) Faujasite-type structures: alumino-silicate framework: positions of cations and molecules: nomenclature. In: Molecular sieves zeolites. I. Am Chem Soc Adv Chem Ser 101:171 – 200

Smith JV (1976) Origin and structures of zeolites. In: Rabo JA (ed) Zeolite chemistry and catalysis. Am Chem Soc Monog 171:1 – 79

Smith JV (1977) Enumeration of 4-connected 3-dimensional nets and classification of framework silicates. I. Perpendicular linkage from simple hexagonal net. Am Miner 62:703 – 709

Smith JV (1978) Enumeration of 4-connected 3-dimensional nets and classification of framework silicates. II. Perpendicular and near-perpendicular linkages from 4.8^2, 3.12^2 and 4.6.12 nets. Am Miner 63:960 – 969

Smith JV (1979) Enumeration of 4-connected 3-dimensional nets and classification of framework silicates. III. Combination of helix, and zig zag, crankshaft and saw chains with simple 2D nets. Am Miner 64:551 – 562

Smith JV (1983) Enumeration of 4-connected 3-dimensional nets and classification of framework silicates: combination of 4-l chain and 2D nets. Z Kristallogr 165:191 – 198

Smith JV, Bennett JM (1981) Enumeration of 4-connected 3-dimensional nets and classification of framework silicates: the infinite set of ABC-6 nets; the Archimedean and 6-related nets. Am Miner 66:777 – 788

Smith JV, Bennett JM (1984) Enumeration of 4-connected 3-dimensional nets and classification of framework silicates: linkages from the two $(5^2.8)_2$ $(5.8^2)_1$ 2D nets. Am Miner 69:104 – 111

Smith JV, Rinaldi F, Dent Glasser LS (1963) Crystal structures with a chabazite framework. II. Hydrated Ca-chabazite at room temperature. Acta Crystallogr 16, 45 – 53

Smith JV, Knowles CR, Rinaldi F (1964) Crystal structures with a chabazite framework. III. Hydrated Ca-chabazite at + 20 and − 150 °C. Acta Crystallogr 17:374 – 384

Smith JV, Pluth JJ, Artioli G, Ross FK (1984) Neutron and X-ray refinements of scolecite. In: Bisio A, Olson DH (eds) Proc 6th Int Zeolite Conf, Butterworths, Guildford, pp 842 – 850

Soloveva LP, Borisov SV, Bakakin VV (1972) New skeletal structure in the crystal structure of barium chloroaluminosilicate. Sov Phys Cryst 16:1035 – 1038

Stadtlaender C (1885) Beiträge zur Kenntnis der am Stempel bei Marburg vorkommenden Mineralien: Analcim, Natrolith und Phillipsit. Neues Jahrb Miner Geol Palaeontol 2:97 – 135

Stamires DN (1973) Properties of the zeolite, faujasite, substitutional series: a review with new data. Clays Clay Minerals 21:379 – 389

Staples LW (1946) Origin of sferoidal clusters of analcime from Benton County, Oregon. Am Miner 31:574 – 581

Staples LW (1957) X-ray study of erionite, a fibrous zeolite. Geol Soc Am Bull 68:1847

Staples LW (1965) Zeolites filling and replacement in fossils. Am Miner 50:1796 – 1801

Staples LW, Gard JA (1959) The fibrous zeolite erionite: its occurrence, unit cell and structure. Miner Mag 32:261 – 281

Steiner A (1955) Wairakite, the calcium analogue of analcime, a new zeolite mineral. Miner Mag 30:691 – 698

Steiner A (1958) Occurrence of wairakite at the Geysers, California. Am Miner 43:781

Steinfink H (1962) The crystal structure of the zeolite phillipsite. Acta Crystallogr 15:644 – 651

Stella Starrabba F (1947) Zeoliti dei Monti Peloritani (analcime e mesolite). Notizie Miner Sicil Calabr 1:49 – 54

Stepisiewiecz M (1978) Hydrothermal laumontite from the Strzelin granitoids. Acta Geol Pol 28:223 – 233

Stewart FH (1941) On sulphatic cancrinite and analcime (eudnophite) from Loch Borolan, Assynt. Miner Mag 26:1 – 8

Stewart RJ (1974) Zeolite facies metamorphism of sandstone in the western Olympic peninsula, Washington. Geol Soc Am Bull 85:1139 – 1142

Stoeffler D, Ewald R, Ostertay R, Reimold WV (1977) Research drilling Nördlinger Ries (1973) composition and texture of polymict impact breccias. Geol Bavarica 75:119 – 132

Stonecipher SA (1976) Origin, distribution and diagenesis of phillipsite and clinoptilolite in deep-sea sediments. Chem Geol 17:307 – 318

Stonecipher SA (1978) Chemistry of Deep-sea phillipsite, clinoptilolite and host sediments. In: Sand LB, Mumpton FA (eds) Natural zeolites. Pergamon, Oxford, pp 221 – 234

Stoppani FS, Curti E (1982) I minerali del Lazio. Olimpia, Firenze

Strauss CA, Truter FC (1950) Post-Bushveld ultrabasic, alkali and carbonatic eruptives at Magnet Heights, Sekukuniland, Eastern Transvaal. Trans Geol Soc S Afr 53:169 – 190

Streckeisen A (1976) To each plutonic rock its proper name. Earth Sci Rev 12:1 – 33

Stringham B (1950) Mordenite from Tinctic, Utah, and the discredited mineral arduinite. Am Miner 35:601 – 604

Sturiale C (1963) Rinvenimento di alcune zeoliti in ialoclastiti presso Palagonia. Rend Soc Miner Ital 19:209 – 212

Sukheswala RN, Avasia RK, Gangopadhyay M (1974) Zeolites and associated secondary minerals in the Deccan Traps of western India. Miner Mag 39:658 – 671

Sumin LG (1955) On laumontite from Dashkesan (in Russian). Trans Miner Mus Acad Sci USSR 7:127 – 131

Suprychev VA (1968) Mordenite from the effusive rocks of Karadag, eastern Crimea (in Ukrainian with English summary). Dopov Akad Nauk Ukr RSR Ser B 30:125 – 128

Suprychev VA (1976) Physico-geographic and geologic aspects of zeolitic deposits of economic significance in Ukraine (in Ukrainian with English summary). Fis Geogr Geomorfol Mishwid Nauk Sbirn 15:90 – 95

Surdam RC, Eugster HP (1976) Mineral reactions in the sedimentary deposits of the Lake Magadi region, Kenya. Geol Soc Am Bull 87:1739 – 1752

Surdam RC, Parker RD (1972) Authigenic aluminosilicate minerals in the tuffaceous rocks of the Green River formation, Wyoming. Geol Soc Am Bull 83:689 – 700

Surdam RC, Sheppard RA (1978) Zeolites in saline, alkaline-lake deposits. In: Sand LB, Mumpton FA (eds) Natural zeolites. Pergamon, Oxford, pp 145 – 174

Sutherland FL (1965) Some new occurrences of zeolites and associated minerals in the Tertiary basalts of Tasmania. Aust J Sci 28:26 – 27

Sutherland FL (1977) Zeolite minerals in the Jurassic dolerites of Tasmania: their use as possible indicators of burial depth. J Geol Soc Aust 24:171 – 178

Suzuki J (1938) On the occurrence of the aegirine-augite in natrolite veins from Nemuro, Hokkaidô. J Fac Sci Hokkaidô Univ Sapporo Ser 4 4:183 – 191

Sweet JM (1961) Tacharanite and other hydrated calcium silicates from Portree, Isle of Skye. Miner Mag 32:745 – 753

Takáts T (1936) Granodiorite from Zsidóvár (in Hungarian). Mat Term Tud Ertesito 54:882–892

Takeshita Y, Yatagai K, Suda K (1975) Zeolites and other minerals from Kuroiwa, Niigata Pref., Japan (in Japanese). Geosci Mag 26:347–359; 27:43–57

Takéuchi Y, Mazzi F, Haga N, Galli E (1979) The crystal structure of wairakite. Am Miner 64:993–1001

Taldykina KS (1958) Stellerite from the ore deposit Savinskoe no. 5 (in Russian). Zap Vses Miner Obshchest 87:108

Tambuyzer E, Bosmans HJ (1976) The crystal structure of synthetic zeolite K-F. Acta Cryst B32:1714–1719

Tamnau F (1836) Monographie des chabasits. Neues Jahrb Miner Geognosie Geol Petrefakten-kunde 1836:635–658

Taylor AM, Roy M (1964) Zeolite studies IV: Na-P zeolites and the ion-exchanged derivatives of tetragonal Na-P. Am Miner 49:656–682

Taylor MW, Surdam RC (1981) Zeolite reactions in the tuffaceous sediments at Teels Marsh, Nevada. Clays Clay Minerals 29:341–352

Taylor WH (1930) The structure of analcite ($NaAlSi_2O_6 \cdot H_2O$). Z Kristallogr 74:1–19

Taylor WH, Jackson R (1933) The structure of edingtonite. Z Kristallogr 86:53–64

Taylor WH, Meek CA, Jackson WW (1933) The structures of the fibrous zeolites. Z Kristallogr 84:373–398

Tchernev DI (1980) The use of zeolites for solar cooling. In: Rees LVC (ed) Proc 5th Int Conf Zeolites. Heyden, London, pp 788–794

Tennyson C (1978) Zeolithe im Basalt des Großen Teichelberges. Lapis 3, no 1:14–17

Teruggi ME (1964) A new and important occurrence of sedimentary analcime. J Sediment Petrol 34:761–767

Thompson AB (1970) Laumontite equilibria and the zeolite facies. Am J Sci 269:267–275

Thompson AB (1971a) P(CO$_2$) in low-grade metamorphism; zeolite, carbonate, clay mineral, prehnite relations in the system $CaO – Al_2O_3 – SiO_2 – CO_2 – H_2O$. Contrib Miner Petrol 33:145–161

Thompson AB (1971b) Analcite-albite equilibria at low temperatures. Am J Sci 271:79–92

Thugutt SJ (1933) Sur la ptilolite de Mydzk en Volhynie (in Polish with French summary). Arch Miner Tow Nauk Warszaw 9:99–104

Tiba T, Matsubara S (1977) Levyne from Dózen (Oki islands), Japan. Can Miner 15:536–539

Tillmanns E, Fischer R (1981) Crystal structure of a new zeolite mineral. Acta Crystallogr Sect A Cryst Phas Diffr Theor Gen Crystallogr 37 (Suppl):C-186

Tomita K, Yamashita H, Oba N (1969) Artificial crystallization of volcanic glass to sodium and potassium form of chabazite at room pressure. J Jpn Assoc Miner Petr Econ Geol 62:80–89

Tomita K, Yamashita H, Oba N (1970) Mordenite in rhyolite at Yoshida area, Kagoshima Pref. J Jpn Assoc Miner Petr Econ Geol 63:16–21

Torii K (1978) Utilization of natural zeolites in Japan. In: Sand LB, Mumpton FA (eds) Natural zeolites. Pergamon, Oxford, pp 441–450

Torii K, Hotta M, Asaka M (1977) Utilization of zeolite tuff. IX. Adsorption properties of cat-ion-exchanged clinoptilolite. (II); Effect of cation exchange in clinoptilolite on chromatograph-ic properties (in Japanese). Nendo Kagaku 17:33–38

Torrie BH, Brown ID, Petch HE (1964) Neutron diffraction determination of the hydrogen posi-tions in natrolite. Can J Phys 42:229–240

Tschernich RW (1972) Zeolites from Skookumchuck Dam, Washington. Miner Record 3:30–34

Tschernich RW, Wise WS (1982) Paulingite: variations in composition. Am Miner 67:799–803

Tyrell GW (1923) Classification and age of the analcite-bearing igneous rocks of Scotland. Geol Mag 60:249–260

Udluft H (1928) Zeolithe als Fossilisationsmaterial. Arkiv Kemi Miner Geol 9, no 33:1–15

Ueda S, Koizumi M (1979) Crystallization of analcime solid solutions from aqueous solutions. Am Miner 64:172–179

Ueda S, Murata H, Koizumi M, Nishimura H (1980) Crystallization of mordenite from aqueous solutions. Am Miner 65:1012–1019

Ulrich F (1930) Contributions to topographic mineralogy of Bohemia II (in Czech with German summary). Věstn Stát Geol Ústavu Českoslov Republ 6:98–110

Ulrych J, Rychly R (1980) The overgrowth crystals of scolecite on mesolite. Cryst Res Technol 15:K91–K93

Ulrych J, Rychly R (1981) Strontian chabazite from Repčice, Bohemia. Chem Erde 40: 68–71

Utada M (1965) Zonal distribution of authigenic zeolites in the Tertiary pyroclastic rocks in Mogami district, Yamagata Prefecture. Sci Pap Coll Gen Educ Univ Tokyo 15:173–216

Utada M (1970) Occurrence and distribution of authigenic zeolites in the Neogene pyroclastic rocks in Japan. Sci Pap Coll Gen Educ Univ Tokyo 20:191–262

Utada M (1971) Zeolitic zoning of the Neogene pyroclastic rocks in Japan. Sci Pap Coll Gen Educ Univ Tokyo 21:189–221

Utada M (1973) The types of alteration in the Neogene sediment relating to the intrusion of volcano-plutonic complexes in Japan. Sci Pap Coll Gen Educ Univ Tokyo 23:167–216

Utada M (1980) Hydrothermal alterations related to igneous activity in Cretaceous and Neogene formations of Japan. Mining Geol Spec Issue 8:67–83

Utada M, Minato H, Ishikawa T, Yoshiro Y (1974) The alteration zones surrounding the Kuroko-type deposits in Nishi-Aizu district, Fukushima Pref., with the emphasis on the analcime zone as an indicator in exploration for ore deposits. Mining Geol Spec Issue 6:291–302

Valkenburg AT Jr Van, Buie BF (1945) Octahedral cristobalite with quartz pseudomorphs from Ellora Caves, at Hyderahad State, India. Am Miner 30:526–535

Vanderstappen R, Verbeek T (1959) Présence d'analcime d'origine sédimentaire dans les Mesozoique du bassin du Congo. Bull Soc Belge Géol Paléontol Hydrol 68:417–421

Varjú Gy (1966) Geology of the Rátka-trass (clinoptilolite bearing rhyolitic tuff). Foldt Kutatas 9(3):21–30

Vaughn PA (1966) The crystal structure of the zeolite ferrierite. Acta Crystallogr 21:983–990

Velde B, Besson JM (1981) Raman spectra of analcime under pressure. Phys Chem Minerals 7:96–99

Venkatarathuam K, Biscaye PE (1973) Deep-sea zeolites: variations in space and time in the sediments of the Indian Ocean. Mar Geol Lett Sect 15:M11–M17

Ventriglia U (1953) Simmetria della heulandite e piezoelettricità di alcune zeoliti. Rend Soc Miner Ital 9:268–269

Venuto PB (1971) Some perspectives on zeolite catalysis. In: Molecular sieve zeolites II. Am Chem Soc Adv Chem Ser 102:260–283

Vernet JP (1961) Concerning the association montmorillonite-analcime in the series of Stanleyville, Congo. J Sediment Petrol 31:293–295

Vezzalini G (1983) A refinement of Elba dachiardite: opposite acentric domains simulating a centric structure. Z Kristallogr 166:63–71

Vezzalini G, Alberti A (1975) Le zeoliti dell'Alpe di Siusi. Rend Soc Ital Miner Petrol 31:711–719

Vezzalini G, Mattioli V (1979) Secondo ritrovamento della roggianite. Period Miner 48:15–20

Vezzalini G, Oberti R (1984) The crystal chemistry of gismondines. Bull Miner 107:805–812

Vince DG (1980) Zeolites and other minerals in vesicles in the Newer Basalt of the Melbourne area, Victoria. Aust Miner 32:155–160

Vuagnat M (1952) Pétrographie, répartition et origine des microbrèches du Flysch nordhelvétique. Mat Carte Géol Suisse Nlle Ser 97:1–109

Walenta K (1974) Zeolithparagenesen aus dem Melilith-Nephelinit des Höwenegg im Hegau. Aufschluß 25:613–626

Walker GPL (1951) The amygdale minerals in the Tertiary lavas of Ireland. I. The distribution of chabazite habits and zeolites in the Garron plateau area, County Antrim. Miner Mag 29:773–791

Walker GPL (1959) The amygdale minerals in the Tertiary lavas of Ireland. II. The distribution of gmelinite. Miner Mag 32:202–217

Walker GPL (1960a) The amygdale minerals in the Tertiary lavas of Ireland. III. Regional distribution. Miner Mag 32:503–527

Walker GPL (1960b) Zeolite zones and dike distribution in relation to the structure of the basalt of eastern Iceland. J Geol 68:515 – 528

Walker GPL (1962a) Garronite, a new zeolite, from Ireland and Iceland. Miner Mag 33: 173 – 186

Walker GPL (1962b) Low-potash gismondine from Ireland and Iceland. Miner Mag 33:187 – 199

Walker TL, Parsons AL (1922) The zeolites of Nova Scotia. Univ Toronto Stud Geol Ser 14:13 – 73

Walton AW (1975) Zeolitic diagenesis in Oligocene volcanic sediments, Trans-Pecos, Texas. Geol Soc Am Bull 86:615 – 624

Weibel M (1963) Chabazit vom Chrüzlistock (Tavetsch). Schweiz Miner Petr Mitt 43:361 – 366

Wells AF (1954) The geometrical basis of crystal chemistry. Part II. Acta Crystallogr 7:545 – 554

Wells AF (1977) Three dimensional nets and polyhedra. Wiley, New York

Wen-Hui H, Shao-Hua T, Kung-Hai W, Chun-Lin C, Cheng-Chih Y (1958) Hsiang-hua-shih, a new beryllium mineral (in Chinese). Ti Chih Yueh K'an 7:35 (see also Am Miner 44:1327 – 1328)

Wetzlar K (1938) Über die Dielektrizitätskonstante wasserhaltiger Kristalle. Zeit angew Miner 1:125 – 133

Wheeler EP (1927) Stellerite from near Juneau, Alaska. Am Miner 12:360 – 364

White JS Jr (1975) Levyne-offretite from Beech Creek, Oregon. Miner Record 6:171 – 173

Whitehouse MJ (1938) The zeolites of Queensland. Univ Queensl Pap Dep Geol 1:1 – 81

Whittermore OJ Jr (1972) Synthesis of siliceous mordenites. Am Miner 57:1146 – 1151

Wilkinson JFG (1958) The petrology of a differentiated teschenite sill near Gunnedah, New South Wales. Am J Sci 256:1 – 39

Wilkinson JFG (1965) Some feldspars, nephelines and analcimes from the Square Top intrusion, Nundle, N. S. W. J Petrol 6:420 – 444

Wilkinson JFG (1968) Analcimes from some potassic igneous rocks and aspects of analcime-rich igneous assemblages. Contrib Miner Petrol 18:252 – 269

Wilkinson JFG (1977) Analcime phenocryst in vitrophyric analcimite – primary or secondary? Contrib Miner Petrol 64:1 – 10

Wilkinson JFG, Whetten JT (1964) Some analcime-bearing pyroclastic and sedimentary rocks from New South Wales. J Sediment Petrol 34:543 – 553

Winchell AN, Winchell H (1951) Elements of optical mineralogy, 4th edn. Part II. Wiley, New York; Chapman and Hall, London

Winkler HGF (1974) Petrogenesis of metamorphic rocks, 3rd edn. Springer, Berlin Heidelberg New York

Wirsching U (1975) Experimente zum Einfluß des Gesteinsglaschemismus auf die Zeolithbildung durch hydrothermale Umwandlung. Contrib Miner Petrol 49:117 – 124

Wirsching U (1981) Experiments on the hydrothermal formation of calcium zeolites. Clays Clay Minerals 29:171 – 183

Wise WS (1959) Occurrence of wairakite in metamorphic rocks of the Pacific Northwest. Am Miner 44:1099 – 1101

Wise WS (1978) Yugawaralite from Bombay, India. Miner Record 9:296

Wise WS (1982) New occurrence of faujasite in southeastern California. Am Miner 67:794 – 798

Wise WS, Tschernich RW (1975) Cowlesite, a new Ca-zeolite. Am Miner 60:951 – 956

Wise WS, Tschernich RW (1976a) Chemical composition of ferrierite. Am Miner 61:60 – 66

Wise WS, Tschernich RW (1976b) The chemical compositions and origin of the zeolites offretite, erionite and levyne. Am Miner 61:853 – 863

Wise WS, Tschernich RW (1978a) Dachiardite-bearing zeolite assemblages in the Pacific Northwest. In: Sand LB, Mumpton FA (eds) Natural zeolites. Pergamon, Oxford, pp 105 – 112

Wise WS, Tschernich RW (1978b) Habits, crystal forms and composition of thomsonite. Can Miner 16:487 – 493

Wise WS, Nokleberg WJ, Kokinos M (1969) Clinoptilolite and ferrierite from Agoura, California. Am Miner 54:887 – 895

Wolf F, Pilchowski K, Georgi K (1980) Zur Selektivität des Ionenaustausches an unterschiedlich dealuminierten Mordeniten. Z Anorg Allg Chem 461:211 – 216

Woolley AK, Symes RF (1976) The analcime-phyric phonolites (blairmorites) and associated analcime kenytes of the Lupata Gorge, Moçambique. Lithos 9:9–15

Wu Dc (1970) Origin of mineral analcite in the Upper Flowerport Shale, northwestern Oklahoma. Trans Kans Acad Sci 73:247–251

Yajima S, Nakamura T, Ishii E (1971) New occurrence of ferrierite. Miner J 6:343–364

Yefimov AF, Ganzeyev AA, Katayeva ZT (1966) Finds of strontian thomsonite in the USSR (in Russian). Dokl Akad Nauk SSSR 169:148–150

Yoder HS Jr, Weir CE (1960) High-pressure form of analcite and free energy change with pressure of analcite reactions. Am J Sci 258-A:420–433

Yoshikawa K, Okada A, Shima M (1982) Phillipsite from a manganese nodule (in Japanese). Rikagaku Kenkyusho Hokoku 58:11–18

Yoshimura T, Wakabayashi S (1977) Na-dachiardite and associated high-silica zeolites from Tsugawa, Northeast Japan. Sci Rep Niigata Univ Ser E Geol Miner 4:49–65

Yoshitani A (1965) Zeolites in the Neogene pyroclastic rocks in the eastern part of Tanzawa Mountainland, Central Japan. Mem Coll Sci Univ Kyoto Ser B 31:199–213

Young M (1965) Stellerite and its genesis in a disseminated molybdenum ore. Sci Geol Sin 3:295–302

Zaporozhtseva AS (1958) Laumontite in the Cretaceous deposits of the Lena coal district (in Russian). Dokl Akad Nauk SSSR 120:384–386

Zaporozhtseva AS, Vishnevskaya TN, Dubar GP (1961) Successive changes in calcium zeolites through vertical sections of sedimentary strata (in Russian). Dokl Akad Nauk SSSR 141:448–451

Zeng Y, Liou JG (1982) Experimental investigation of yugawaralite-wairakite equilibrium. Am Miner 67:937–943

Zirkl EJ (1973) Ferrierit im Basalt von Weitendorf, Steiermark. Neues Jahrb Miner Monatsh 1973:524–528

Zoch I (1915) Über den Basenaustausch kristallisierter Zeolithe gegen neutrale Salzlösungen. Chem Erde 1:219–269

Zoltai T (1960) Classification of silicates and other minerals with tetrahedral structures. Am Miner 45:960–973

Zyla M, Zabinski W (1978) Sorption properties of Mg-form of stilbite. Miner Pol 9:71–75

Mineral and Synthetic Zeolite Index

Numbers in **bold** *type refer to the main section devoted to the minerals; X-ray = X-ray powder pattern; IR = Infrared spectrum*

Locality Index

F.J.Sawkins

Metal Deposits in Relation to Plate Tectonics

1984. 173 figures. XIV, 325 pages
(Minerals and Rocks, Volume 17). ISBN 3-540-12752-6

Contents: Introduction: Plate Tectonics and Geology. – Convergent Plate Boundary Environments: Principal Arcs and Their Associated Metal Deposits. Metal Deposits on the Inner Sides of Principal Arcs. Metal Deposits of Arc-Related Rifts. Additional Aspects of Arc-Related Metallogeny. – Divergent Plate Boundary Environments: Metallogeny of Oceanic-Type Crust. Intracontinental Hotspots, Anorogenic Magmatism, and Associated Metal Deposits. Metal Deposits Associated with the Early Stages of Continental Rifting. Metal Deposits Related to Advanced Stages of Rifting. – Collisional Environments and Other Matters: Metal Deposits in Relation to Collision Events. Metal Deposits and Plate Tectonics – an Attempt at Perspective. – Afterword. – References. – Subject Index.

The author of this work has written a valuable overview of the geology and settings of metal deposits providing an understanding of the controls of their time-space distribution, and insights into some of the problems that remain to be resolved regarding their genesis. In it, he shows how the generation of most types of metal deposits can be meaningfully related to specific plate tectonic environments, and how tectonic settings provide a viable framework within which ore-generating systems can be studied. The most important features of the book are a careful assessment of various plate tectonic environments in terms of structure and lithology and the types of metal deposits that can be ascribed to each. Furthermore, details of specific ore deposit types are included to provide the reader with an understanding of characteristics of a significant number of categories of metal deposits.

This approach leads to a clearer understanding of the systematics of metal deposits and their distribution patterns. These insights, combined with the suggestions for exploration strategy included in many sections, will be of particular interest to industry geologists involved in metals exploration. Although some recent books on the relationship between ore deposits and tectonics have appeared, they do not investigate the causative relationships between plate tectonics and metal deposits as deeply and extensively as is true of the present work.

Springer-Verlag
Berlin
Heidelberg
New York
Tokyo

J. Gill

Orogenic Andesites and Plate Tectonics

1981. 109 figures. XIV, 390 pages
(Minerals and Rocks, Volume 16). ISBN 3-540-10666-9

Contents: What is "Typical Calcalkaline Andesite"? –
The Plate Tectonic Connection. – Geophysical Setting
of Volcanism at Convergent Plate Boundaries. – Ande-
site Magmas, Ejecta, Eruptions, and Volcanoes. – Bulk
Chemical Composition of Orogenic Andesites. – Mine-
ralogy and Mineral Stabilities. – Spatial and Temporal
Variations in the Composition of Orogenic Andesites. –
The Role of Subducted Ocean Crust in the Genesis of
Orogenic Andesites. – The Role of the Mantle Wedge. –
The Role of the Crust. – The Role of Basalt Differentia-
tion. – Conclusions. – Appendix. – References. –
Subject Index.

The solution to the question of andesite genesis is a
major, multidisciplinary undertaking facing geoscientists
in the 1980's. *Orogenic Andesites and Plate Tectonics* was
written in response to the growing need of researchers in
this area for a clarification of the long-standing problem
and identification of profitable areas for future investiga-
tions.
This book critically summarizes the vast relevant literature on the tectonics, geophysics, volcanology, geology,
geochemistry, and mineralogy of andesites and their
volcanoes. The author cites over 1100 references in
these specialties and includes information on the location and rock composition of more than 300 recently
active volcanoes. In addition, he provides a systematic
and original evaluation of genetic hypotheses. Numerous
cross references enhance the integrated subject development which consistently emphasizes the implications of
data for theories of magma genesis.
Orogenic Andesites and Plate Tectonics will prove an
invaluable reference source to researchers and graduate
students in the geosciences seeking a careful evaluation
of genetic hypotheses and key data, arguments or gaps
in this promising field.

Springer-Verlag
Berlin
Heidelberg
New York
Tokyo